Recent Advances in III-Nitride Semiconductors

Recent Advances in III-Nitride Semiconductors

Editors

Peng Chen
Zhizhong Chen

MDPI

Basel • Beijing • Wuhan • Barcelona • Belgrade • Novi Sad • Cluj • Manchester

Editors

Peng Chen
Nanjing University
Nanjing, China

Zhizhong Chen
Peking University
Beijing, China

Editorial Office
MDPI
St. Alban-Anlage 66
4052 Basel, Switzerland

This is a reprint of articles from the Special Issue published online in the open access journal *Crystals* (ISSN 2073-4352) (available at: https://www.mdpi.com/journal/crystals/special_issues/nitride_semiconductors_2).

For citation purposes, cite each article independently as indicated on the article page online and as indicated below:

Lastname, A.A.; Lastname, B.B. Article Title. *Journal Name* **Year**, *Volume Number*, Page Range.

ISBN 978-3-0365-8624-3 (Hbk)
ISBN 978-3-0365-8625-0 (PDF)
doi.org/10.3390/books978-3-0365-8625-0

Contents

About the Editors

Peng Chen

Dr. Chen received his Ph.D. in Physics from Nanjing University in December 2000. From 2001 to 2007, he served as a Research Scientist and Principal Investigator at the Institute of Materials Research, Singapore S&T Research Agency, and a doctoral supervisor at the Department of Electrical and Computer Engineering, National University of Singapore. In 2007, he joined the Department of Physics at Nanjing University as a professor. He won the title of "Young and Middle-Aged Experts with Outstanding Contributions" in Jiangsu Province in 2010. In 2011, he won the "Outstanding Contribution Award" of "333 Talents Project" and the "Innovation Individual Award of China Industry-University Research". He has been in charge of several international collaboration projects. As the Principal Investigator, he was in charge of the national "863" key project "Third Generation Semiconductor Devices and Evaluation Technology", and also in charge of or participated in a number of National Basic Research Program, the National Natural Science Foundation of China and other major research projects. He has applied for more than 90 international and national patents. A total of more than 230 research papers have been published, of which more than 200 are cited in the SCI, including more than 40 high-level papers published by the first author or corresponding author in the fields of semiconductor optoelectronics, including in Advanced Materials, Advanced Optical Materials, Small, IEEE Electron Device Letters, Applied Physics Letters and IEEE Photonics Journal. He also participated in the compilation of four academic books.

Zhizhong Chen

Dr. Zhizhong Chen received his Ph.D. in Physics from Nanjing University in 2000. From 2000 to 2002, he was a postdoctoral fellow at the postdoctoral flow station in condensed matter physics, Peking University, and as a Senior Engineer in LED Chip Department, Shanghai Blue Light Technology Co., Ltd. From 2002, he worked at the Institute of Condensed Matter Physics, School of Physics, Peking University and as an Associate Professor since 2004. In February 2005, he was a short-term visiting scholar at the Institute of High Energy Particles at the University of Leuven in Belgium. He also was a visiting scholar at the Institute of Optoelectronics, Strathclyde University, UK, from August 2009 to August 2010. He has been in charge of or participated in a number of national "973" and "863" projects, as well as the National Natural Science Foundation of China, the key research and development projects of Beijing and other major research projects. He has applied for more than 50 national patents. A total of more than 180 research papers have been published, including high-level papers published by the first author or corresponding author in the fields of semiconductor optoelectronics, including in Applied Physics Letters, Optical Express, Scientific Reports and IEEE Photonics Journal, which have been cited more than 2000 times.

Preface

Dear colleagues,

Group-III nitrides, viz, GaN, InN, AlN and their compounds, are typical direct bandgap semiconductor materials. Interest in group-III nitrides stems from their irreplaceable and efficient blue UV luminescence capability. By adjusting the material composition of the alloy, the bandgap of the compound semiconductors formed by the group-III nitrides can be continuously changed from 0.7eV (InN) to 6.2eV (AlN), and its emission wavelength can cover from the near-infrared to the deep ultraviolet (UV) band wavelength range. Therefore, III-nitride semiconductors are considered the most suitable materials for green to deep ultraviolet (DUV) and white light sources. These solid-state light sources have spurred new developments in various fields including high-density optical storage, water treatment and biochemical detection. At present, many studies have been carried out on the development of light-emitting diodes and laser devices in the deep ultraviolet DUV spectral range below a 280 nm wavelength. Realizing laser emission at such a short wavelength requires high-quality materials with good internal quantum efficiency (IQE) and an appropriate cavity structure. Recent progress in GaN-based material quality and device design relies on well-mastered techniques of material growth and the formation of desired structures with other elements. This offers a significant possibility of creating high-quality materials and diverse functional devices.

GaN is also a promising candidate for next-generation power electronic applications because of its outstanding material properties, but its potential is far from being realized. Regarding whether GaN can be competent in the field of ultra-high voltage (UHV, >10 kV) applications, there is currently a huge controversy in the community, especially under the strong background of SiC in this field, although GaN has a farther limit than SiC based on the nature of the materials.

Therefore, this Special Issue includes 16 excellent works, covering a broad spectrum of topics from the study of materials, micro/nano structures and novel functional devices, to new applications in frontier fields. The topics include, but are not limited to:

Growth of GaN-based materials and micro/nanostructures;

Characterization of the materials and heterostructures;

GaN-based novel devices, including emission, detection and power devices;

Application and integration of other wide-bandgap materials and novel devices in novel electronics and photonics.

We believe that this Special Issue will contribute to the learning and communication of scientists, graduate students and engineers working in this field.

We deeply appreciate the outstanding research work of all authors. We are also very grateful to Mr. Mars Tan from the Editorial Office for his continuous assistance and support, which enabled our Special Issue to be successful.

Peng Chen and Zhizhong Chen
Editors

crystals

MDPI

Article

Proposal for Deep-UV Emission from a Near-Infrared AlN/GaN-Based Quantum Cascade Device Using Multiple Photon Up-Conversion

Daniel Hofstetter [1,*], David P. Bour [2] and Hans Beck [3]

1 Chemin du Château 5, 2068 Hauterive, Switzerland
2 Apple, Inc., San Jose, CA 94086, USA
3 Rue des Peupliers 6, 2014 Bôle, Switzerland
* Correspondence: dani_hofstetter@outlook.com; Tel.: +41-32-753-78-25

Abstract: We propose the use of an n-doped periodic AlN/GaN quantum cascade structure for the optical up-conversion of multiple near-infrared (near-IR) photons into deep-ultraviolet (deep-UV) radiation. Without applying an external bias voltage, the active region of such a device will (similar to an un-biased quantum cascade laser) resemble a sawtooth-shaped inter-subband structure. A carefully adjusted bias voltage then converts this sawtooth pattern into a 'quantum-stair'. Illumination with $\lambda = 1.55$ µm radiation results in photon absorption thereby lifting electrons from the ground state of each main well into the first excited state. Three additional GaN quantum wells per period then provide by LO-phonon-assisted tunneling a diagonal transfer of these electrons towards the ground level of the neighboring period. From there, the next near-infrared (near-IR) photon absorption, electron excitation, and partial relaxation takes place. After 12 such absorption, transfer, and relaxation processes, the excited electrons have gained a sufficiently high amount of energy to undergo in the final AlN-based p-type contact layer an electron-hole band-to-band recombination. By employing this procedure, multiple near-IR photons will be up-converted to produce deep-UV radiation. Since for a wavelength of 1.55 µm very powerful near-IR pump lasers are readily available, such an up-conversion device will (even at a moderate overall conversion efficiency) potentially result in an equal or even higher output power than the one of an AlN-based p-n-junction light-emitting diode. The proposed structures are therefore very interesting for applications such as ultra-high-resolution photolithography or printing, water purification, medical equipment disinfection, white light generation, or the automotive industry.

Keywords: nitride semiconductors; AlN/GaN superlattice; optical pumping; near-IR absorption; optically induced up-conversion; deep-UV emission

Citation: Hofstetter, D.; Bour, D.P.; Beck, H. Proposal for Deep-UV Emission from a Near-Infrared AlN/GaN-Based Quantum Cascade Device Using Multiple Photon Up-Conversion. *Crystals* **2023**, *13*, 494. https://doi.org/10.3390/cryst13030494

Academic Editors: Peng Chen and Zhizhong Chen

Received: 9 February 2023
Revised: 7 March 2023
Accepted: 8 March 2023
Published: 13 March 2023

1. Introduction

Deep ultraviolet (deep-UV) light (also called UV-C radiation) is of crucial importance for many applications in environmental and life sciences. Up to now, different approaches to generate such short-wavelength radiation have been reported. Among them are frequency-doubling of blue InGaN-based laser diodes [1], cathodoluminescence on Y_2O_3 [2], or an electric discharge through [Ar]-gas striking a graphene/hexagonal BN/graphene heterostructure [3]. Since the direct generation of spectrally narrow deep-UV radiation using semiconductor-based light-emitting diodes (LEDs) still suffers from issues such as low doping, high contact resistance, or poor efficiency, alternative methods such as up-conversion of multiple near-IR photons for the subsequent emission of a single deep-UV photon are being actively investigated [4–6]. Such up-conversion processes have already enabled a variety of interesting applications ranging from bio-imaging [7] or solar energy conversion [8] to high-density optical storage [9].

Very promising semiconductor materials for such deep-UV applications are AlN and GaN. This is not only because of the 2014 Physics Nobel Prize in this area [10,11] or the III-nitrides' great radiation hardness, chemical inertness, and mechanical stability. Rather, it is also due to their potential as opto-electronic materials of choice for the entire visible wavelength range [12]. By using inter-subband (ISB) transitions in AlN/GaN-based multi-quantum well structures, it is even possible to access the near-IR wavelength range [13,14]. For these reasons, III-nitride semiconductors have become very interesting for ultrafast detector applications in both the mid- and the near-IR or in free-space optical telecommunications.

In the present publication, however, we will discuss a completely new device idea—taking advantage of optically up-converting near-IR light into deep-UV light. The resulting short-wavelength radiation can—for instance—be used for the sterilization of medical equipment [15], water purification [16], deep-UV lithography [17], or high-density optical data storage [18].

2. Experimental Setup

For the fabrication of the envisaged devices, the following layer structure is proposed. On top of a high-quality AlN/sapphire template [19], a fully relaxed 2 μm-thick n-doped AlN:Si layer ($N_D = 5 \times 10^{17}$ cm^{-3}) will be grown. The latter serves both as a quasi-substrate and a lower contact material. It will be overgrown by the active region which consists of a 12-period superlattice containing relaxed AlN barriers and strained GaN quantum wells (QWs). The layer structure of one period—starting with a barrier layer—corresponds to the following sequence: **15**/15/**15**/15/**15**/15/**20**/15 Å. Boldface numbers denote AlN barriers, roman numbers stand for GaN QWs; and n-type modulation-doped layers ([Si], $N_D = 2 \times 10^{18}$ cm^{-3}, $E_A = 25$ meV) are marked by underlining. The final top contact layer consists of 200 nm fully relaxed, p-doped AlN:Be ($N_A = 3 \times 10^{17}$ cm^{-3}, $E_A = 37$ meV), and it will be prepared by metal modulated epitaxy (MME) [20]. The entire SL will thus essentially have the lattice constant of relaxed AlN.

Using an ultraviolet HeCd-laser (P = 100 mW, $\lambda_{HeCd} = 325$ nm), a second-order diffraction grating with a period of $\Lambda = (1.55\ \mu m/2.3) = 674$ nm is holographically defined by two-beam interference. After photoresist exposure and development, the resulting pattern is metalized (Pt/Pd/Au 10/10/200 nm)—followed by metal lift-off. Using the grating's metal stripes as an etch mask, the semiconductor surface is then dry-etched using Cl_2/H_2-based reactive ion etching with an inductively coupled plasma (ICP-RIE) [21]. The etch depth will be 100 nm—i.e., about halfway through the p-type contact layer. A small portion of such a grating looks typically like the example in the SEM image shown in Figure 1. In order to achieve a very strong diffraction, an enhanced coupling strength grating with a low-index liner and a high-index cover layer may be used [22].

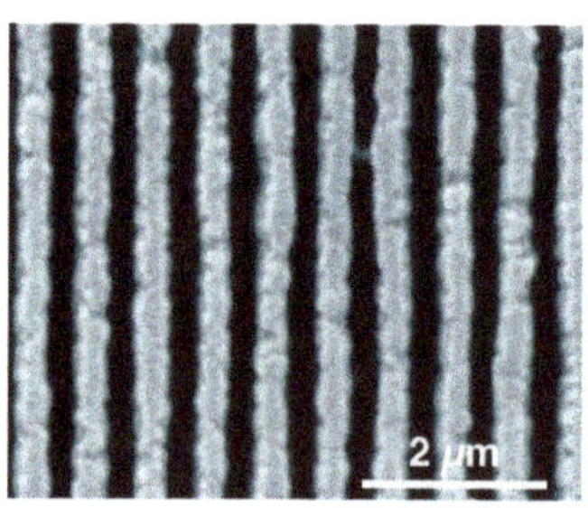

Figure 1. Scanning electron microscopy (SEM) image of a second-order diffraction grating for vertical coupling. The period of this grating is $\Lambda = 674$ nm, as shown by the scale bar. The metal layers are identical to the p-type metallization used for the contacts.

In a next step, square-shaped mesa structures with a side length of 200 µm and a height of 1.2 µm are defined using, again, ICP-RIE. Such a mesa is schematically presented in Figure 2.

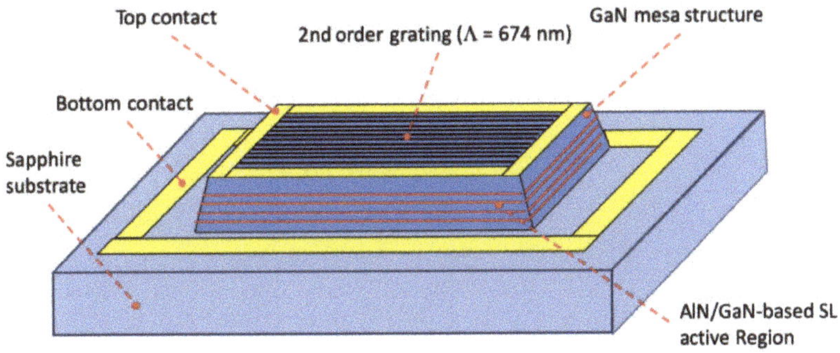

Figure 2. Schematic drawing of the square-shaped AlN/GaN-based inter-subband mesa with a diffraction grating on top. Metallization allows electrical contact at the top and the bottom of the mesa.

P-type contact metals (Pt/Pd/Au 10/10/200 nm) will then be lift-off deposited on the shoulders of the square mesas. The resulting 'contact contour' provides a common electrical connection to all grating lines of a given mesa structure and assures a sufficiently large top contact area. The relatively high p-type doping of the upper contact layer guarantees a good Ohmic contact. The n-type lower contact metal layers—to be deposited on the dry-etched areas outside the mesas—consist of Ti/Al/Ni/Au (10/10/10/200 nm). These contacts will then be annealed at 825 °C for 1 min.

3. Proposed Operation of the Device

When applying zero bias voltage, the conduction band structure schematically appears sawtooth-shaped. There, the label 'g' denotes the first main well's ground state while 'e' is its first excited state. Labels with primes (such as g', g'', …) indicate levels in the second, third, … main QW. In order to operate this device, a total voltage of 6.2 V (corresponding to an electric field of 413 kV/cm) will be applied between the top and the bottom contact layers of the mesa. This results in a stair-like conduction band structure; as shown in Figure 3. The above voltage will be carefully adjusted in order to result in a sufficiently small forward current (<1 mA)—indicating a slightly incomplete level alignment and a missing anti-crossing of the relevant energy levels 'e' and '1' (see Figure 3). In order to provide the small amount of remaining electric field, the mesa under investigation will be illuminated vertically from above using light with a wavelength of 1.55 µm. The second-order grating diffracts this vertically incoming radiation into the directions parallel to the mesa surface. Due to subsequent light propagation in the horizontal direction, strong inter-subband absorption will take place in the 1.5 nm wide, n-doped GaN ([Si], $N_D = 2 \times 10^{18}$ cm^{-3}) QW layers.

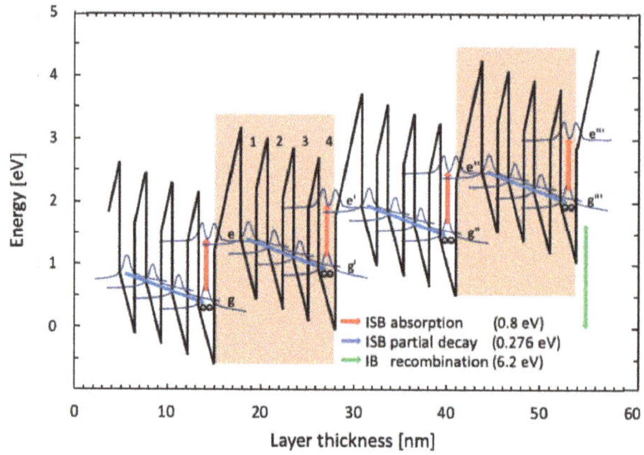

Figure 3. Conduction band structure under an applied bias voltage of 6.2 V on 12 periods—corresponding to an electric field of 413 kV/cm. QWs/barriers 1, 2, 3, 4 correspond to one SL period. Excitation through photon absorption (vertical red arrows) and partial relaxation (oblique blue arrows) are schematically shown. The vertical transition energy in the main QW measures 800 meV (λ = 1.55 μm). The green arrow marks the interband transition in the deep-UV.

As shown schematically in Figure 3, the absorbed near-IR radiation efficiently lifts electrons from the ground level (labelled level 'g') into the excited QW state (labelled level 'e'). From there, the electrons have two possibilities to proceed. They will either drop back vertically into level 'g' or transfer—in a second step—towards the neighboring QW (labelled level '1'). Due to the specifically designed multiple LO-phonon resonances of the quantum stair, this process is extremely efficient and assumed to have a higher probability than the vertical one back to level 'g'.

Therefore, those electrons which perform the above two-step process with absorption in the main QW and LO-phonon-induced transfer into well '1' are very likely to further transit through the adjacent 'quantum stair'. In this case, they will finally end up at the ground level of the next active region period (g'), from where they will be excited again. This way, they travel through the entire structure and they gain a certain amount of energy for each traversed period. At this point, the probability of such a multiple excitation process will be estimated.

After each optical excitation of 800 meV, the electrons will lose some of their acquired energy by undergoing three subsequent relaxations. This energy loss corresponds to three GaN-based LO-phonon energies (3 × 92 meV = 276 meV), thus roughly one-third of the initial energy of 800 meV. Since these optical pumping and decaying processes take place simultaneously in each of the 12 active region periods, a successive climbing across the entire quantum stair — at least of a certain fraction of the entire electron population—will occur. Obviously, we note that 12 × (0.8 − 0.276) eV = 6.288 eV is just slightly above the AlN bandgap of 6.2 eV. As this climbing constitutes the exact opposite of the 'normal' quantum cascade current flow, we will—as a net effect—see a slight reduction in the injection current.

How are we going to prevent the electrons from making a 'too early' recombination—i.e., already after two or three instead of 12 periods? Simply by growing only one single p-doped layer: namely the final layer of the structure. Therefore, the electrons have to reach the last (thin) QW layer of the structure. From there, the 12th optical absorption lifts the electrons onto a sufficiently high level to undergo afterward the desired inter-band recombination in AlN.

Concerning the efficiency of such a device, the following reasoning applies: an optical input power of 1 W will first be diffracted and lead to an ideal diffraction efficiency of 50% (i.e., 500 mW). This amount of power will be available for inter-subband excitation and carrier transport. By further admitting that—by a specific design exploiting a LO-phonon resonance as presented in reference [23]—a 'two-step excitation/transfer' towards level '1' is (due to its shorter lifetime of τ_{e1} = 0.3 ps) about three times as probable as a vertical transition back to level 'g' with τ_{eg} = 0.8 ps [24], we estimate a power of 375 mW going towards the 'forward direction' while only 125 mW will go towards the 'backward direction'. This situation is equivalent to the one of a quantum cascade detector: While both the ('inefficient') vertical and the ('desired') diagonal excitation will be designed to have roughly the same probability of 50%, the diagonally excited electrons will be rapidly transferred by LO-phonon-assisted tunneling into the next well of the quantum stair. The carriers then swiftly reach the ground state of the following period, where they will be excited again. After 12 such excitations, we therefore obtain 1 W × (0.375 W/1 W)^{12} = 7.7 µW reaching the p-contact layer of the structure. With a roughly 80% conversion efficiency towards the deep-UV wavelength range, we estimate that a total power of 6 µW of deep-UV radiation at a wavelength as short as 200 nm will be generated in such a device. This value needs to be compared for instance with an AlN LED at λ = 210 nm emitting 0.02 mW (Taniyasu et al. [25]), an AlGaN LED at λ = 231 nm emitting 3.5 mW (Knauer et al. [26]), or an AlGaN LED at λ = 304 nm emitting 40 mW (Khan et al. [27]). From this short list, it becomes clear that going to an even shorter wavelength results in a rapidly decreasing output power.

Finally, as is quite generally the case for light-emitting diodes, the output will at first follow the torus-shaped radiation pattern typically seen in AlN (mainly horizontal radiation). However, it will then be efficiently diffracted towards a roughly vertical direction by the diffraction grating already used for the optical input.

4. Prospective Results

In order to measure this power with a sufficiently high signal-to-noise ratio, the incoming infrared radiation will be chopped at a frequency of 10 kHz. Although this could be done mechanically by a rotating chopper wheel, it is more efficient to perform this task electrically by directly applying a pulsed input signal between the top and the bottom contacts of the mesa. Between two electrical pulses, the conduction band structure is thus unbiased. During the electrical pulses, however, it looks biased like in Figure 3. Therefore, this structure changes periodically between its flat and its stair-like configuration. The resulting small difference in the injection current can be detected best using a sensitive lock-in technique. The latter typically allows the measurement of 1 % of the mean measurement values. In this case, this would be on the order of 1% × 6 µW = 60 nW.

Given the previously calculated 6 µW of deep-UV power for a near-IR input power of 1 W, a linear efficiency curve can be assumed in a first approximation. Figure 4 therefore presents the deep-UV output intensity as a function of near-IR optical input power.

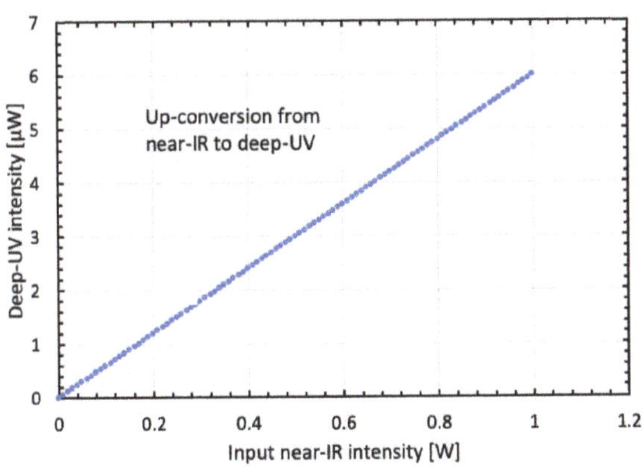

Figure 4. Calculated output intensity at an energy corresponding to the AlN bandgap of 6.2 eV (λ = 200 nm) as a function of optical illumination at λ = 1.55 µm. For the available range of input intensities, no pronounced saturation effects are to be expected.

Although the achievable output power of the proposed device with 6 μW seems to be relatively modest, one sees easily that this up-conversion has the potential for more interesting power levels. In addition, despite the fact that with AlN-based p-n-junction devices output powers approaching the 1 mW range have been achieved at a wavelength of 230 nm, going to an even shorter wavelength range will be fundamentally difficult. It is clear, however, that many interesting applications crucially rely on the additionally available photon energy corresponding to a wavelength of 200 nm instead of 230 nm. In any case, the device proposed here constitutes a useful intermediate approach until p-n-junctions in AlN are up to the point. Already at a modest device performance, more than 6 μW of deep-UV optical output power can be generated. In addition, the presented up-conversion approach is particularly interesting due to the fact that very powerful $\lambda = 1.55$ μm pump lasers are easily available. Finally, the optical excitation and relaxation scheme shown here is relatively simple and straightforward to implement.

5. Conclusions

In the present publication, we have outlined, conceived, and analyzed a new type of optical up-conversion device for the generation of deep-UV radiation. The proposed III-nitride semiconductor structure is based on carefully designed inter-subband transitions in a regular, chirped AlN/GaN-based superlattice. Using this structure, it will be possible to up-convert near-IR radiation into the deep-UV wavelength range. Although such multi-stage up-conversion processes are in general limited in their efficiency, the generated output power will in this case take advantage of the specific design. It uses LO-phonon-induced, mainly diagonal transfer of the excited charge carriers toward the ground state of the next period, thereby prolonging their lifetime quite considerably. In addition, the generated output power will obviously profit from readily available, watt-level optical input intensities in this important telecom wavelength range. Finally, thanks to better crystal growth, fabrication, and processing techniques—along with a higher level of characterization technology—considerable improvements are to be expected in the very near future.

Author Contributions: Conceptualization, D.H. and H.B.; methodology, D.H.; software, D.H. and H.B.; validation, D.P.B. and D.H.; formal analysis, D.H.; writing—original draft preparation, D.H.; writing—review and editing, D.H., H.B. and D.P.B. All authors have read and agreed to the published version of the manuscript.

Funding: This research received no external funding.

Data Availability Statement: The original data can be obtained from the first author of this article.

Acknowledgments: The authors would like to thank all of the involved colleagues and institutions—in particular, Lutz Kirste from the Institute of Applied Solid State Physics of the German Fraunhofer Gesellschaft and Cynthia Aku-Leh from ISCIENCES LLC. in Ann Arbor, MI, USA for their precious contributions to this work. We also greatly acknowledge scientific advice from John E. Epler from Lumileds, Inc., San Jose, CA, USA.

Conflicts of Interest: The authors declare no conflict of interest.

References

1. Ruhnke, N.; Müller, A.; Eppich, B.; Maiwald, M.; Sumpf, N.; Erbert, G.; Tränkle, G. Compact deep UV system at 222.5 nm based on frequency doubling of GaN laser diode emission. *IEEE Photonics Technol. Lett.* **2018**, *30*, 289–292. [CrossRef]
2. Den Engelsen, D.; Fern, G.R.; Ireland, T.G.; Harris, P.G.; Hobson, P.R.; Lipman, A.; Dhillon, R.; Marsh, P.J.; Silver, J. Ultraviolet and blue cathodoluminescence from cubic Y_2O_3 and Y_2O_3:Eu^{3+} generated in a transmission electron microscope. *J. Mater. Chem. C* **2016**, *4*, 7026. [CrossRef]
3. Song, S.B.; Yoon, S.; Kim, S.Y.; Yang, S.; Seo, S.-Y.; Cha, S.; Jeong, H.-W.; Watanabe, K.; Taniguchi, T.; Lee, G.-H.; et al. Deep-ultra-violet electroluminescence and photocurrent generation in graphene/hBN/graphene heterostructures. *Nat. Commun.* **2021**, *12*, 7134. [CrossRef]

4. Milleville, C.C.; Chen, E.Y.; Lennon, K.R.; Cleveland, J.M.; Kumar, A.; Zhang, J.; Bork, J.A.; Tessier, A.; LeBeau, J.M.; Chase, D.B.; et al. Engineering efficient photon upconversion in semiconductor heterostructures. *ACS Nano* **2019**, *13*, 489–497. [CrossRef] [PubMed]
5. Sun, T.; Chen, B.; Guo, Y.; Zhu, Q.; Zhao, J.; Li, Y.; Chen, X.; Wu, Y.; Gao, Y.; Jin, L.; et al. Ultralarge anti-Stokes lasing through tandem upconversion. *Nat. Commun.* **2022**, *13*, 1032. [CrossRef]
6. Motmaen, A.; Rostami, A.; Matloub, S. Ultra-high efficiency integrated mid infrared to visible up-conversion system. *Sci. Rep.* **2020**, *10*, 9325. [CrossRef] [PubMed]
7. Kumamoto, Y.; Fujita, K.; Smith, N.I.; Kawata, S. Deep-UV biological imaging by lanthanide ion molecular protection. *Biomed. Opt. Express* **2016**, *7*, 158–170. [CrossRef]
8. Mahboub, M.; Huang, Z.; Tang, M.L. Efficient infrared to visible upconversion with subsolar irradiance. *Nano Lett.* **2016**, *16*, 7169–7175. [CrossRef]
9. Kallepalli, D.L.N.; Alshehri, A.L.; Marquez, D.T.; Andrzejewski, L.; Scaiano, J.C.; Bhardwaj, R. Ultra-high density optical data storage in common transparent plastics. *Nat. Sci. Rep.* **2016**, *6*, 26163. [CrossRef]
10. Nakamura, S.; Fasol, G. *The Blue Laser Diode: GaN Based Light Emitters and Lasers*; Springer: Berlin/Heidelberg, Germany, 1997.
11. Akasaki, I.; Amano, H.; Nakamura, S. The Nobel Prize in Physics. 2014. Available online: https://www.nobelprize.org/prizes/physics/2014/press-release (accessed on 14 April 2015).
12. Murayama, M.; Nakayama, Y.; Yamazaki, K.; Hoshina, Y.; Watanabe, H.; Fuutagawa, N.; Kawanishi, H.; Uemura, T.; Narui, H. Watt-class green (530 nm) and blue (465 nm) laser diodes. *Phys. Status Solidi A* **2018**, *215*, 1700513. [CrossRef]
13. Hofstetter, D.; Baumann, E.; Giorgetta, F.R.; Théron, R.; Wu, H.; Schaff, W.J.; Dawlaty, J.; George, P.A.; Eastman, L.F.; Rana, F.; et al. Photodetector based on intersubband transitions using III-nitride superlattice structures. *J. Phys. Condens. Matter* **2009**, *21*, 174208. [CrossRef] [PubMed]
14. Giorgetta, F.R.; Baumann, E.; Guillot, F.; Monroy, E.; Hofstetter, D. High frequency (f = 2.37 GHz) room temperature operation of 1.55 μm AlN/GaN intersubband detector. *Electron. Lett.* **2007**, *43*, 185–187. [CrossRef]
15. Shur, M.S.; Gaska, R. Deep-ultraviolet light-emitting diodes. *IEEE Trans. Electron Devices* **2010**, *57*, 12–25. [CrossRef]
16. Song, K.; Mohseni, M.; Taghipour, F. Application of ultraviolet light-emitting diodes (UV-LEDs) for water disinfection: A review. *Water Res.* **2016**, *94*, 341–349. [CrossRef] [PubMed]
17. SaifAddin, B.K.; Almogbel, A.S.; Zollner, C.J.; Wu, F.; Bonef, B.; Iza, M.; Nakamura, S.; Speck, J.S. AlGaN deep-ultraviolet light-emitting diodes grown on SiC substrates. *ACS Photonics* **2020**, *7*, 554–561. [CrossRef]
18. Ichimura, I.; Maeda, F.; Osato, K.; Yamamoto, K.; Kasami, Y. Optical disk recording using a GaN blue-violet laser diode. *Jpn. J. Appl. Phys.* **2000**, *39*, 937. [CrossRef]
19. Wang, D.; Uesugi, K.; Xiao, S.; Norimatsu, K.; Miyake, H. High-quality AlN/sapphire templates prepared by thermal cycle annealing for high-performance ultraviolet light-emitting. *Appl. Phys. Express* **2021**, *14*, 035505. [CrossRef]
20. Ahmad, H.; Lindemuth, J.; Engel, Z.; Matthews, C.M.; McCrone, T.M.; Doolittle, W.A. Substantial p-type conductivity of AlN achieved via Beryllium doping. *Adv. Mater.* **2021**, *33*, 2104497. [CrossRef]
21. Shul, R.J.; McClellan, G.B.; Casalnuovo, S.A.; Rieger, D.J.; Pearton, S.J.; Constantine, C.; Barratt, C.; Karlicek, R.F., Jr.; Tran, C.; Schurman, M. Inductively coupled plasma etching of GaN. *Appl. Phys. Lett.* **1996**, *69*, 1119–1121. [CrossRef]
22. Evans, G.A.; Butler, J.K.; He, R.-H.; Kirk, J.B.; Yao, J.; Zheng, X.; Krishnamoorthy, A.V. Enhanced coupling strength gratings for couplers in III-V-compound dielectric waveguides. *IEEE J. Light. Technol.* **2017**, *35*, 2234–2242. [CrossRef]
23. Graf, M.; Scalari, G.; Hofstetter, D.; Faist, J.; Beere, H.; Linfield, E.; Ritchie, D.; Davies, G. Terahertz range quantum-well infrared photodetector. *Appl. Phys. Lett.* **2004**, *84*, 475–477. [CrossRef]
24. Heber, J.D.; Gmachl, C.; Ng, N.M.; Cho, A.Y. Comparative study of ultrafast intersubband electron scattering times at ~1.55 μm wavelength in GaN/AlGaN heterostructures. *Appl. Phys. Lett.* **2002**, *81*, 1237–1239. [CrossRef]
25. Taniyasu, Y.; Kasu, M.; Makimoto, T. An aluminium nitride light-emitting diode with a wavelength of 210 nanometres. *Nature* **2006**, *441*, 325–328. [CrossRef] [PubMed]
26. Knauer, A.; Kolbe, T.; Hagedorn, S.; Hoepfner, J.; Guttmann, M.; Cho, H.K.; Rass, J.; Ruschel, J.; Einfeldt, S.; Kneissl, M.; et al. Strain induced power enhancement of far-UVC LEDs on high temperature annelaed AlN templates. *Appl. Phys. Lett.* **2023**, *122*, 011102. [CrossRef]
27. Khan, M.A.; Noritoshi, M.; Joosun, Y.; Masafumi, J.; Yoichi, Y.; Hirayama, H. Achieving 9.6% efficiency in 304 nm p-AlGaN UVB LED via increasing the holes injection and light reflectance. *Sci. Rep.* **2022**, *12*, 2591. [CrossRef] [PubMed]

crystals

MDPI

Article

MOCVD of InGaN on $ScAlMgO_4$ on Al_2O_3 Substrates with Improved Surface Morphology and Crystallinity

Guangying Wang [1,*], Yuting Li [1], Jeremy Kirch [1], Yizhou Han [1], Jiahao Chen [1], Samuel Marks [2], Swarnav Mukhopadhyay [1], Rui Liu [2], Cheng Liu [1], Paul G. Evans [2] and Shubhra S. Pasayat [1]

1 Department of Electrical & Computer Engineering, University of Wisconsin–Madison, Madison, WI 53706, USA
2 Department of Materials Science and Engineering, University of Wisconsin–Madison, Madison, WI 53706, USA
* Correspondence: gwang265@wisc.edu

Abstract: $ScAlMgO_4$ (SAM) is a promising substrate material for group III-nitride semiconductors. SAM has a lower lattice mismatch with III-nitride materials compared to conventionally used sapphire (Al_2O_3) and silicon substrates. Bulk SAM substrate has the issues of high cost and lack of large area substrates. Utilizing solid-phase epitaxy to transform an amorphous SAM on a sapphire substrate into a crystalline form is a cost-efficient and scalable approach. Amorphous SAM layers were deposited on 0001-oriented Al_2O_3 by sputtering and crystallized by annealing at a temperature greater than 850 °C. Annealing under suboptimal annealing conditions results in a larger volume fraction of a competing spinel phase ($MgAl_2O_4$) exhibiting themselves as crystal facets on the subsequently grown InGaN layers during MOCVD growth. InGaN on SAM layers demonstrated both a higher intensity and emission redshift compared to the co-loaded InGaN on GaN on sapphire samples, providing a promising prospect for achieving efficient longer-wavelength emitters.

Keywords: MOCVD; epitaxy; III-nitrides; crystallization; InGaN

Citation: Wang, G.; Li, Y.; Kirch, J.; Han, Y.; Chen, J.; Marks, S.; Mukhopadhyay, S.; Liu, R.; Liu, C.; Evans, P.G.; et al. MOCVD of InGaN on $ScAlMgO_4$ on Al_2O_3 Substrates with Improved Surface Morphology and Crystallinity. *Crystals* **2023**, *13*, 446. https://doi.org/10.3390/cryst13030446

Academic Editors: Peng Chen and Zhizhong Chen

Received: 11 February 2023
Revised: 24 February 2023
Accepted: 2 March 2023
Published: 4 March 2023

1. Introduction

The group III-nitride material system has enabled numerous technologies including electronic and optoelectronic devices. InGaN-based LEDs can be used for emitters of different wavelengths by varying the indium content in the film. A higher indium concentration in the quantum well is required for longer emission wavelengths. An increasing indium content changes their bandgap from 0.7 eV to 3.4 eV, spanning the near IR to near UV range, respectively [1]. One of the most difficult aspects of the InGaN-based material synthesis is its lattice mismatch with the conventionally used GaN buffer layer, which is approximately 11% between GaN and InN [2–5]. Such a high lattice mismatch results in a high strain, causing dislocations in the crystalline structure and defect generation [1,2]. The use of a relaxed InGaN buffer layer reduces the lattice mismatch between the buffer layers and the active region of the emitters, enhancing the efficiency of the devices [6]. AlInGaP-based red emitters are conventionally used; however, size reduction results in a severe efficiency reduction [7–9], which is not a critical concern for nitride-based emitters [10–12]. Therefore, there is a critical need for a relaxed InGaN substrate technology. One of the methods to achieve relaxed InGaN is to grow an InGaN buffer layer on a suitable lattice-matched substrate such as ZnO (lattice-matched to $In_{0.22}Ga_{0.78}N$) [13–17] or $ScAlMgO_4$ (SAM) (lattice-matched to $In_{0.17}Ga_{0.83}N$) [18–20]. Using a modified flexible underlayer to achieve relaxed InGaN has been studied, such as the porosification of Si-doped GaN [21–24] and thermal decomposition of underlying high-composition InGaN [25,26]. These methods have successfully demonstrated relaxed InGaN buffer layers and achieved different extents of emission redshifts, however, they also had different constraints. Deposition on ZnO requires a very low deposition temperature (<600 °C), porous GaN based substrates require additional fabrication steps, and thermal decomposition of high composition InGaN layer results in an increased dislocation density [14,18,26]. These constraints motivated us to pursue the study of accessible SAM substrates for high-composition InGaN synthesis. SAM has a lattice mismatch of 0% with $In_{0.17}Ga_{0.83}N$ and ~2% with GaN, making it a suitable candidate for hetero-epitaxial growth of III-nitride materials [18–20]. The high substrate cost of bulk SAM substrates—USD8900 for 2 in diameter substrates—is quite limiting for academic research [27]. The solid-phase epitaxy (SPE) of amorphous SAM on sapphire followed by crystallization is a cost-efficient way to solve this issue. A problem associated with the creation of thin layers of SAM is that the crystallization of the SAM layer with unoptimized annealing conditions leads to the appearance of scandium-deficient regions on the surface, which lead to the formation of undesirable spinel-phase $MgAl_2O_4$ material [28]. The surface roughness of SPE-grown SAM is considerably higher than the other substrate technologies cited before due to its multi-crystallinity. In a previous study, Chen et al. demonstrated the synthesis of epitaxial SAM on a sapphire substrate with a preferred lower temperature of crystallization to constrain the formation of spinel-phase $MgAl_2O_4$ [28].

In this work, we present an improved crystallization method for SAM layers grown on a sapphire substrate and present Metal-Organic Chemical Vapor Deposition (MOCVD) growth of InGaN layers on crystalline SAM. The surface roughness and crystallinity were optimized by annealing SAM at different temperatures for various durations. The InGaN layers grown on SAM on sapphire demonstrated a higher photoluminescence (PL) intensity and a longer peak wavelength of emission compared to the same layers grown on a co-loaded reference GaN on sapphire template.

2. Materials and Methods

InGaN on SAM on sapphire was achieved following a three-step process. Initially, an amorphous SAM was sputtered onto a sapphire substrate, which was then subjected to crystallization through annealing. Finally, InGaN films were deposited on the crystalized SAM film on sapphire using MOCVD (Figure 1). The amorphous SAM layer was deposited using on-axis radio frequency (RF) magnetron sputtering with a chamber pressure of 2.4 Pa and RF power of 45 W applied on the substrate at room temperature. The crystallization of the SAM layer was accomplished by annealing in a pre-heated furnace in air at different temperatures (1123 K–1223 K) for different time ranges (20–30 h). After annealing, the samples were cooled outside the furnace in ambient air.

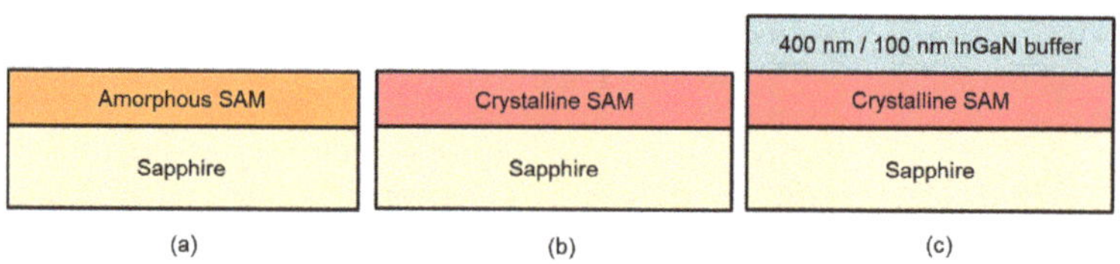

Figure 1. Cross-sectional schematics of (**a**) amorphous SAM deposited with RF magnetron sputtering, (**b**) crystallization of the SAM by annealing in a tube furnace, (**c**) subsequent MOCVD growth of InGaN.

In the first set of experiments, the effect of pre-annealing sapphire was studied. The pre-annealing step improved the surface morphology and atomic step uniformity of the sapphire substrate [29,30]. Moreover, high-temperature treatment will result in a uniform oxygen-terminated surface that can improve the deposition of the SAM [31]. Pre-annealing of the sapphire substrate was performed at 1673 K for 10 h. A 40 nm thick amorphous SAM layer was deposited on the pre-annealed sapphire and unannealed sapphire. Both samples were then annealed for 10 h at 1173 K followed by 15 h at 1223 K for the layer crystallization. Further series of crystallization experiments were performed. The first was a time series where three samples with nominally the same amorphous SAM layer thickness of 75 nm were deposited on sapphire substrates. These samples were annealed at 1223 K for 20 h, 25 h, and 30 h. In the second set of crystallization experiments, three different thicknesses of the SAM layers were studied: 25 nm, 75 nm, and 90 nm. All the samples in the second series were annealed at 1223 K for 25 h. In the third series, the temperature of one of the two annealing steps was varied as follows. A thickness of 40 nm was chosen for two samples that were annealed for a total of 25 h at different temperatures for the first 10 h and 1223 K for the remaining 15 h. The second annealing step of 1223 K for 15 h was designed to ensure crystallization, and by varying the temperature of the first 10 h of annealing, the volumetric ratio of spinel to SAM phases was compared to determine the ideal crystallization annealing condition.

After SAM crystallization, InGaN layers were grown using MOCVD with triethylgallium (TEGa), trimethylindium (TMI), and ammonia (NH_3). The 100 nm and 400 nm thick InGaN layers were grown with a calibrated composition of x_{In} = 0.14 on both SAM on sapphire as well as a reference GaN template on sapphire. These layers were grown with TMI and TEGa flows of 2.6 and 4.9 μmol min^{-1} at 1138K, respectively, with an NH_3 flow of 89 mmol/min.

An atomic force microscopy (AFM) scan was performed to compare the surface morphology of the layers using a Bruker Icon AFM tool. The roughnesses of the films were evaluated and compared through the root-mean-square (RMS) Ra parameter. The thickness of the SAM layer was measured with X-ray reflectivity (XRR) using the Panalytical Empyrean X-ray diffractometer. The same equipment was utilized for $\omega-2\theta$ and ($\omega-2\theta$)—ω reciprocal space map (RSM) scans to evaluate the film crystallinity as well as the layer composition. SAM- and spinel-phase $MgAl_2O_4$ peaks were identified with the indexes from previously reported structures [32,33]. The quality of the crystallized SAM layer was evaluated by comparing the spinel (311) to SAM (0009) peak-to-background intensity ratio, the SAM (0009) peak's full width of half maximum range (FWHM), and the spinel (311) peak's FWHM from the $\omega-2\theta$ scans. Transmission electron microscope (TEM) studies are currently underway to investigate crystalline quality. Due to the thickness restriction on acquiring high-quality RSM on thin SAM films, RSM scans were performed on the reference InGaN on GaN template on sapphire to determine the calibrated indium content in 100 nm and 400 nm thick InGaN layers. The approximate indium content on different substrates was quantified with photoluminescence (PL) spectroscopy using a Horiba LabRAM HR Evolution Raman Spectrometer, using a blue 405 nm laser with an acquisition time of 0.5 s and 125 W/cm^2 power. The systematic variation of the experimental parameters is summarized in Table 1. Sputtered amorphous SAM film on Si along with a bulk SAM substrate were externally characterized using X-ray fluorescence (XRF). XRF was performed using a Rigaku Primus II WDXRF X-ray source with a rhodium X-ray tube in vacuum conditions across a 10 mm diameter measurement area.

Table 1. Phase content, $\omega-2\theta$ FWHM, and roughness for crystallized SAM layers. The annealing temperature for part one (T1) and annealing time for part two (t2) were varied in different experiments. The time of annealing part one was 10 h (t1) and the temperature for annealing part two was 1223 K (T2).

Sample	Pre-Annealing	Thickness (nm)	T1 (K)	t2 (h)	Spinel-to-SAM Ratio	SAM FWHM (Arc-Seconds)	Spinel FWHM (Arc-Seconds)	RMS Roughness (nm) in 0.5 μm × 0.5 μm Scan	In Content
A	Yes	40	1173	15	0.220	1307	901	1.57	
B	No	40	1173	15	Semi-crystalized				
C	Yes	25	1223	15	0.203	1512	1188	1.25	
D	Yes	75	1223	15	0.521	1033	900	1.75	
E	Yes	90	1223	15	0.408	1008	936	2.3	
F	Yes	75	1223	10	0.453	720	959	1.41	20.8%
G	Yes	75	1223	15	0.446	953	991	1.7	20.8%
H	Yes	75	1223	20	0.177	953	929	2	25.9%
I	Yes	40	1123	15	0.263	1059	994	1.48	20.9%

3. Results and Discussions

3.1. XRF Results

The results of the film compositions (Wt%) and atomic percentages (At%) of sputtered amorphous SAM film on Si and bulk SAM substrate are summarized in Table 2. These results indicate that the stoichiometry of both the sputtered SAM film and the SAM substrate were fairly comparable, pointing to the possibility of using a crystalline thin SAM film as a substitute for SAM substrate.

Table 2. Stoichiometry of amorphous SAM film on Si and SAM Substrate.

	Amorphous SAM on Si Wt%	Amorphous SAM on Si At%	Bulk SAM Substrate Wt%	Bulk SAM Substrate At%
Sc	26.8	13.6	28.2	14.4
Al	17.6	14.8	16.3	13.8
Mg	15.5	14.5	15.7	14.8
O	40.1	57.1	39.8	57.0

3.2. Effect of Pre-Annealing

The pre-annealing experiment was performed to improve the sapphire substrate morphology and study its effect on the crystallization of sputtered SAM.

Sample A and B (Table 1) have the same thickness (40 nm) with the same deposition and annealing condition (1173 K for 10 h and 1223 K for 15 h). Sample A used pre-annealed sapphire substrate and Sample B used unannealed sapphire substrate.

Sample A had a SAM reflection with a high peak-to-background ratio in the thin-layer XRD measurements (Figure 2a). Sample B remained amorphous or partially crystallized (Figure 2b). The surface roughness of A increased from 1.16 nm to 1.57 nm after crystallization, as shown in the (500 nm)^2 size AFM scans (Figure 2d,c). Sample B showed an RMS roughness of 1 nm (Figure 2e) after annealing, which was similar to the RMS roughness of amorphous Sample A, 1.16 nm (Figure 2d). The crystalized samples tended to have a higher roughness than amorphous layers. As mentioned before, the pre-annealing step improved the surface morphology, and the oxygen-terminated surface aided the formation of crystalline SAM film above the sapphire substrate [29–31]. An improved substrate morphology of sapphire resulted in an efficient crystallization of Sample A compared to Sample B.

3.3. Effect of SAM Thickness

Samples C, D, and E were deposited with thicknesses of 25 nm, 75 nm, and 90 nm, respectively, on a pre-annealed sapphire substrate. The crystallization for all three samples was conducted at 1223 K for 25 h. The results are compared using XRD and AFM (Table 1). As the thickness varied from 25 nm to 90 nm, the RMS roughness changed from 1.25 nm to 2.3 nm. The spinel-to-SAM ratio also increased from 0.203 to 0.408. The reason for this behavior could be that the spinel was crystallizing faster than SAM for a given thickness, and the higher the thickness of the layer, the longer the crystallization duration [34]. For thinner 25 nm SAM layers (Sample C), the spinel and SAM phase both crystallized for the annealing duration of 25 h, whereas for the 90 nm SAM layer (Sample E), the spinel phase crystallized; however, the SAM phase did not crystallize efficiently. The high spinel-to-SAM volumetric ratio resulted in the formation of facets, which increased the surface roughness, degrading the morphology. Increasing the layer thickness from 25 nm to 75 nm resulted in a decrease in the FWHM of the $\omega-2\theta$ of the SAM 0009 reflection from 1188 to 900 arcseconds, which then nominally remained the same from 75 nm to 90 nm thick SAM layers. This trend could be related to the interfacial reaction between SAM and sapphire during the annealing process resulting in a lower quality of the SAM layer near the interface [28]. With the increasing thickness of the SAM layer on sapphire, the contribution of this poor-quality interfacial layer on the peak intensity of the SAM layer in the XRD scan was reduced, resulting in a lower FWHM of the SAM layer.

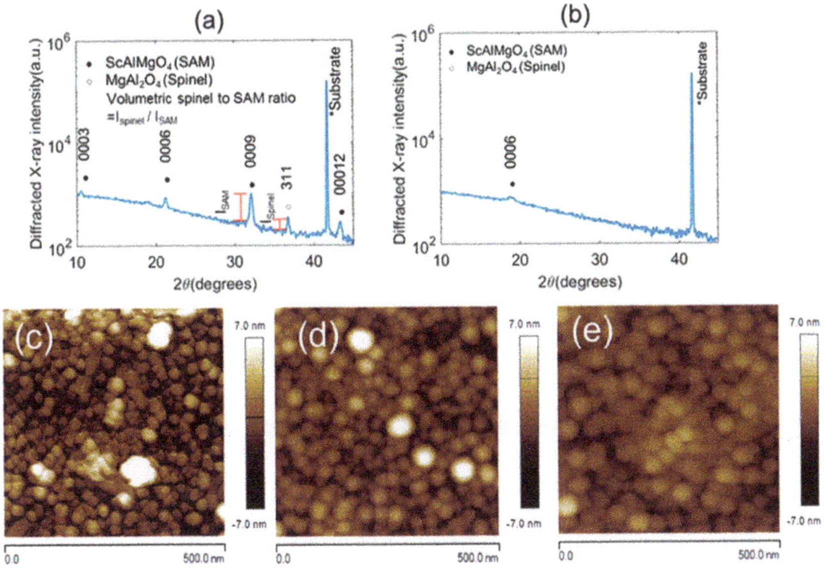

Figure 2. X-ray diffraction ω–2θ scans of annealed SAM layer on (**a**) pre-annealed sapphire (Sample A) and (**b**) unannealed sapphire (Sample B). AFM images of (**c**) crystallized Sample A, with RMS roughness of 1.57 nm, (**d**) amorphous Sample A, with RMS roughness of 1.16 nm, and (**e**) Sample B (not fully crystallized), with RMS roughness of 1 nm.

3.4. Effect of Annealing Time

The effect of annealing at a fixed temperature (1123 K) for different durations was studied using 75 nm thick crystallized SAM on sapphire samples. Samples F, G, and H were annealed with durations of 20 h, 25 h, and 30 h, respectively. The RMS roughness increased from 1.41 nm to 2 nm with increasing annealing duration. The spinel-to-SAM volumetric ratio decreased from 0.453 to 0.177 with increasing duration. Although the spinel-to-SAM ratio decreased, the results are difficult to interpret due to the decrease in the overall intensity of both spinel (311) and SAM (0009) peaks, the reason for which is not well understood at this point. With a longer annealing duration, we speculate that the SAM crystallizes more effectively compared to a shorter annealing time due to a lower crystallization rate. The spinel-phase crystallization rate was faster than the SAM phase and the spinel X-ray reflections appeared with a relatively higher intensity than SAM when the sample was annealed for a shorter duration. Subsequent annealing steps resulted in the formation of more SAM-phase material, enhancing its relative intensity [34,35].

3.5. Effect of Annealing Temperature

A 25 h total annealing duration was chosen to proceed with the annealing temperature experiment. A high SAM thickness may increase the roughness of the film; however, an insufficient SAM thickness might fail to serve as a lattice-matched layer for the subsequently grown InGaN films. Thus, an intermediate-thickness 40 nm SAM layer was chosen in the following series, as surface roughness reduction was one of the main goals of this study. The temperature series was divided into a two-step annealing experiment for Samples A and I. The annealing temperature of the first 10 h was 1123 K (Sample A) and 1173 K (Sample I). In both cases, the first annealing step was followed by 15 h at 1223 K to ensure crystallization. Samples A and I were both pre-annealed and used the same thickness of SAM layer. After crystallization, Samples A and I had similar RMS roughness. Sample I had a slightly higher volumetric spinel-to-SAM ratio than Sample A, because the spinel phase may have formed at a faster rate compared to that of SAM at 1123 K (Table 1). The

SAM may require a higher annealing temperature to crystallize sooner; however, beyond an optimized temperature, Sc-deficient regions may start to appear. More data points are required to determine a clear trend for the annealing temperature. The optimized crystallization temperature for SAM will be investigated in future work.

3.6. *MOCVD Growth of Thick InGaN Layers*

Samples F and H were placed in the MOCVD reactor to deposit a 100 nm thick InGaN layer. In addition, Samples G and I were utilized to deposit a 400 nm thick InGaN layer (Table 1). In both these experiments, reference GaN on sapphire templates were co-loaded to determine the indium content and InGaN relaxation extent on these samples with RSM (−1 −1 4) reflection scans. The SAM layers had low thickness compared with the sapphire substrate and the InGaN layer, resulting in a relatively low intensity in the XRD RSM scans. SAM was required as a reference for the lattice constant to measure the InGaN lattice constant and its relative degree of relaxation. However, the low intensity of SAM introduced challenges in obtaining high-resolution RSM measurements on InGaN on SAM on sapphire samples. RSM scans were performed on InGaN on GaN on sapphires samples and PL was performed on both samples to provide a quantitative comparison of the indium concentration between the InGaN on GaN template and the InGaN on SAM on sapphire. The grown structure included a targeted 3–5 nm of GaN interlayer to smooth the morphology, as reported in reference [36]. The corresponding ω–2θ diffraction patterns of the SAM Sample H before and after the MOCVD growth are shown in Figure 3a,b. The growth of the 100 nm InGaN layer on GaN on sapphire indicated the indium content of 0.145 with a degree of relaxation of 8.7%. The 400 nm thick layer showed two InGaN peaks: one corresponding to 14.2% indium content with a degree of relaxation of 15% and the second one with 14.9% indium content and a 67% degree of relaxation (Figure 3c,d). With a 400 nm InGaN thickness, the layer had exceeded its critical thickness on GaN on sapphire; hence, the second peak appeared with higher relaxation [37]. Both peaks of 400 nm InGaN had a higher degree of relaxation compared to the 100 nm InGaN layer owing to higher strain in the structure [38,39]. Comparing the surface morphology using RMS roughness for Samples F and H from Table 1, prior to the InGaN deposition, Sample H showed a significant reduction in the spinel-phase facets although with a slightly higher surface roughness, the reason for which is not well understood at this point (Table 1). However, Sample H demonstrated a lower spinel-to-SAM volumetric ratio of ~0.18 compared to ~0.45 for Sample F, which might be the reason for fewer formations of facets after the InGaN growth (Figure 4a,b). For the second MOCVD growth of the thicker 400 nm InGaN layer, prior to the MOCVD deposition, Samples G and I had a spinel-to-SAM volumetric ratio of ~0.45 and ~0.22, respectively. After the InGaN deposition, the AFM scan showed similar roughness for Samples G and I; however, Sample I showed a smaller grain size than Sample G, which we speculate may be due to the lower spinel-to-SAM volumetric ratio (Figure 4c,d). This suggests that further lowering of the spinel-to-SAM volumetric ratio is necessary to suppress the spinel facets in SAM and improve the homogeneity in the subsequently grown InGaN layer.

The InGaN layer grown on SAM and GaN showed different PL wavelengths of emission (Figure 5a,b) suggesting different degrees of strain in the structure and different indium incorporation. Based on the strain information achieved from the RSM scans and PL emission wavelength, the indium compositions of InGaN on GaN on sapphire samples were estimated and compared to the indium composition measured in RSM scans for validation. Equations (1) and (2) were utilized [40,41], where E_g is the bandgap energy, R is the degree of relaxation, and x is the indium composition. The PL peak emission wavelength was converted to bandgap energy E_g using Equation (2) [41], where λ is the wavelength and $h\vartheta$ is the photon energy, interpreted here as the bandgap energy of the InGaN layer. With this method, 100 nm InGaN on GaN on sapphire with a relaxation of 8% (from RSM) luminescing at 432 nm was determined to have an indium content of 14.5%, which is the same as the indium concentration measured with the XRD RSM scan shown in

Figure 3c. However, the RSM scan of 400 nm InGaN on GaN on sapphire layer (Figure 3d), as discussed above, demonstrated two separate indium peaks, thereby making the indium content estimation challenging. Due to a ~10% larger lattice constant of InN compared to GaN [2], when thicker InGaN films are grown on GaN, the layer exceeds its critical thickness resulting in strain-relaxed films. The appearance of two separate peaks may be an indication of unintentionally relaxation-graded films, with a higher degree of relaxation for layers closer to the film surface, with RSM scans showing an averaged indium content and degree of relaxation across a range of thickness. Assuming the layers at the surface of the 400 nm thick InGaN on GaN on sapphire sample were 100 % relaxed, a luminescence at 475 nm as depicted in Figure 5b should have ~16 % indium content, using Equations (1) and (2), which is nominally similar to that observed from the two InGaN peaks in the Figure 3d RSM scan. As the PL characterization was performed on the top surface of the wafer, consideration of full relaxation of the films close to the surface to determine the indium content accurately can be justified.

$$E_g(R, x) = 3.42 + (0.69R + 3.42)x^2 - (1.57R + 4.07)x. \tag{1}$$

$$\lambda(\mu m) = \frac{1.2398}{h\vartheta(eV)} \tag{2}$$

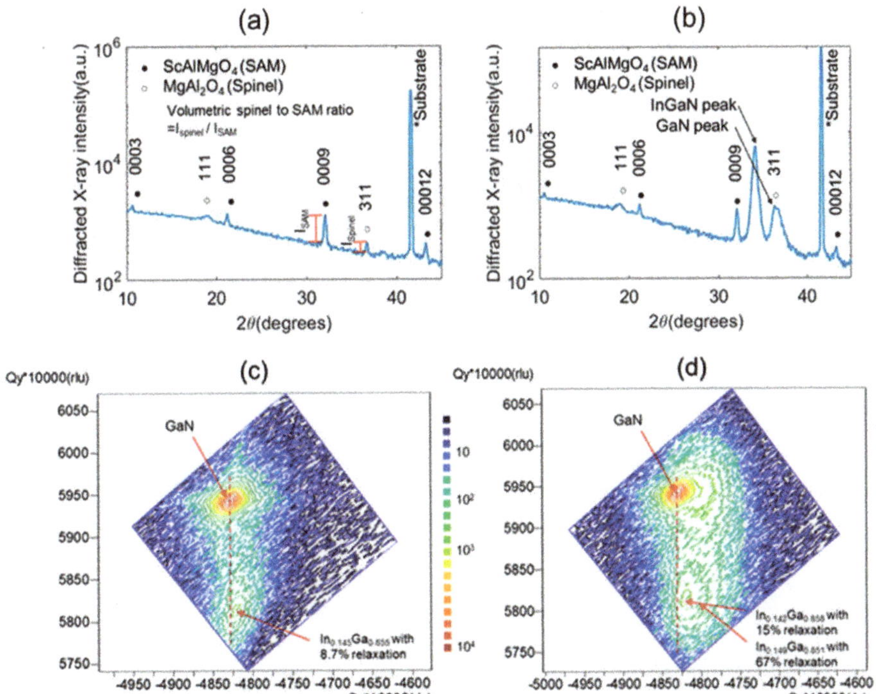

Figure 3. → XRD ω–2θ of (**a**) Sample H, (**b**) Sample H with 100 nm InGaN on 75 nm SAM layer on Al_2O_3; RSM (−1 −1 4) of (**c**) 100 nm InGaN on GaN template; RSM (−1 −1 4) of (**d**) 400 nm InGaN on GaN template.

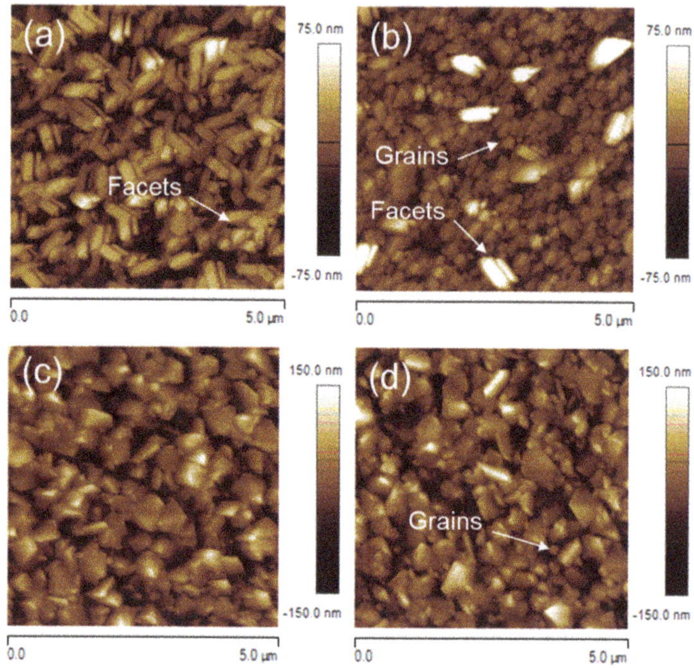

Figure 4. AFM size 5 µm × 5 µm scan of (**a**) Sample F (100 nm InGaN layer), RMS roughness: 18.3 nm, (**b**) Sample H (100 nm InGaN layer), RMS roughness: 14.5 nm, (**c**) Sample G (400 nm InGaN layer), RMS roughness: 30.8 nm, (**d**) Sample I (400 nm) InGaN layer, RMS roughness: 31.3 nm.

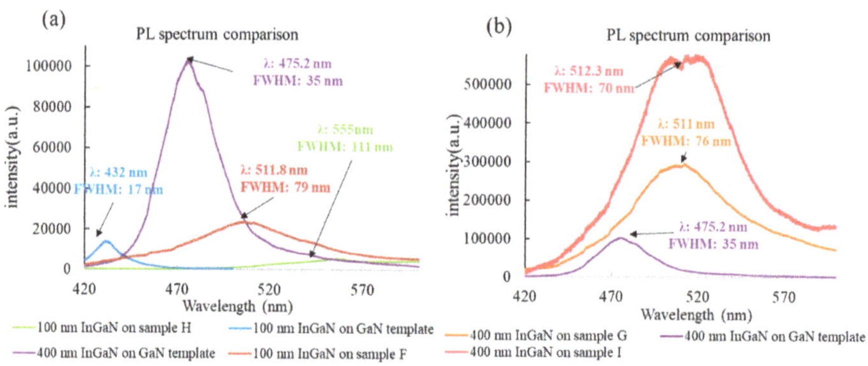

Figure 5. PL spectrum of (**a**) 100 nm InGaN on SAM and GaN template with an intensity range of 0 to ~100,000. (**b**) 400 nm InGaN on SAM and GaN template with an intensity range of 0 to ~600,000 (400 nm InGaN on GaN template was the common spectrum for both plots).

For the InGaN on SAM on sapphire, as mentioned above, without a GaN buffer layer, utilization of a thin crystalline SAM layer posed challenges in acquiring the exact film composition and degree of relaxation with XRD RSM scans. Therefore, some assumptions were made for the degree of relaxation parameter to estimate the indium content using Equations (1) and (2) while using the PL wavelength as the only measured entity. For the InGaN on SAM on sapphire samples, applying 8.7% relaxation to Equation (1) to calculate the indium content of the 100 nm thick InGaN films for Samples F and H, the results were 32.7% and 46.1%, which are impractical. In addition, SAM films are lattice-matched to

$In_{0.17}Ga_{0.83}N$ [18–20]; therefore, these films can be considered as grown 100 % relaxed. Therefore, the indium composition of Samples F and H were recalculated considering full relaxation (R = 1 in Equation (1)), resulting in an indium composition of 20.8% for Sample F and 25.9% for Sample H. For the 400 nm thick InGaN on SAM on sapphire samples, Sample G and Sample I, the same assumption of 100% relaxation yielded indium concentrations of 20.8% and 20.9%, respectively.

The PL results of InGaN film on SAM on sapphire and the reference GaN on sapphire samples are summarized in Table 3. With a 100 nm InGaN layer grown on GaN on sapphire, a peak wavelength of 432 nm with a PL intensity of around 11,000 arbitrary units (a.u.) was measured, which will be considered as the baseline for comparison of the PL intensity of 100 nm InGaN samples grown on SAM. An increased thickness (400 nm) of the InGaN layer on GaN on sapphire depicted higher indium incorporation, thereby emitting a longer wavelength of 475 nm compared to 432 nm, with an improvement in the intensity (100,000 a.u) of almost ten times. The thick (400 nm) InGaN on GaN on sapphire was the second reference sample for the comparison with thick (400 nm) InGaN grown on SAM. Thick (400 nm) InGaN layer grown on 40 nm SAM (Sample I) produced a two times higher PL intensity, a 1.3 nm redshift in the peak emission wavelength, and a 6 nm lower FWHM compared to Sample G (75 nm of SAM), as shown in Figure 5b. This improvement in the PL emission characteristics could be due to the lower volumetric spinel-to-SAM ratio of Sample I (0.263) compared to Sample G (0.446). Although these two samples showed similar surface roughness measured with AFM, more small-sized grains were formed in Sample I whereas Sample G had larger facets, which might be the reason for the improved crystalline quality and lower FWHM of Sample I compared to Sample G (Figure 4c,d).

Table 3. PL results of InGaN film on SAM on sapphire substrates along with reference InGaN on GaN on sapphire samples.

Sample Name	SAM Film Thickness (nm)	InGaN Film Thickness (nm)	Spinel-to-SAM Ratio	Peak Wavelength (nm)	PL FWHM (nm)
F	75	100	0.453	512	79
H	75	100	0.177	555	111
100 nm InGaN on GaN on sapphire	-	100	-	432	17
G	75	400	0.446	511	76
I	40	400	0.263	512	70
400 nm InGaN on GaN on sapphire	-	400	-	475	35

Although the PL emission of 400 nm InGaN on SAM on sapphire samples demonstrated a higher peak emission intensity, 100 nm InGaN grown on SAM showed a longer redshift, which is the intended target. The 100 nm InGaN layers on SAM on sapphire (Sample H) demonstrated the longest wavelength among all the samples with an indium composition of ~26%, but it also demonstrated a two times lower PL intensity compared with reference InGaN on GaN on sapphire (Figure 5a). Sample H was the only sample that demonstrated a lower intensity among all InGaN on SAM on sapphire samples compared with InGaN on GaN template. Further investigations are needed to understand this behavior. Comparing the 100 nm InGaN layers grown on SAM samples, Samples F (75 nm) and H (75 nm) emitted at different wavelengths with different intensities. While Sample F demonstrated a lower redshift of ~80 nm, a higher intensity of emission by approximately two times was observed, even with a poorer surface morphology. Owing to the higher indium incorporation, a drop in the emission intensity by half is still promising, considering the ~120 nm (555 nm compared to 432 nm) redshift for InGaN on Sample H compared to the GaN on sapphire. A more controlled experiment on Sample H needs to be performed to emit in the same range as Sample F to draw further conclusions about the lower emission intensity. The key difference between F and H was that Sample H had a lower spinel-to-SAM ratio and less formation of spinel facets (Figure 4a,b). The SAM samples with improved homogeneity result in an overall smoother surface morphology, critical for the subsequently grown InGaN layers. This PL result further strengthens our assertion that a lower spinel-to-SAM volumetric ratio is important to achieve high-quality InGaN layers on the optimized SAM on sapphire substrates that we have developed, but in this specific case it resulted in a higher indium content, thereby making the quality comparison inconclusive.

The FWHM of the PL for the InGaN on SAM layers was higher than that for the InGaN on GaN templates. This might be caused by the high surface roughness of the underlying SAM layers, resulting in a stronger indium surface segregation, which leads to a non-uniform indium incorporation [42]. The FWHM of the PL for the InGaN on SAM layers on sapphire obtained in our experiments were comparable to the InGaN layers grown on ZnO substrates [43]. Both thicknesses of the InGaN layers on the SAM on sapphire samples demonstrated higher intensities (two times higher for 100nm InGaN on Sample F and six times higher for 400 nm InGaN on Sample I) and longer emission wavelengths (around 123 nm for 100 nm InGaN and around 37 nm for 400 nm InGaN) than those grown on GaN on sapphire templates. Although the high surface roughness of the InGaN on SAM on sapphire samples may contribute to the increase in the PL intensity, it still cannot account for the high level of intensity enhancement observed in our experiments. Prior reports have shown that roughening of the surface on the nanometer scale has conventionally either degraded the PL intensity or led to minor enhancements [44,45]. However, when the surface roughness increased from the nanometer scale to the micrometer scale, the PL intensity was enhanced by almost twice. In our experiment, compared to InGaN on GaN on sapphire, the InGaN on SAM on sapphire samples had a surface roughness degradation on the scale of tens of nanometers, changing from ~30 nm to ~10 nm, respectively, for a 5 μm × 5 μm AFM scan. A two to six times higher PL intensity observed in this work, therefore, cannot be accounted for by this nanometer-scale change in surface roughness. Therefore, we speculate that the high InGaN material quality may be the reason for the higher PL intensity observed for the InGaN on SAM on sapphire samples.

This is one of the highest reported redshifts for any relaxed InGaN pseudo-substrate technology so far and one of the few wherein the redshift is accompanied by an increased emission intensity [21]. Further investigations will be pursued to achieve device results on these substrates. These results demonstrate the potential of SAM on sapphire substrates as a viable candidate to achieve thick, high-indium-concentration InGaN layers for long-wavelength emitters.

4. Conclusions

In this work, further optimization of SAM crystallization was investigated to suppress the spinel phase to obtain an improved surface morphology. Compared to the previous work [28], both the surface roughness and the volumetric ratio of the spinel to SAM phase have improved. From the thick InGaN layer growth, we identified that the spinel-to-SAM ratio was one of the key parameters to suppress spinel-phase facets that transfer onto the MOCVD-grown layer surface, increasing the overall surface roughness value. The higher PL intensity along with the redshift demonstration of the InGaN layers on SAM demonstrate the potential of sputtered and crystallized SAM as a promising substrate for longer-wavelength III-nitride emitters. The growth condition for the InGaN layers was not extensively optimized in this work. In the future, we will further investigate the optimized annealing temperature for SAM crystallization and design better growth conditions for the epitaxial growth of InGaN on SAM to grow the light-emitter structures.

Author Contributions: Investigation, G.W., S.S.P. and P.G.E.; Formal analysis, G.W. and Y.L.; Data curation, G.W., Y.L., J.K., Y.H. and J.C.; Methodology, G.W., Y.L., J.K., S.M. (Samuel Marks), R.L., C.L., S.S.P. and P.G.E.; Visualization, G.W.; Writing—original draft, G.W.; Writing—review and editing, G.W., S.M. (Swarnav Mukhopadhyay), S.S.P. and P.G.E.; Project administration, S.S.P. and P.G.E.;

Supervision, S.S.P. and P.G.E.; Validation, S.S.P. and P.G.E.; Funding acquisition, S.S.P. All authors have read and agreed to the published version of the manuscript.

Funding: This work was financially supported by the National Science Foundation Materials Research Science and Engineering Center (DMR-1720415).

Institutional Review Board Statement: Not applicable.

Informed Consent Statement: Not applicable.

Data Availability Statement: The data that support the findings of this study are available from the corresponding author upon reasonable request.

Acknowledgments: We would like to thank Brandon Erickson for helping with manuscript proofreading. The authors gratefully acknowledge the use of facilities and instrumentation supported by NSF through the University of Wisconsin Materials Research Science and Engineering Center (DMR-1720415).

Conflicts of Interest: The authors declare that they have no conflict of interest.

References

1. Kour, R.; Arya, S.; Verma, S.; Singh, A.; Mahajan, P.; Khosla, A. Review—Recent Advances and Challenges in Indium Gallium Nitride ($In_x Ga_{1-x} N$) Materials for Solid State Lighting. *ECS J. Solid State Sci. Technol.* **2020**, *9*, 015011. [CrossRef]
2. Rinke, P.; Scheffler, M.; Qteish, A.; Winkelnkemper, M.; Bimberg, D.; Neugebauer, J. Band Gap and Band Parameters of InN and GaN from Quasiparticle Energy Calculations Based on Exact-Exchange Density-Functional Theory. *Appl. Phys. Lett.* **2006**, *89*, 161919. [CrossRef]
3. Yamaguchi, S.; Kariya, M.; Nitta, S.; Takeuchi, T.; Wetzel, C.; Amano, H.; Akasaki, I. Structural Properties of InN on GaN Grown by Metalorganic Vapor-Phase Epitaxy. *J. Appl. Phys.* **1999**, *85*, 7682–7688. [CrossRef]
4. Kim, Y.H.; Park, H.J.; Kim, K.; Kim, C.S.; Yun, W.S.; Lee, J.W.; Kim, M.D. Strain Distribution and Interface Modulation of Highly Lattice-Mismatched InN/GaN Heterostructure Nanowires. *Appl. Phys. Lett.* **2009**, *95*, 033112. [CrossRef]
5. Matsuoka, T. Progress in Nitride Semiconductors from GaN to InN—MOVPE Growth and Characteristics. *Superlattices Microstruct.* **2005**, *37*, 19–32. [CrossRef]
6. Tan, A.K.; Hamzah, N.A.; Ahmad, M.A.; Ng, S.S.; Hassan, Z. Recent Advances and Challenges in the MOCVD Growth of Indium Gallium Nitride: A Brief Review. *Mater. Sci. Semicond. Process.* **2022**, *143*, 106545. [CrossRef]
7. Fan, K.; Tao, J.; Zhao, Y.; Li, P.; Sun, W.; Zhu, L.; Lv, J.; Qin, Y.; Wang, Q.; Liang, J.; et al. Size Effects of AlGaInP Red Vertical Micro-LEDs on Silicon Substrate. *Results Phys.* **2022**, *36*, 105449. [CrossRef]
8. Wong, M.S.; Kearns, J.A.; Lee, C.; Smith, J.M.; Lynsky, C.; Lheureux, G.; Choi, H.; Kim, J.; Kim, C.; Nakamura, S.; et al. Improved Performance of AlGaInP Red Micro-Light-Emitting Diodes with Sidewall Treatments. *Opt. Express* **2020**, *28*, 5787–5793. [CrossRef]
9. Oh, J.-T.; Lee, S.-Y.; Moon, Y.-T.; Moon, J.H.; Park, S.; Hong, K.Y.; Song, K.Y.; Oh, C.; Shim, J.-I.; Jeong, H.-H.; et al. Light Output Performance of Red AlGaInP-Based Light Emitting Diodes with Different Chip Geometries and Structures. *Opt. Express* **2018**, *26*, 11194–11200. [CrossRef]
10. Wasisto, H.S.; Prades, J.D.; Gülink, J.; Waag, A. Beyond Solid-State Lighting: Miniaturization, Hybrid Integration, and Applications of GaN Nano- and Micro-LEDs. *Appl. Phys. Rev.* **2019**, *6*, 041315. [CrossRef]
11. Horng, R.-H.; Ye, C.-X.; Chen, P.-W.; Iida, D.; Ohkawa, K.; Wu, Y.-R.; Wuu, D.-S. Study on the Effect of Size on InGaN Red Micro-LEDs. *Sci. Rep.* **2022**, *12*, 1324. [CrossRef] [PubMed]
12. Huang, Y.-M.; Peng, C.-Y.; Miao, W.-C.; Chiang, H.; Lee, T.-Y.; Chang, Y.-H.; Singh, K.J.; Iida, Z.D.; Horng, R.-H.; Chow, C.-W.; et al. High-Efficiency InGaN Red Micro-LEDs for Visible Light Communication. *Photon. Res.* **2022**, *10*, 1978–1986. [CrossRef]
13. Matsuoka, T.; Yoshimoto, N.; Sasaki, T.; Katsui, A. Wide-Gap Semiconductor InGaN and InGaAln Grown by MOVPE. *J. Electron. Mater.* **1992**, *21*, 157–163. [CrossRef]
14. Wang, S.-J.; Li, N.; Park, E.-H.; Feng, Z.C.; Valencia, A.; Nause, J.; Kane, M.; Summers, C.; Ferguson, I. MOCVD Growth of GaN-Based Materials on ZnO Substrates. *Phys. Status Solidi C* **2008**, *5*, 1736–1739. [CrossRef]
15. Lei, Y.; Xu, J.; Zhu, K.; He, M.; Zhou, J.; Gao, Y.; Zhang, L.; Chen, Y. A GaN-Based LED With Perpendicular Structure Fabricated on a ZnO Substrate by MOCVD. *J. Display Technol.* **2013**, *9*, 377–381. [CrossRef]
16. Wang, S.-J.; Li, N.; Park, E.-H.; Lien, S.-C.; Feng, Z.C.; Valencia, A.; Nause, J.; Ferguson, I. Metalorganic Chemical Vapor Deposition of InGaN Layers on ZnO Substrates. *J. Appl. Phys.* **2007**, *102*, 106105. [CrossRef]
17. Li, N.; Wang, S.-J.; Huang, C.-L.; Feng, Z.C.; Valencia, A.; Nause, J.; Summers, C.; Ferguson, I. Effect of an Al_2O_3 Transition Layer on InGaN on ZnO Substrates by Organometallic Vapor-Phase Epitaxy. *J. Cryst. Growth* **2008**, *310*, 4908–4912. [CrossRef]
18. Ozaki, T.; Takagi, Y.; Nishinaka, J.; Funato, M.; Kawakami, Y. Metalorganic Vapor Phase Epitaxy of GaN and Lattice-Matched InGaN on $ScAlMgO_4$(0001) Substrates. *Appl. Phys. Express* **2014**, *7*, 091001. [CrossRef]
19. Ozaki, T.; Funato, M.; Kawakami, Y. InGaN-Based Visible Light-Emitting Diodes on $ScAlMgO_4$(0001) Substrates. *Appl. Phys. Express* **2015**, *8*, 062101. [CrossRef]

20. Takahashi, R.; Fujiki, R.; Hozo, K.; Hiramatsu, R.; Matsukura, M.; Kojima, T.; Han, D.-P.; Iwaya, M.; Takeuchi, T.; Kamiyama, S. Improvement of 650-Nm Red-Emitting GaIn0.17N/GaIn0.38N Multiple Quantum Wells on $ScAlMgO_4$ (0001) Substrate by Suppressing Impurity Diffusion/Penetration. *Appl. Phys. Lett.* **2022**, *120*, 142102. [CrossRef]
21. Pasayat, S.S.; Gupta, C.; Wong, M.S.; Wang, Y.; Nakamura, S.; Denbaars, S.P.; Keller, S.; Mishra, U.K. Growth of Strain-Relaxed InGaN on Micrometer-Sized Patterned Compliant GaN Pseudo-Substrates. *Appl. Phys. Lett.* **2020**, *116*, 111101. [CrossRef]
22. Mynbaeva, M.; Titkov, A.; Kryganovskii, A.; Ratnikov, V.; Mynbaev, K.; Huhtinen, H.; Laiho, R.; Dmitriev, V. Structural Characterization and Strain Relaxation in Porous GaN Layers. *Appl. Phys. Lett.* **2000**, *76*, 1113–1115. [CrossRef]
23. Pasayat, S.S.; Gupta, C.; Acker-James, D.; Cohen, D.A.; DenBaars, S.P.; Nakamura, S.; Keller, S.; Mishra, U.K.; Fellow, I. Fabrication of Relaxed InGaN Pseudo-Substrates Composed of Micron-Sized Pattern Arrays with High Fill Factors Using Porous GaN. *Semicond. Sci. Technol.* **2019**, *34*, 115020. [CrossRef]
24. Pasayat, S.S.; Gupta, C.; Wong, M.S.; Ley, R.; Gordon, M.J.; DenBaars, S.P.; Nakamura, S.; Keller, S.; Mishra, U.K. Demonstration of Ultra-Small (<10 μm) 632 nm Red InGaN Micro-LEDs with Useful on-Wafer External Quantum Efficiency (>0.2%) for Mini-Displays. *Appl. Phys. Express* **2020**, *14*, 011004.
25. Chan, P.; DenBaars, S.P.; Nakamura, S. Growth of Highly Relaxed InGaN Pseudo-Substrates over Full 2-in. Wafers. *Appl. Phys. Lett.* **2021**, *119*, 131106. [CrossRef]
26. Chan, P.; Rienzi, V.; Lim, N.; Chang, H.-M.; Gordon, M.; DenBaars, S.P.; Nakamura, S. Demonstration of Relaxed InGaN-Based Red LEDs Grown with High Active Region Temperature. *Appl. Phys. Express* **2021**, *14*, 101002. [CrossRef]
27. Quote from Aemdeposition. Available online: www.aemdeposition.com (accessed on 5 July 2022).
28. Chen, Y.; Zuo, P.; Guan, Y.; Yusuf, M.H.; Babcock, S.E.; Kuech, T.F.; Evans, P.G. Reduction of Interface Reactions in the Low-Temperature Solid-Phase Epitaxy of $ScAlMgO_4$ on Al_2O_3(0001). *Cryst. Growth Des.* **2020**, *20*, 6001–6007. [CrossRef]
29. Yoshimoto, M.; Maeda, T.; Ohnishi, T.; Koinuma, H.; Ishiyama, O.; Shinohara, M.; Kubo, M.; Miura, R.; Miyamoto, A. Atomic-scale Formation of Ultrasmooth Surfaces on Sapphire Substrates for High-quality Thin-film Fabrication. *Appl. Phys. Lett.* **1995**, *67*, 2615–2617. [CrossRef]
30. Shan, J.; Sun, J.; Liu, Z. Chemical Vapor Deposition Synthesis of Graphene over Sapphire Substrates. *ChemNanoMat* **2021**, *7*, 515–525. [CrossRef]
31. Tsuda, M.; Watanabe, K.; Kamiyama, S.; Amano, H.; Akasaki, I.; Liu, R.; Bell, A.; Ponce, F.A. Mechanism of H2 Pre-Annealing on the Growth of GaN on Sapphire by MOVPE. *Appl. Surf. Sci.* **2003**, *216*, 585–589. [CrossRef]
32. Kimizuka, N.; Mohri, T. Structural Classification of $RAO_3(MO)_n$ Compounds (R = Sc, In, Y, or Lanthanides; A = Fe(III), Ga, Cr, or Al; M = Divalent Cation; $n = 1-11$). *J. Solid State Chem.* **1989**, *78*, 98–107. [CrossRef]
33. Zorina, N.G.; Kvitka, S.S. Refinement of the Structure of the Spinel Al_2MgO_4. *Sov. Phys. Crystallogr.* **1969**, *13*, 599–600.
34. Mani, G.K.; Rayappan, J.B.B. Impact of Annealing Duration on Spray Pyrolysis Deposited Nanostructured Zinc Oxide Thin Films. *Superlattices Microstruct.* **2014**, *67*, 82–87. [CrossRef]
35. Kužel, R.; Nichtová, L.; Matěj, Z.; Musil, J. In-Situ X-Ray Diffraction Studies of Time and Thickness Dependence of Crystallization of Amorphous TiO2 Thin Films and Stress Evolution. *Thin Solid Film.* **2010**, *519*, 1649–1654. [CrossRef]
36. White, R.C.; Khoury, M.; Wu, F.; Keller, S.; Rozhavskaia, M.; Sotta, D.; Nakamura, S.; DenBaars, S.P. MOCVD Growth of Thick V-Pit-Free InGaN Films on Semi-Relaxed InGaN Substrates. Semicond. *Sci. Technol.* **2021**, *36*, 015011.
37. Holec, D.; Costa, P.M.F.J.; Kappers, M.J.; Humphreys, C.J. Critical Thickness Calculations for InGaN/GaN. *J. Cryst. Growth* **2007**, *303*, 314–317. [CrossRef]
38. Chaudhuri, R.; Zhang, Z.; Xing, H.G.; Jena, D. Very High Density (>10^{14} cm^{-2}) Polarization-Induced 2D Hole Gases Observed in Undoped Pseudomorphic InGaN/AlN Heterostructures. *Adv. Elect. Mater.* **2022**, *8*, 2101120. [CrossRef]
39. Pasayat, S.S.; Gupta, C.; Wang, Y.; DenBaars, S.P.; Nakamura, S.; Keller, S.; Mishra, U.K. Compliant Micron-Sized Patterned InGaN Pseudo-Substrates Utilizing Porous GaN. *Materials* **2020**, *13*, 213. [CrossRef]
40. Parker, C.A.; Roberts, J.C.; Bedair, S.M.; Reed, M.J.; Liu, S.X.; El-Masry, N.A.; Robins, L.H. Optical Band Gap Dependence on Composition and Thickness of $In_xGa_{1-x}N$. *Appl. Phys. Lett.* **1999**, *75*, 2566–2568. [CrossRef]
41. Soga, T. Chapter 1-Fundamentals of Solar Cell. In *Nanostructured Materials for Solar Energy Conversion*; Soga, T., Ed.; Elsevier: Amsterdam, The Netherlands, 2006; pp. 3–43.
42. Chen, H.; Feenstra, R.M.; Northrup, J.E.; Neugebauer, J.; Greve, D.W. Indium Incorporation and Surface Segregation during InGaN Growth by Molecular Beam Epitaxy: Experiment and Theory. *MRS Internet J. Nitride Semicond. Res.* **2001**, *6*, e11. [CrossRef]
43. Khan, K.; Biswas, M.; Ahmadi, E. Growth of High Quality (In,Ga)N Films on O-Face ZnO Substrates by Plasma-Assisted Molecular Beam Epitaxy. *AIP Adv.* **2020**, *10*, 075120. [CrossRef]
44. Wagner, A.; Ratzker, B.; Kalabukhov, S.; Frage, N. Enhanced External Luminescence Quantum Efficiency of Ceramic Phosphors by Surface Roughening. *J. Lumin.* **2019**, *213*, 454–458. [CrossRef]
45. López, J.; Morales, A.; Aceves, M.; Yu, Z.; Domínguez, C. Analysis of Surface Roughness and Its Relationship with Photoluminescence Properties of Silicon-Rich Oxide Films. *J. Vac. Sci. Technol. A Vac. Surf. Film.* **2008**, *27*, 57. [CrossRef]

crystals

MDPI

Article

Plasma-Assisted Halide Vapor Phase Epitaxy for Low Temperature Growth of III-Nitrides

Galia Pozina *, Chih-Wei Hsu, Natalia Abrikossova and Carl Hemmingsson

Department of Physics, Chemistry and Biology, Linköping University, 58183 Linköping, Sweden
* Correspondence: galia.pozina@liu.se

Abstract: Developing growth techniques for the manufacture of wide band gap III-nitrides semiconductors is important for the further improvement of optoelectronic applications. A plasma-assisted halide phase vapor epitaxy (PA-HVPE) approach is demonstrated for the manufacture of undoped and In-doped GaN layers at ~600 °C. A dielectric barrier discharge (DBD) plasma source is utilized for the low-temperature activation of ammonia. The use of the plasma source at a growth temperature of ~600 °C increases the growth rate from ~1.2 to ~4–5 μm/h. Furthermore, the possibility for the growth of InGaN at ~600 °C has been studied. Precursors of GaCl and InCl/$InCl_3$ are formed in situ in the reactor by flowing HCl gas over a melt of metallic Ga and In, respectively. The In concentration was low, in the order of a few percent, as the incorporation of In is reduced by plasma due to the activation of chlorine-containing species that etch the relatively poorly bonded In atoms. Nevertheless, the approach of using plasma for ammonia activation is a very promising approach to growing epitaxial III-nitrides at low temperatures.

Keywords: PA-HVPE; GaN; InGaN; plasma activated ammonia; low-temperature growth

Citation: Pozina, G.; Hsu, C.-W.; Abrikossova, N.; Hemmingsson, C. Plasma-Assisted Halide Vapor Phase Epitaxy for Low Temperature Growth of III-Nitrides. *Crystals* **2023**, *13*, 373. https://doi.org/10.3390/cryst13030373

Academic Editors: Peng Chen and Zhizhong Chen

Received: 31 January 2023
Revised: 16 February 2023
Accepted: 20 February 2023
Published: 22 February 2023

1. Introduction

III-N wide bandgap semiconductors such as GaN and its alloys with indium or aluminum are widely used for manufacturing LEDs, laser diodes, photovoltaics, and high-frequency high-power devices [1–3]. GaN has the potential for high pressure, high temperature, and high voltage applications due to its high critical electric field, good thermal conductivity, and high electron mobility [4,5].

Today, most white LED lamps are based on down conversion when phosphors absorb light from UV or blue LEDs and then re-emit photons at lower energies resulting in a white spectrum. However, the energy losses using down conversion can be about 20–30% depending on the wavelength due to the Stokes shift [6]. These losses can be avoided if a multi-chip design with red, green, and blue LEDs is used with color-mixing optics. Currently, there are red LEDs based on alloys of GaAsP and blue LEDs based on InGaN with relatively low In content demonstrating high quantum efficiencies exceeding 80% [7]. However, efficient green LEDs emitting around 550–600 nm are still difficult to manufacture. This phenomenon is known as the green gap [8]. The origin of the green gap is related to the degradation in crystalline quality and piezoelectric fields in strained InGaN quantum wells (QWs) due to an increase in lattice mismatch between $In_{1-x}Ga_xN$ and the underlying GaN, which is roughly 1% for every 0.1 indium mole fraction "*x*" [9,10]. This gives rise to the poor structural quality of the material, especially for thicker layers and it will result in reduced efficiency of the InGaN-based LEDs. The problem is further complicated when growth is performed on foreign substrates such as on the commonly used sapphire (Al_2O_3) or silicon carbide (SiC) that have a significant lattice mismatch with GaN of ~16 and 3%, respectively [11,12]. The lattice mismatch results in an increased density of threading dislocations that act as non-radiative recombination channels, which is a critical issue in the In-rich InGaN-based LEDs [13]. To reduce stress, it is advantageous to fabricate thick lattice-matched InGaN layers of high crystalline quality that can be used as initial templates for further growth of InGaN-based device structures.

To overcome the abovementioned problems several different techniques to fabricate LEDs by growth on a relaxed InGaN layer have been suggested [14]. For example, highly relaxed thick layers of InGaN templates [15], growth on thermally decomposed InGaN layers [16], growth of InGaN on lattice-matched $ScAlMgO_4$ substrates [17,18], growth of relaxed InGaN on InGaN pseudo-substrates [19,20], and relaxed InGaN templates grown on porous GaN [21].

Molecular beam epitaxy (MBE) and metal-organic vapor phase epitaxy (MOVPE) are common methods for fabricating epitaxial layers and heterostructures for various III-N applications. However, due to low growth rates (<1 µm/h) these methods are not suitable for growing thick III-N layers that can be used as templates. Another problem is related to the tendency of $In_{1-x}Ga_xN$ to phase-segregate into In-rich and Ga-rich alloy regions during the growth, especially as the indium fraction increases [22]. This is partly caused by a high surface mobility of the In atoms compared to the Ga atoms and the large disparity in the vapor pressures of these two metals. While halide vapor phase epitaxy (HVPE) is a more promising technique to produce thick GaN layers at ~1000 °C due to high growth rates in the range of 10–100 µm/h [23,24], it remains challenging to reduce the growth temperature to ~500–600 °C, i.e., to temperatures necessary to prevent thermal segregation of InGaN. This is because the cracking efficiency of ammonia, which is a typical precursor of nitrogen, sharply reduces at temperatures below 1000 °C, which results in low growth rates and poor crystal quality [25].

Thus, in this work, we report the development of a proof-of-concept method to grow GaN layers using a plasma-assisted HVPE (PA-HVPE) process. The implementation of the plasma source in the reactor chamber before the sample holder allows ammonia to be activated even at lower growth temperatures of ~570 °C, which results in an increase in growth rates by a factor of 2 up to 5 µm/h. We have also used metallic indium to form in situ precursors of $InCl/InCl_3$ with the aim of studying the possibility of In incorporation during HVPE growth.

2. Materials and Methods

For the experiments, we used a horizontal homemade HVPE reactor schematically shown in Figure 1. The HVPE system consists of a resistively heated oven with three zones. Each temperature zone can be controlled and heated up to 1050 °C. The system has two boats for in situ generation of GaCl and $InCl_3/InCl$ by flowing HCl over metal melts. The Ga and In metals have a purity of 7 N. Nitrogen was used as carrier gas and the pressure in the reactor during growth was 500 mbar. The HCl flow over the Ga melt was kept constant at 5 mL/min, and, for the In melt, the flow was varied in the range of 2–10 mL/min. The NH_3 and N_2 gas flows were kept constant at 80 mL/min and 1000 mL/min, respectively. The temperature of the metal melts can be varied by changing the position of the boats in the reactor. In this study, the temperature of the Ga melt was 700 °C, while the temperature of the In melt was varied in the range of 260 to 460 °C. As a substrate for the growth, both *c*-axis Al_2O_3 with a thickness of 450 µm and GaN templates were used. The GaN templates were commercially available templates consisting of a 3 µm thick MOVPE GaN layer grown on *c*-axis Al_2O_3 wafers with a thickness of 430 µm. Since the growth of high-quality GaN on foreign substrates, such as sapphire, requires several elaborate steps to form a nucleation layer [26], we have mainly carried out homoepitaxial growth on GaN templates in this work. The size of the samples was 1 cm × 1 cm.

For plasma generation, dielectric barrier discharge (DBD) was used, where the plasma was generated between two graphite electrodes with a 430 µm thick Al_2O_3 wafer acting as a dielectric (Figure 2a). The gap distance d was 5 mm. The electronics for the plasma generation control consists of four 15 kV high-voltage switches connected in a so-called H-bridge, in which the polarity between the electrodes can be changed by alternately turning on and off each pair of switches (S1, S4) and (S2, S3). Two 15 kV MOSFET push-pull high-voltage switches (BEHLKE HTS 151-03-GSM) and a high-voltage unit (Heinzinger EVO10000-200rev) were used. Plasma generation was carried out using a voltage of ~3–5 kV. The pulse width was limited to 10 μs since longer pulses resulted in discharging and an unstable plasma. The pulse repetition frequency was limited to 5 kHz by the MOSFET push-pull high-voltage switches. Figure 2b schematically shows a typical pulse sequence with "on" and "off" times. The HVPE growth of GaN is commonly performed at atmospheric pressure. At such pressures, it was difficult to obtain a stable nitrogen plasma even at high temperatures. Therefore, the growth of the studied layers was done at a lower pressure of 500 mbar. To prevent arc-discharge due to parasitic deposition on the electrodes, the growth time was limited to 15 min.

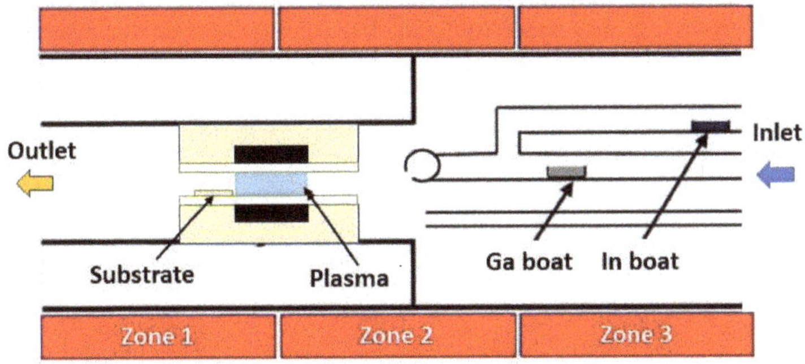

Figure 1. Schematic drawing of the HVPE reactor.

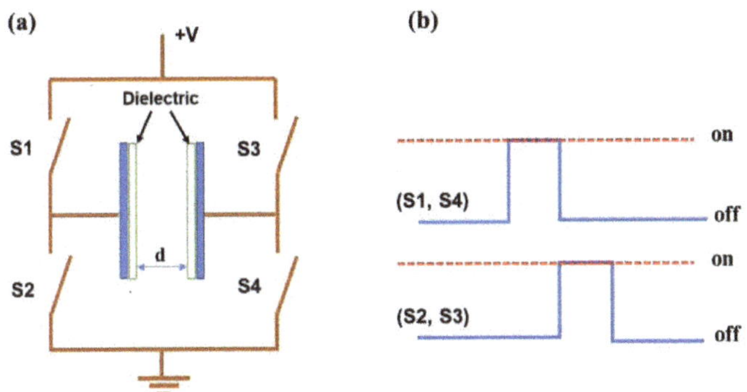

Figure 2. Schematic drawing of plasma generation DBD. (**a**) High-voltage H-bridge to switch polarity between electrodes with the electrode distance d. (**b**) The pulse sequence used to control H-bridge.

The thickness of the layers was estimated from the weight of the samples before and after growth to give quick feedback to the growth runs, but it was also confirmed by measurements using scanning electron microscopy (SEM) and white light reflectance spectroscopy (WLRS).

Optical characterization was performed by cathodoluminescence (CL) measurements at different temperatures from ~5 to 295 K using a MonoCL4 setup in situ SEM instrument LEO1550. For plasma and transmission spectra measurements, we used a Thorlabs fiber-based compact Czerny-Turner CCD spectrometer, model CCS200/M. Atomic force microscopy (AFM) measurements were performed using a Dimension 3100 SPM microscope with the VT-102 vibration isolation table. X-ray diffraction (XRD) measurements were done using a PANalytical X'Pert Pro diffractometer with Cu-Kα radiation.

3. Results and Discussion

Figure 3 shows optical spectra of the nitrogen plasma generated between the electrodes inside the reactor with parameters used during the growth of samples. Measurements are shown for 20 and 200 °C for pure nitrogen (blue and red lines) and when ammonia with a flow of 80 mL/min was added in the chamber (green lines). At higher temperatures, the chamber was protected by an isolating shield, which prevented optical measurements. The plasma spectrum shows peaks in the range of 300–500 nm related to the transitions of the N_2 s positive system (2PS) overlapping with N_2^+ first negative system (1NS) originating from the transition between the excited states and the ground state of molecular ion N_2^+ [27]. Electronic transitions in the region of 500–900 nm are related to the first positive system (1PS) of N_2 [28]. The 1PS lines become more intense at 200 °C. We also observed at this temperature a weak sharp line at ~746.6 nm, which might correspond to the excited nitrogen atom transition at 746.83 nm.

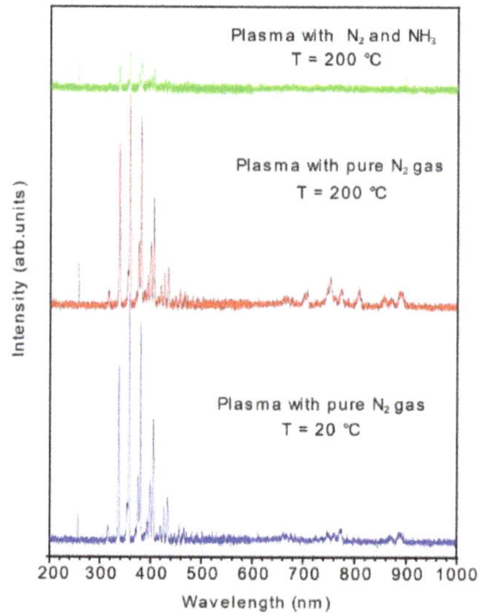

Figure 3. Spectra of nitrogen plasma at 20 °C (blue line) and 200 °C (red line). The top graph (green line) shows the plasma spectrum at 200 °C when ammonia gas was added to the nitrogen.

When the ammonia gas was added, the intensities of the transition peaks decrease as can be seen in Figure 3 for the plasma spectrum shown by the green line. In this case, the spectrum was dominated mainly by the 2PS transitions, while no additional lines were observed, although the transition at 746.6 nm was still present in the spectrum. We also observed a sharp peak at ~256.6 nm in all measured plasma spectra. The origin of this peak has not yet been identified; however, it is not related to the response function of the instrument, but to some unknown species in the plasma.

First, the HVPE growth of GaN layers at a low temperature of ~600 °C was performed without and with plasma generation. The thickness of the layers increased from 0.17 μm to 0.5 μm, which means a significant increase in the growth rate from ~0.6 μm/h to 2 μm/h by activation of ammonia using the plasma source. The deposition in this particular case was done directly on a sapphire substrate to avoid any influence on the layer properties from the MOVPE GaN templates. When growth is performed on sapphire, the growth rate is not constant and is lower before coalescence.

Using the GaN templates, the growth rate at ~600 °C increases to ~4–5 μm/h for the samples produced with ammonia activated by plasma, while attempts to grow GaN layers without plasma generation resulted in much lower growth rates of ~1.2 μm/h.

SEM measurements confirmed that the estimation of thickness from the weight gives a reasonable average value. For the sample in Figure 4a, the HVPE layer thickness determined from the weight measurements was 0.96 µm, and from the SEM image was 0.95 µm. Figure 4b shows a thickness map of a layer grown at 620 °C on a GaN template. For this layer, the average thickness of ~1.6 µm was determined from the weight measurements, which correlates with the thickness map, where the variation in thickness was obtained in the range of 1.3–1.7 µm. There is a gradient in the thickness over the sample (~25%) in the direction of the gas flow. Thus, for scaling up the process and deposition of larger areas, the system requires rotation of the sample and/or higher carrier gas flows. The roughness is ~14 nm for the root mean square (RMS) as estimated from AFM measurements in Figure 4c.

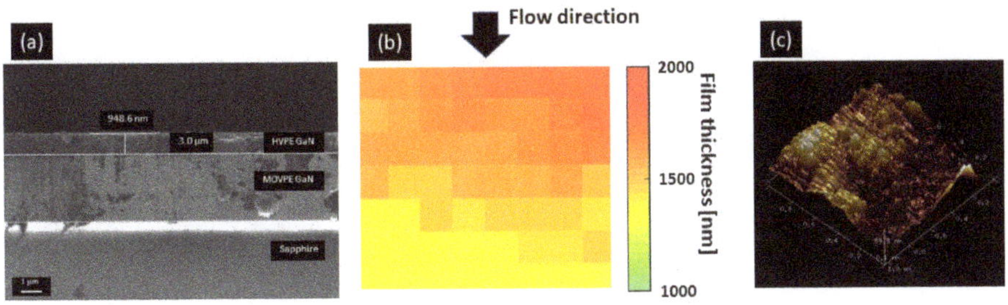

Figure 4. (**a**) Cross-section SEM image shows the thickness of the HVPE layer. (**b**) WLRS thickness map. The thickness varies in the region 1300–1740 nm. The top of the figure is closest to the plasma source. (**c**) AFM image taken over 1 µm × 1 µm area.

The structural quality of the GaN layers grown directly on sapphire was examined by XRD (see Figure 5). The GaN layer grown with plasma generation shows a reflection peak from the (0002) plane and, thus, an alignment to the c-axis, while the sample grown without ammonia activation shows only a sapphire-related peak.

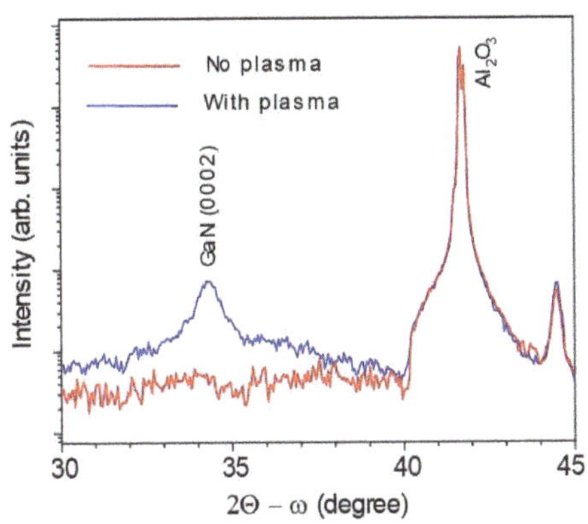

Figure 5. XRD spectra measured for the GaN layers grown at 620 °C directly on sapphire (0001) substrates with plasma (blue line) and without plasma (red line) generation of nitrogen precursor. The peak at ~44.3 degrees is related to the holder.

We have investigated the possibility of indium incorporation in GaN using the PA-HVPE process with different temperatures of the boat with metallic indium, with different HCl flows over the boat with indium. The process was also studied using growth temperatures of 570 and 620 °C. The XRD patterns measured for the symmetric (0002) reflection are shown in Figure 6a for samples grown at 620 °C at three different temperatures of the metallic indium. The dominating diffraction peak is related to GaN, which also involves a reflection from the GaN template. Therefore, to distinguish signals from the HVPE layers and the MOVPE GaN template, grazing incidence XRD (GIXRD) 2θ scans with an incident angle of 0.5 degrees were acquired (Figure 6b). Clearly, for the samples grown at different temperatures of the metallic indium, the XRD peak position shifts towards lower angles, which indicates the incorporation of indium in GaN. Note that the thickness of all layers was ~1 μm, which means that the layers can be considered to be fully relaxed [25]. In this case, the lattice constant *c* can be determined from the X-ray (0002) peak position according to the Bragg law and the indium molar fraction x can be calculated from Vegard's law:

$$c_{InGaN} = x c_{InN} + (1 - x) c_{GaN}, \tag{1}$$

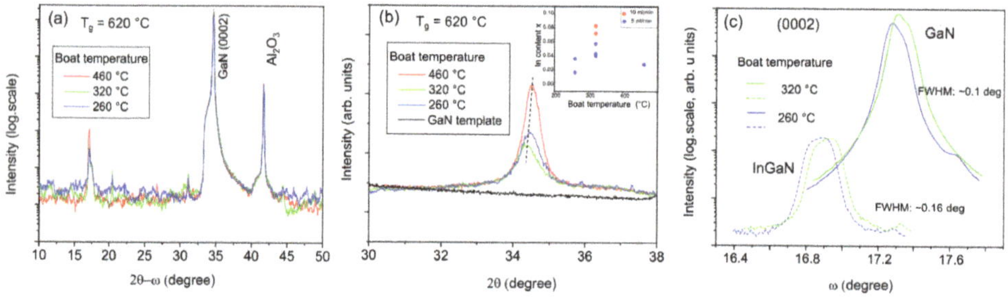

Figure 6. (**a**) XRD patterns measured for the In-doped GaN layers grown at 620 °C on GaN templates using plasma activation of ammonia and different temperatures of the boat with metallic In melt of 460 °C (red line), 320 °C (green line) and 260 °C (blue line). (**b**) Grazing incidence 2 theta scans performed with a small incident angle of 0.5 degrees shown for the same samples as in (**a**) and for the GaN template for reference. The inset shows estimated the indium content vs. The boat temperature. (**c**) XRD omega rocking curves for InGaN and GaN (0002) reflections for samples grown with In boat temperatures of 260 °C (blue lines) and 320 °C (green line).

To estimate the indium concentration in the layers from XRD data, we have used values of 5.702 and 5.185 Å for the lattice parameter *c* of InN and GaN, respectively [29]. The obtained x values are plotted in the inset of Figure 6b for different growth parameters. The highest indium concentration of ~8% was obtained for the medium boat temperature of ~320 °C and HCl flow of 10 mL/min. However, the XRD method to assess indium composition in InGaN depends on the lattice parameters, which have a spread of values according to the literature [29], which can arise to inaccuracies. The rocking curves are rather broad with full widths at half maximum (FWHM) of ~0.1 and ~0.16 degrees for (0002) GaN and InGaN reflections, respectively (see Figure 6c), which indicates structural imperfections. This is an additional problem contributing to the determination of lattice parameters and, thus, to correctly estimating In content from XRD measurements. Nevertheless, GIXRD has been shown as a feasible technique to study InGaN films with various In% on Si substrate [30].

Furthermore, SIMS analysis shows that, although indium indeed incorporates in the HVPE layers at 620 °C, the estimated concentration was lower than that obtained from XRD data. The depth profiles of indium content for samples grown at 620 °C with HCl flows of 5 and 10 mL/min and with the ammonia activated by plasma are shown in Figure 7a in the linear scale. For comparison, the indium concentration profile is also plotted for the layer grown with the same parameters but without plasma generation in Figure 7b. Clearly, the growth rate is twice as high with plasma-assisted growth. The activated ammonia concentration is higher after the plasma source and that gives rise to a higher growth rate of GaN. At the same time, plasma has a negative effect on indium concentration because plasma simultaneously activates chlorine-containing species. The relatively poor bonding ability of In compared to Ga in InGaN means that In can be more easily etched by the highly reactive plasma-activated chlorine-containing species and that will result in a lower In concentration [31,32]. In addition, it is also known that the growth of InGaN does not occur when InCl is used since the equilibrium constant of the reaction between InCl and NH_3 is small [33]. Since the $InCl_3$ in our system is passed through the plasma, we cannot rule out that some of the In precursors are decomposed to InCl. Thus, to improve the incorporation of In in the layers, the precursors $InCl_3$ and GaCl should bypass the plasma.

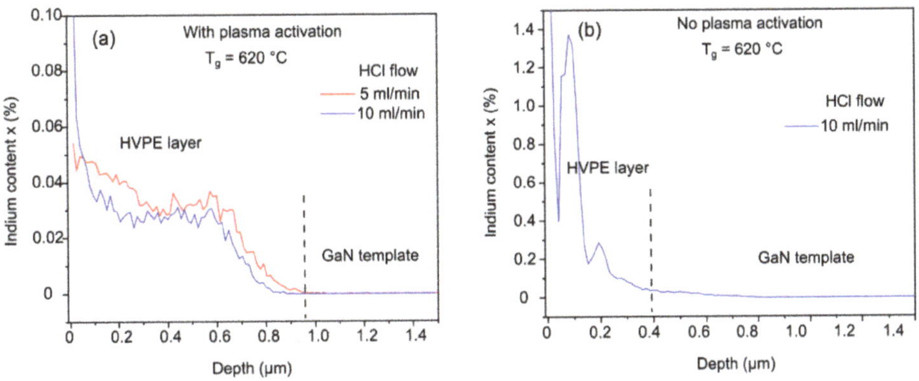

Figure 7. (**a**) SIMS depth profile for indium concentration for samples grown with ammonia activated by plasma. (**b**) SIMS profile for the sample grown with the same parameters as in (**a**) but without activation of ammonia by plasma. The indium boat temperature was 320 °C. The dashed lines are a guide to the eye.

The band gap shift caused by different concentrations of indium in the InGaN layers grown using plasma-activated ammonia can be observed in transmission measurements. Then, the optical band gap can be estimated from the absorption coefficient α. For the direct band gap semiconductors α is related to the band gap energy E_g as follows:

$$\alpha^2 E^2 \sim (E - E_g), \quad (2)$$

where E is the photon energy. Graphs of $\alpha^2 E^2$ as a function of photon energy in the vicinity of the band edge are shown in Figure 8a for samples grown with different HCl flows through the boat with indium. The reference spectrum for the GaN template is also shown in the inset. The shift in the absorption edge is clearly seen in Figure 8b for the samples grown using different temperatures of the indium melt. The energy of the band gap can be estimated by extrapolating the linear part of $\alpha^2 E^2$ to zero as indicated in Figure 8a by dashed lines. The extracted values of E_g are plotted in Figure 9a,b for the same samples as in Figure 8. Note, the optical band energy of 3.396 eV estimated for the GaN template is slightly higher than the 3.39 eV for relaxed GaN, which is common for layers under compression when grown on sapphire [34].

The energy of the bandgap for alloy $In_xGa_{1-x}N$ can be approximated by Vegard's law:

$$E_g(In_xGa_{1-x}N) = xE_g(InN) + (1 - x)E_g(GaN), \quad (3)$$

which allows the calculation of the indium composition x. Using the bandgap energy of 3.39 and 0.7 eV for GaN and InN [25], respectively, and the InGaN bandgap energy from the absorption data, we have determined that the indium concentration does not exceed 3% in the layers grown with plasma-activated ammonia. Taking into account nonlinearity in the InGaN band gap with bowing parameter in the range between ~1.3 and 3.8 eV [35,36], the indium concentration can be lower. This result correlates well with the data obtained by the SIMS analysis. Indium concentrations determined by different techniques are collected in Table 1. The discrepancy in the determination of indium content has several causes. First, the estimation of indium fraction by SIMS can give an error due to the matrix effect as

SIMS is mainly used for the determination of the concentrations of impurities [37]. Second, the extraction of lattice constant *c* from the XRD patterns can be also inaccurate due to broad XRD peaks and material imperfection. Thus, there is a moderate overestimation in In content determined by GIXRD compared to absorption spectra. Similar observations have also been reported when comparing In% determined by various techniques [30].

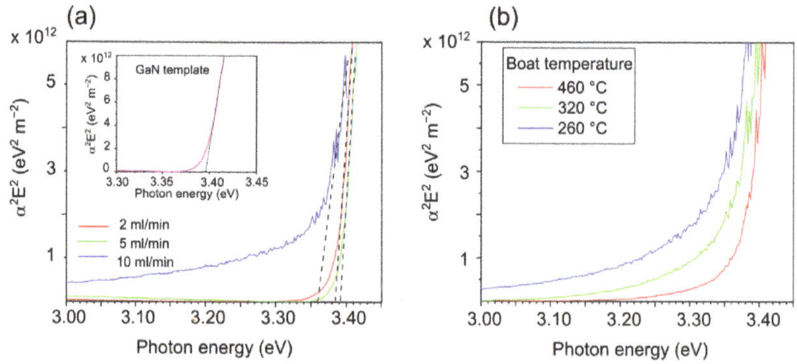

Figure 8. (**a**) Squared absorption coefficient plotted as a function of photon energy for layers grown at 570 °C at three different flows of HCl through the boat with indium. Temperature of the boat kept constant at 320 °C. The inset shows the spectrum for the GaN template for reference. The dashed lines are a guide to the eye. (**b**) Graphs for layers grown at 620 °C for different temperatures of the boat with metallic indium. HCl flow over indium melt was kept to 5 mL/min.

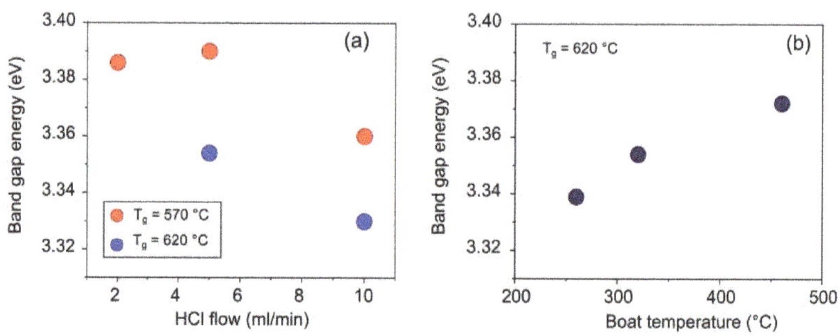

Figure 9. The band gap energy estimated from the absorption measurements for samples grown (**a**) with different HCl flows through the boat with indium and (**b**) at different temperatures of the boat with indium.

Table 1. In content estimated by different techniques for layers grown with different process parameters.

In Boat Temperature (°C)	In Content % (XRD)	In Content % (Absorption)
260	3.7	2.3
320	5.8	1.7
460	2.8	1.0
HCl (mL/min), T_g = 570 °C		
2	-	0.5
5	8.3	0.4
10	6.3 (SIMS 0.03)	1.5
HCl (mL/min), T_g = 620 °C		
5	5.8 (SIMS 0.04)	1.7
10	7.2	2.6

The emission properties have been studied by low-temperature CL simultaneously with SEM measurements. The morphology of the studied layers produced with plasma generation shows a rougher surface compared to the MOVPE GaN layer (template) (see Figure 10a,c). At the same time, panchromatic CL (PCL) images at 5 K demonstrate that threading dislocations observed as dark spots in the image for the GaN template (Figure 10b) are likely overgrown during the HVPE process judging from the PCL image, which demonstrates uniform contrast (Figure 10d). A similar tendency to overgrow threading dislocations has been observed previously in HVPE GaN layers [38]. However, the uniform contrast in Figure 10d can be also caused by the presence of strong nonradiative centers over the entire InGaN layer, which will prevent the decoration of threading dislocations.

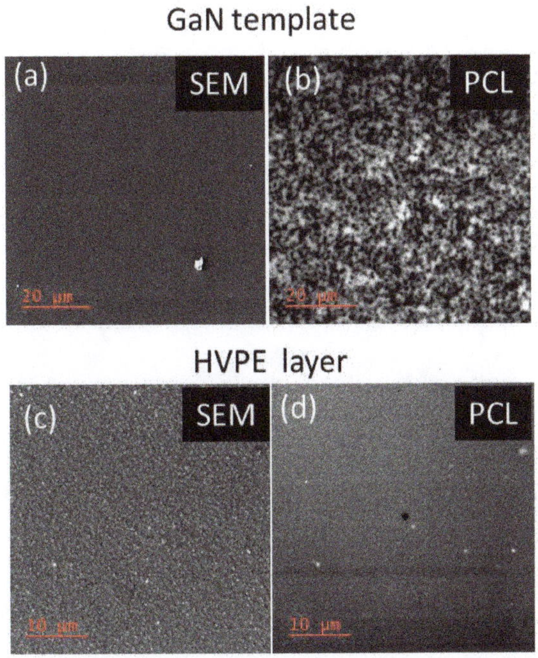

Figure 10. (**a**) SEM and (**b**) panchromatic CL images for the GaN template. (**c**) SEM and (**d**) panchromatic CL images for the InGaN layer grown with plasma-activated ammonia at T_g = 620 °C with HCl flow of 5 mL/min and at the indium melt temperature of 320 °C.

Examples of CL spectra measured at 5 K for the samples grown with three different process parameters are shown in Figure 11. The reference CL spectrum for the GaN template is shown by the dashed line. CL spectra for all studied samples grown by using plasma-activated ammonia are dominated by defect luminescence. Increasing the HCl flow over the indium melt at a boat temperature of 320 °C results in the shift of the maximum of the defect luminescence to the lower photon energies for the samples grown at 570 °C (Figure 11a). The influence of the indium boat temperature on the luminescence is also noticeable as shown for the layer grown at 620 °C with a constant HCl flow of 5 mL/min (Figure 11b). CL spectra measured at different temperature for the layer grown with a constant HCl flow of 10 mL/min through the indium boat is presented in Figure 11c. The ratio of the integrated intensities for the defect emission at 5 K and room temperature is ~2.5. The layer shows rather uniform emission properties with the same peak position independently of the point of acquisition as shown in Figure 11d. Conversely, for different samples, the shift of the CL band maximum correlates with the increased concentration of indium precursor and with indium content determined by XRD. Similar luminescence was observed for plasma-assisted MBE InGaN samples, where the emission shift was assigned to the increasing indium content [39]. In our case, however, the observed CL shift can hardly be attributed to the narrowing of the band gap of $In_xGa_{1-x}N$ with increasing x, as the variation of the optical band gap does not exceed 30 meV, while the shift of the CL bands is more than 300 meV and ranged between 2.2 eV and 1.85 eV. The observed emissions will be, thus, considered as related to defect luminescence in GaN, known as yellow (YL) and red (RL) bands. Similar RL has been observed in HVPE GaN samples but it is not so common for MOVPE GaN [40]. Although the position of the RL bands is around 1.8 eV, it can be caused by different defects with different properties, and the nature of the RL band is yet to be revealed. The YL1 band with the maximum at 2.2 eV is better studied and associated recently with a carbon-related defect (C_N) rather than defects involving Ga vacancy [40]. YL1 is typical for MOVPE GaN layers where metal-organics are used as Ga precursor. In HVPE samples, the carbon concentration is usually very low. While YL1 is not pronounced in the undoped HVPE GaN, it appears under doping with carbon impurity [41]. Here, using graphite plasma electrodes for ammonia activation, carbon species can be produced, which explains the strong YL1 emission. The observed shift in the emission to red wavelengths and the enhancement in the RL band are correlated with an increased concentration of indium precursor (i.e., InCl or $InCl_3$). Considering that RL is typical for HVPE-grown GaN, our observation may indicate that the origin of the RL band is related to the chlorine-containing impurities.

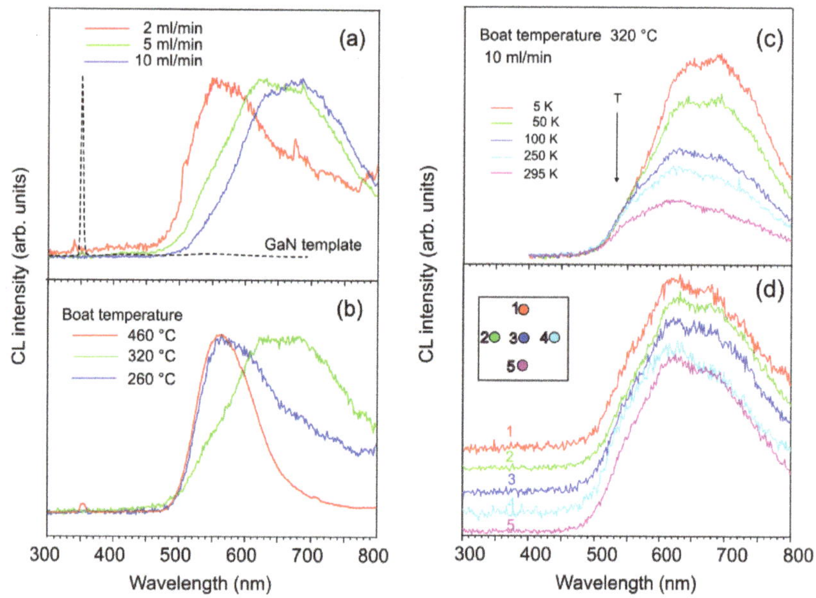

Figure 11. (**a**) Low-temperature CL spectra for samples grown with different HCl flow rates through the boat with In. CL spectrum for the GaN template is shown by the dashed line for reference. (**b**) CL spectra at 5 K for layers grown at different temperatures of the indium boat. (**c**) CL spectra measured at different temperatures for the layer grown with the indium boat temperature of 320 °C and with a constant HCl flow rate of 10 mL/min. (**d**) CL spectra at room temperature measured at different points as schematically indicated in the inset for the same sample as in (**c**).

4. Conclusions

We have demonstrated a PA-HVPE method for the growth of GaN layers at a low temperature of ~600 °C. Plasma at a pressure of 500 mbar inside the growth chamber was generated using dielectric barrier discharge. The use of plasma allows ammonia to be activated even at low temperatures, which resulted in increased growth rates of up to ~4–5 µm/h at ~600 °C compared to ~1–1.5 µm/h for the conventional HVPE process. We have investigated the possibility of using metallic indium to form $InCl/InCl_3$ precursors directly inside the growth chamber. The effect of various process parameters on the In concentration in the InGaN layers has been studied. The incorporation of indium was confirmed by XRD, SIMS, and absorption measurements, but does not exceed a few percent. We have also found that, although plasma is very efficient to activate ammonia and to increase the growth rates at low temperatures, the concentration of indium significantly reduces due to the activation of reactive chlorine-containing species that etch In at the growth surface. To increase the efficiency of the indium precursor, further modifications of the gas inlets and the plasma source should be done. To avoid the negative effect of chlorine species on the structural quality and the incorporation of indium into InGaN layers, the supply of GaCl and $InCl/InCl_3$ precursors should approach the growth surface without passing through the plasma source. Nevertheless, the suggested PA-HVPE process is very promising for activating ammonia and promoting the growth of thick III-nitrides layers at low temperatures.

Author Contributions: Conceptualization, G.P.; methodology, C.-W.H. and N.A.; software, C.H.; validation, G.P., C.-W.H. and C.H.; investigation, G.P., C.-W.H., N.A. and C.H.; data curation, C.H.; writing—original draft preparation, G.P.; writing—review and editing, C.H.; visualization, G.P. and C.H., project administration, G.P. and C.H.; funding acquisition, G.P. and C.H. All authors have read and agreed to the published version of the manuscript.

Funding: This research was funded by the Swedish Research Council (grant nos. 2018-04552, 2019-05154) and the Swedish Energy Agency (grant no. 46563-1).

Data Availability Statement: The data presented in this study are available on reasonable request from the corresponding author.

Acknowledgments: The authors are grateful for the technical support from Linköping University.

Conflicts of Interest: The authors declare no conflict of interest.

References

1. Ayari, T.; Sundaram, S.; Li, X.; Alam, S.; Bishop, C.; El Huni, W.; Jordan, M.B.; Halfaya, Y.; Gautier, S.; Voss, P.L.; et al. Heterogeneous integration of thin-film InGaN-based solar cells on foreign substrates with enhanced performance. *ACS Photonics* **2018**, *5*, 3003–3008. [CrossRef]
2. Li, K.H.; Fu, W.Y.; Choi, H.W. Chip-scale GaN integration. *Prog. Quantum Electron.* **2020**, *70*, 100247. [CrossRef]
3. Bottaro, E.; Rizzo, S.A.; Salerno, N. Circuit models of power MOSFETs leading the way of GaN HEMT modelling—A review. *Energies* **2022**, *15*, 3415. [CrossRef]
4. Kumar, A.; Moradpour, M.; Losito, M.; Franke, W.T.; Ramasamy, S.; Baccoli, R.; Gatto, G. Wide band gap devices and their application in power electronics. *Energies* **2022**, *15*, 9172. [CrossRef]
5. Xu, J.; Gu, L.; Ye, Z.; Kargarrazi, S.; Rivas-Davila, J.M. Cascode GaN/SiC: A wide-bandgap heterogenous power device for high-frequency applications. *IEEE Trans. Power Electron.* **2020**, *35*, 6340–6349. [CrossRef]
6. Yin, S.; Wang, C.; Zhu, W.; Yao, J.; Zou, J.; Lin, X.; Luo, C. Single chip super broadband InGaN/GaN LED enabled by nanostructured substrate. *Opt. Express* **2014**, *22*, A1380–A1388. [CrossRef]
7. Narukawa, Y.; Ichikawa, M.; Sanga, D.; Sano, M.; Mukai, T. White light emitting diodes with super-high luminous efficacy. *J. Phys. D Appl. Phys.* **2010**, *43*, 354002. [CrossRef]
8. Auf der Maur, M.; Pecchia, A.; Penazzi, G.; Rodrigues, W.; Di Carlo, A. Efficiency drop in green InGaN/GaN light emitting diodes: The role of random alloy fluctuations. *Phys. Rev. Lett.* **2016**, *116*, 027401. [CrossRef]
9. Wakahara, A.; Tokuda, T.; Dang, X.Z.; Noda, S.; Sasaki, A. Compositional inhomogeneity and immiscibility of a GaInN ternary alloy. *Appl. Phys. Lett.* **1997**, *71*, 906–908. [CrossRef]
10. Hiramatsu, K.; Kawaguchi, Y.; Shimizu, M.; Sawaki, N.; Zheleva, T.; Davis, R.F.; Tsuda, H.; Taki, W.; Kuwano, N.; Oki, K. The composition pulling effect in MOVPE grown InGaN on GaN and AlGaN and its TEM characterization. *MRS Internet J. Nitride Semicond. Res.* **1997**, *2*, 11. [CrossRef]
11. Li, G.; Wang, W.; Yang, W.; Lin, Y.; Wang, H.; Lin, Z.; Zhou, S. GaN-based light-emitting diodes on various substrates: A critical review. *Rep. Prog. Phys.* **2016**, *79*, 056501. [CrossRef]
12. Cordier, Y.; Portail, M.; Chenot, S.; Tottereau, O.; Zielinski, M.; Chassagne, T. AlGaN/GaN high electron mobility transistors grown on 3C-SiC/Si(111). *J. Cryst. Growth* **2008**, *310*, 4417–4423. [CrossRef]

13. Kaneta, A.; Funato, M.; Kawakami, Y. Nanoscopic recombination processes in InGaN/GaN quantum wells emitting violet, blue, and green spectra. *Phys. Rev. B* **2008**, *78*, 125317. [CrossRef]
14. Iida, D.; Ohkawa, K. Recent progress in red light-emitting diodes by III-nitride materials. *Semicond. Sci. Technol.* **2022**, *37*, 013001. [CrossRef]
15. Abdelhamid, M.; Reynolds, J.G.; El-Masry, N.A.; Bedair, S.M. Growth and characterization of $In_xGa_{1-x}N$ ($0 < x < 0.16$) templates for controlled emissions from MQW. *J. Cryst. Growth* **2019**, *520*, 18–26.
16. Chan, P.; Rienzi, V.; Lim, N.; Chang, H.M.; Gordon, M.; DenBaars, S.P.; Nakamura, S. Demonstration of relaxed InGaN-based red LEDs grown with high active region temperature. *Appl. Phys. Express* **2021**, *14*, 101002. [CrossRef]
17. Ozaki, T.; Funato, M.; Kawakami, Y. InGaN-based visible light-emitting diodes on ScAlMgO4(0001) substrates. *Appl. Phys. Express* **2015**, *8*, 062101. [CrossRef]
18. Ozaki, T.; Funato, M.; Kawakami, Y. Red-emitting $In_xGa_{1-x}N/In_yGa_{1-y}N$ quantum wells grown on lattice-matched $In_yGa_{1-y}N$/ScAlMgO4(0001) templates. *Appl. Phys. Express* **2019**, *12*, 011007. [CrossRef]
19. Dussaigne, A.; Barbier, F.; Damilano, B.; Chenot, S.; Grenier, A.; Papon, A.M.; Samuel, B.; Ben Bakir, B.; Vaufrey, D.; Pillet, J.C.; et al. Full InGaN red light emitting diodes. *J. Appl. Phys.* **2020**, *128*, 135704. [CrossRef]
20. Dussaigne, A.; Maitre, P.L.; Haas, H.; Pillet, J.C.; Barbier, F.; Grenier, A.; Michit, N.; Jannaud, A.; Templier, R.; Vaufrey, D.; et al. Full InGaN red (625 nm) micro-LED (10 lm) demonstration on a relaxed pseudo-substrate. *Appl. Phys. Express* **2021**, *14*, 092011. [CrossRef]
21. Pasayat, S.S.; Ley, R.; Gupta, C.; Wong, M.S.; Lynsky, C.; Wang, Y.; Gordon, M.J.; Nakamura, S.; Denbaars, S.P.; Keller, S.; et al. Color-tunable <10 lm square InGaN micro-LEDs on compliant GaN-on-porous-GaN pseudo-substrates. *Appl. Phys. Lett.* **2020**, *117*, 061105.
22. Singh, R.; Doppalapudi, D.; Moustakas, T.D.; Romano, L.T. Phase separation in InGaN thick films and formation of InGaN/GaN double heterostructures in the entire alloy composition. *Appl. Phys. Lett.* **1997**, *70*, 1089. [CrossRef]
23. Hemmingsson, C.; Paskov, P.P.; Pozina, G.; Heuken, M.; Schineller, B.; Monemar, B. Hydride vapour phase epitaxy growth and characterization of thick GaN using a vertical HVPE reactor. *J. Cryst. Growth* **2007**, *300*, 32–36. [CrossRef]
24. Hemmingsson, C.; Pozina, G. Optimization of low temperature GaN buffer layers for halide vapor phase epitaxy growth of bulk GaN. *J. Cryst. Growth* **2013**, *366*, 61–66. [CrossRef]
25. Yam, F.K.; Hassan, Z. InGaN: An overview of the growth kinetics, physical properties and emission mechanisms. *Superlattices Microstruct.* **2008**, *43*, 1–23. [CrossRef]
26. Martin, D.; Napierala, J.; Ilegems, M.; Butté, R.; Grandjean, N. High quality thin GaN templates grown by hydride vapor phase epitaxy on sapphire substrates. *Appl. Phys. Lett.* **2006**, *88*, 241914. [CrossRef]
27. Akatsuka, H. Progresses in experimental study of N_2 plasma diagnostics by optical emission spectroscopy. In *Chemical Kinetics*; Patel, V., Ed.; InTech: Vienna, Austria, 2012; p. 297. [CrossRef]
28. Camacho, J.; Poyato, J.M.L.; Diaz, L.; Santos, M. Optical emission studies of nitrogen plasma generated by IR CO_2 laser pulses. *J. Phys. B At. Mol. Opt. Phys.* **2007**, *40*, 4573–4590. [CrossRef]
29. Vickers, M.E.; Kappers, M.J.; Smeeton, T.M.; Thrush, E.J.; Barnard, J.S.; Humphreys, C.J. Determination of the indium content and layer thicknesses in InGaN/GaN quantum wells by X-ray scattering. *J. Appl. Phys.* **2003**, *94*, 1565–1574. [CrossRef]
30. Hsu, C.W.; Martinovic, I.; Magnusson, R.; Bakhit, B.; Palisaitis, J.; Persson, P.O.Å.; Rouf, P.; Pedersen, H. Homogeneous high In content $In_xGa_{1-x}N$ films by supercycle atomic layer deposition. *J. Vac. Sci. Technol. A* **2022**, *40*, 060402. [CrossRef]
31. Chung, C.W.; Hong, F.C.N. Chlorine-assisted growth of epitaxial InGaN and AlGaN crystals at low temperatures using plasma-enhanced chemical vapor deposition. *Cryst. Growth Des.* **2023**, *23*, 37–48. [CrossRef]
32. Efremov, A.; Svettsov, V.; Sitanov, D.; Balashov, D. A comparative study of plasma parameters and gas phase compositions in Cl_2 and HCl direct current glow discharges. *Thin Solid Films* **2008**, *516*, 3020–3027. [CrossRef]
33. Kumagai, Y.; Takemoto, K.; Koukitu, A.; Seki, H. Thermodynamics on halide vapor-phase epitaxy of InN using InCl and $InCl_3$. *J. Cryst. Growth* **2001**, *222*, 118–124. [CrossRef]
34. Pozina, G.; Edwards, N.V.; Bergman, J.P.; Paskova, T.; Monemar, B.; Bremser, M.D.; Davis, R.F. Time-resolved spectroscopy of strained GaN/AlN/6H–SiC heterostructures grown by metalorganic chemical vapor deposition. *Appl. Phys. Lett.* **2001**, *78*, 1062–1064. [CrossRef]
35. Mosesa, P.G.; Van de Walle, C.G. Band bowing and band alignment in InGaN alloys. *Appl. Phys. Lett.* **2010**, *96*, 021908. [CrossRef]
36. Wetzel, C.; Takeuchi, T.; Yamaguchi, S.; Katoh, H.; Amano, H.; Akasaki, I. Optical band gap in $Ga_{1-x}In_xN$ ($0 < x < 0.2$) on GaN by photoreflection spectroscopy. *Appl. Phys. Lett.* **1998**, *73*, 1994–1996.
37. Deline, V.R.; Katz, W.; Evans, C.A.; Williams, P. Mechanism of the SIMS matrix effect. *Appl. Phys. Lett.* **1978**, *33*, 832–835. [CrossRef]
38. Pozina, G.; Khromov, S.; Hemmingsson, C.; Hultman, L.; Monemar, B. Effect of silicon and oxygen doping on donor bound excitons in bulk GaN. *Phys. Rev. B* **2011**, *84*, 165213. [CrossRef]
39. Fabien, C.A.M.; Gunning, B.P.; Doolittle, W.A.; Fischer, A.M.; Wei, Y.O.; Xie, H.; Ponce, F.A. Low-temperature growth of InGaN films over the entire composition range by MBE. *J. Cryst. Growth* **2015**, *425*, 15–118. [CrossRef]

40. Reshchikov, M.A.; McNamara, J.D.; Helava, H.; Usikov, A.; Makarov, Y. Two yellow luminescence bands in undoped GaN. *Sci. Rep.* **2018**, *8*, 8091. [CrossRef]
41. Khromov, S.; Hemmingsson, C.; Monemar, B.; Hultman, L.; Pozina, G. Optical properties of C-doped bulk GaN wafers grown by halide vapor phase epitaxy. *J. Appl. Phys.* **2014**, *116*, 223503. [CrossRef]

crystals

MDPI

Article

Micro-Nanoarchitectonics of Ga_2O_3/GaN Core-Shell Rod Arrays for High-Performance Broadband Ultraviolet Photodetection

Ruifan Tang [1,2], Guanqi Li [2], Xun Hu [2], Na Gao [2], Jinchai Li [2,*], Kai Huang [2,*], Junyong Kang [2] and Rong Zhang [2]

1 School of Physics and Electronic Science, Hunan Institute of Science and Technology, Yueyang 414006, China
2 Fujian Key Laboratory of Semiconductor Materials and Applications, CI Center for OSED Collaborative Innovation Center for Optoelectronic Semiconductors and Efficient Devices, Department of Physics, Xiamen University, Xiamen 361005, China
* Correspondence: jinchaili@xmu.edu.cn (J.L.); k_huang@xmu.edu.cn (K.H.)

Abstract: This study presents broadband ultraviolet photodetectors (BUV PDs) based on Ga_2O_3/GaN core-shell micro-nanorod arrays with excellent performance. Micro-Nanoarchitectonics of Ga_2O_3/GaN core-shell rod arrays were fabricated with high-temperature oxidization of GaN micro-nanorod arrays. The PD based on the microrod arrays exhibited an ultrahigh responsivity of 2300 A/W for 280 nm at 7 V, the peak responsivity was approximately 400 times larger than those of the PD based on the planar Ga_2O_3/GaN film. The responsivity was over 1500 A/W for the 270–360 nm band at 7 V. The external quantum efficiency was up to 1.02×10^6% for 280 nm. Moreover, the responsivity was further increased to 2.65×10^4 A/W for 365 nm and over 1.5×10^4 A/W for 270–360 nm using the nanorod arrays. The physical mechanism may have been attributed to the large surface area of the micro-nanorods coupled with the Ga_2O_3/GaN heterostructure, which excited more photogenerated holes to be blocked at the Ga_2O_3 surface and Ga_2O_3/GaN interface, resulting in a larger internal gain. The overall high performance coupled with large-scale production makes it a promising candidate for practical BUV PD.

Keywords: ultraviolet photodetectors; Ga_2O_3/GaN heterostructure; core-shell micro-nanorod arrays

Citation: Tang, R.; Li, G.; Hu, X.; Gao, N.; Li, J.; Huang, K.; Kang, J.; Zhang, R. Micro-Nanoarchitectonics of Ga_2O_3/GaN Core-Shell Rod Arrays for High-Performance Broadband Ultraviolet Photodetection. *Crystals* **2023**, *13*, 366. https://doi.org/10.3390/cryst13020366

Academic Editors: Dmitri Donetski, Peng Chen and Zhizhong Chen

Received: 3 January 2023
Revised: 2 February 2023
Accepted: 15 February 2023
Published: 20 February 2023

1. Introduction

Broadband ultraviolet photodetectors (BUV PDs) have attracted significant attention as an emerging technology, and they have been widely used in land invasion, forest fire and ozone hole security monitoring, satellite and security communications, and other fields [1–4]. Over the past few decades, significant developments in fabricating BUV PDs have been achieved based on wide-band-gap semiconductors, such as MgZnO/ZnO, AlGaN/GaN [5–10]. However, high Al/Mg-content AlGaN/MgZnO layers with high crystal quality are difficult to achieve by epitaxial growth, as the process of alloying makes the fabrication complex and introduces high defect density, thereby increasing the dark current and limiting the performance of PDs [11–17]. Alternatively, Ga_2O_3 can avoid the complex and unmanageable alloying process owing to its intrinsic solar-blind region band gap. Furthermore, Ga_2O_3 exhibits high thermal stability (MP 1730 °C), chemical stability and excellent thermal conductivity. Accordingly, Ga_2O_3-based PDs gained vast attention in the past few years [18–28]. Moreover, Ga_2O_3/GaN heterostructures have a broadband ultraviolet bandgap. More importantly, the intrinsic conduction band difference between Ga_2O_3 and GaN is 0.1 eV, which facilitates the flow of electrons. The intrinsic valence-band barrier between Ga_2O_3 and GaN is approximately 1.4 eV, which can effectively block hole transport, resulting in a large internal gain. Therefore, Ga_2O_3/GaN heterostructures are considered to be promising materials for BUV PDs. Guo et al. fabricated a super-high-performance self-powered UV PD based on the GaN/Sn:Ga_2O_3 p–n junction. The responsivity at 254 nm reached up to 3.05 A/W without consuming external power [29].

Lin et al. proposed a typical UV PD based on the "sandwich" structure (graphene/β-Ga_2O_3/GaN heterojunction). Using rectifying effect of the p-GaN/β-Ga_2O_3 diode, the photodetector showed an extremely low dark current density of 1.25×10^{-8} A/cm^2, and the responsivity is up to 12.8 A/W by the hot carrier multiplication in graphene under Solar-blind ultraviolet illumination [30]. Kalra et al. demonstrated epitaxial β-Ga_2O_3/GaN-based vertical metal–heterojunction-metal (MHM) broadband UV-A/UV-C PDs, the PD exhibited the responsivity (R) of 3.7 A/W at 256 nm, and the UV-to-visible rejection >10^3, and a photo-to-dark current ratio of >100 [31].

Although there are many reports on BUV PD based on Ga_2O_3/GaN heterostructures, they are mostly thin-film structures [29–33]. The Micro-Nanoarchitectonics of Ga_2O_3/GaN rod arrays are featured with a higher surface-to-volume ratio display superior optical absorption ability and higher recovery efficiency. Specifically, the higher surface area induces more photogenerated carriers and enhances recombination of nonequilibrium carrier via the high surface states, leading to excellent response characteristics. He et al. [34] grew Ga_2O_3/GaN nanowires by partially thermally oxidizing GaN nanowires grown using molecular beam epitaxy (MBE). The fabricated PD based on vertical Ga_2O_3/GaN nanowire arrays showed a broadband spectral response from 276 to 366 nm with a high responsivity exceeding 550 A/W at −5 V bias and a fast response speed in the millisecond range. Wang et al. [35] fabricated monoclinic β-Ga_2O_3 vertically aligned nanorod arrays (NRAs) by hydrothermal and post-annealing methods. The fabricated solar-blind PD based on β-Ga_2O_3 NRAs exhibited a high responsivity of 550 A/W under a bias of 5 V. Moreover, the PD exhibited self-powered characteristics, that is, a responsivity of 10.80 mA/W at 0 V bias and response time of 0.38 s. Although significant developments have been achieved based on Micro-Nanoarchitectonics Ga_2O_3/GaN rod arrays, there are still many areas that could be further optimized. For example, most of the reported structures are based on relative complex vertical structures, and the preparation methods of Ga_2O_3/GaN heterostructures are mainly MBE [36–39], metal-organic chemical vapor deposition (MOCVD) [40–42], pulsed laser deposition (PLD) [43–45], atomic layer deposition (ALD) [46–48] and so on. These methods require precise control. It is worth noting that thermal oxidation, that is, the high-temperature thermal oxidation of GaN to obtain Ga_2O_3, is relatively simple and suitable for mass production [49–61]. Furthermore, the metal-semiconductor-metal (MSM) structure takes advantage of the high light–electron conversion of nano/micro-rod arrays and possesses the unique advantage of easier fabrication compared to other vertical sandwich structures. More importantly, such an MSM geometry constructed two electrodes on the nano/microarrays to increase the electrode effective area and thus effectively collect photogenerated carriers, further improving the performance of PDs.

In this study, Micro-Nanoarchitectonics of Ga_2O_3/GaN core-shell rod arrays were prepared by high-temperature oxidization of GaN M-NRAs. MSM PD based on Ga_2O_3/GaN core-shell microrod arrays (MRAs) was fabricated. The PD showed excellent performance; an ultrahigh responsivity of 2300 A/W for 280 nm and over 1500 A/W for the 270–360 nm band was obtained at a 7 V bias. The external quantum efficiency was approximately 1.02×10^6% at 280 nm, respectively. The physical mechanism was attributable to the large surface area of the MRA arrays and Ga_2O_3/GaN heterojunction band structure, which blocked more holes at the surface of the M-NRA and Ga_2O_3/GaN heterojunction interface, resulting in a larger internal gain compared to the planar Ga_2O_3/GaN film-based PD. This mechanism is consistent with the results of PD fabricated using NRAs. It is worth highlighting that the features of inexpensive manufacturing and easy scalability are particularly attractive for mass production.

2. Materials and Methods

To take advantage of the internal gain brought by the special band structure of Ga_2O_3/GaN heterojunction, the Micro-architectonics of Ga_2O_3/GaN with core-shell structure were fabricated. Figure 1 is a schematic of the Ga_2O_3/GaN core-shell MRAs fabrication. GaN/sapphire samples with approximately 5 μm-thick unintentionally doped GaN epitaxial layer were used in this study. The GaN/sapphire wafers were ultrasonically cleaned in acetone and isopropyl alcohol for 15 min and then dipped into a diluted hydrochloric acid water solution (HCL:H_2O = 1:1) for 5 min to eliminate the native oxide. A SiO_2 film with a thickness of 200 nm was deposited on GaN using plasma-enhanced chemical vapor deposition (SENTECH SI500D). Subsequently, the polystyrene microsphere solution and alcohol were mixed in a 1:1 ratio, and then the polystyrene microspheres were transferred to the surface of the sample. A highly ordered self-assembled monolayer of microspheres with a diameter of 1.7 μm was arranged on the sample. To reduce the size of the microspheres and increase the distance between them, the samples were treated with oxygen plasma for 10 min. The polystyrene microsphere template was then transferred to the SiO_2 film by etching the SiO_2 film using inductively coupled plasma (ICP) with CF_4 gas for 120 s. After the ICP etching process, the polystyrene nanospheres were eliminated by dipping the sample in tetrahydrofuran for 2 h and sonicating for 30 s. Subsequently, the SiO_2 template was transferred to the GaN film through etching of the GaN film by ICP with Cl_2 and BCl_3 gases. Then, the sample was dipped in the BOE solution to eliminate the SiO_2 mask, and GaN MRAs with a cone morphology were fabricated. Subsequently, the sample was placed in an oxidizing furnace. The heating time from room temperature to 1000 °C was approximately 55 min in an oxygen-gas environment. GaN was oxidized in a quartz tube furnace injected with 50 sccm of oxygen gas at 1000 °C for 45 min at atmospheric pressure. Finally, the samples were removed from the oxidizing furnace when the temperature was reduced to 200 °C in a nitrogen protection environment. An interdigitated metal contact (Ti/Al) with a thickness of 30/100 nm was patterned using standard photolithography (Karlsuss MA6/BA6), DC magnetron sputtering, and lift-off processes. Each device comprised 10 pairs of interdigital electrodes with lengths, widths, and spacings of 300, 6, and 6 μm, respectively. Consequently, the effective illumination area was 3.42×10^{-4} cm^2. The fabricated PD was finally annealed by rapid thermal annealing under an N_2 atmosphere at 200 °C for 120 s. The contrast PD based on the planar Ga_2O_3/GaN film was fabricated using a direct high-temperature oxidization process from a GaN film, as described above.

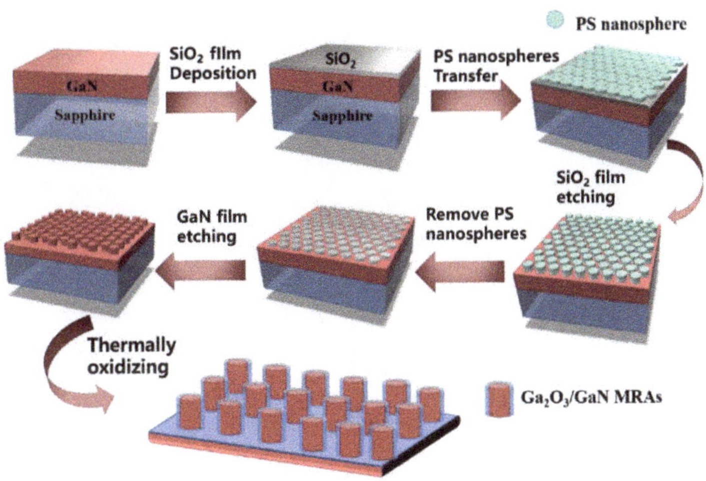

Figure 1. Schematic of the Ga_2O_3/GaN MRAs fabrication.

To characterize the material properties, the surface and cross-sectional morphology, microstructure, and the elemental mapping were determined using a ZEISS SIGMA high-resolution field emission scanning electron microscope (FE-SEM). Phase formation and crystallinity were monitored using a Helios Nano-Lab 460HP high-resolution transmission electron microscope (HR-TEM). The I–V characteristics were measured using a Keithley 2410 source meter and Keithley 6514 programmable electrometer. A 450 W Xe arc lamp, mechanical chopper, and lock-in amplifier were applied to measure the photocurrent response spectra. The system was calibrated using a standard silicon detector. The devices were all illuminated perpendicularly to the front metal/semiconductor contact side. All measurements were performed at room temperature.

3. Results

Figure 2a shows the surface morphology of the polystyrene microspheres. A highly ordered self-assembled monolayer of microspheres with a diameter of 1.7 μm was arranged on the sample. Other sizes of the polystyrene microspheres can also be used to fabricate the GaN rod arrays with different sizes. The diameter of the microspheres was reduced to 1.3 μm and the spacing of the microspheres increased after oxygen plasma treatment, as shown in Figure 2b. GaN rod arrays with different sizes and distances can also be prepared flexibly by adjusting the power and duration of oxygen plasma processing. Figure 2c shows the surface morphology of the GaN MRAs, and the highly ordered GaN MRAs with a diameter of about 1.3 μm was obtained. This size is similar to the size of the polystyrene microspheres after plasma treatment. Figure 2d shows the surface morphology of the oxidized Ga_2O_3/GaN MRAs. The diameter of the Ga_2O_3/GaN MRAs was approximately 1.35 μm. The height of the Ga_2O_3/GaN MRAs was approximately 750 nm, as shown in the inset of Figure 2d. Clearly, the Ga_2O_3/GaN MRAs show a core-shell structure, in which the outer shell is Ga_2O_3 with a 200–300 nm thickness and the core is GaN with an 850–1050 nm thickness. Some Ga_2O_3 grains with several hundred nanometers in size were observed on the top surface of the MRAs, which was also confirmed in our previous study. The X-ray diffraction (XRD) spectrum of the sample and morphology of the planar Ga_2O_3/GaN film were similar to those previously reported and are not shown here [54,55,59–61].

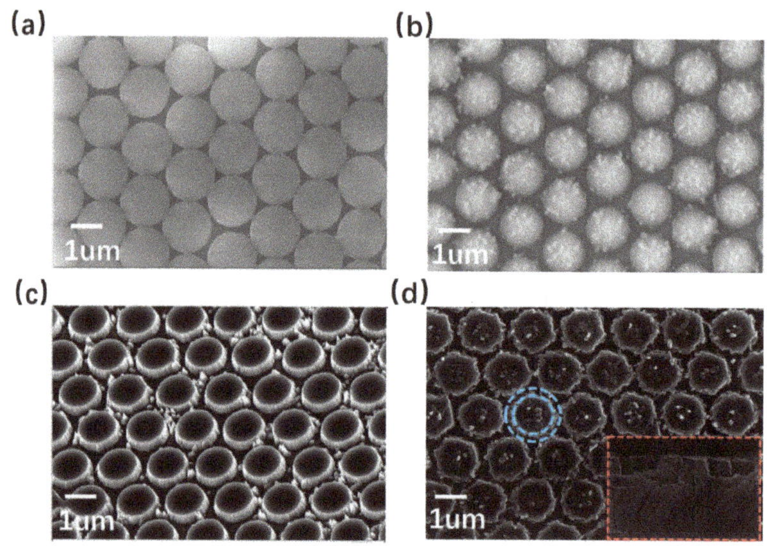

Figure 2. (**a**) Surface SEM image of microspheres. (**b**) Surface SEM image of microsphere after oxygen plasma treatment. (**c**) Surface morphology of the GaN MRAs after etching the GaN film. (**d**) Surface morphology of the oxidized Ga_2O_3/GaN MRAs. The inset shows the cross-sectional SEM image of Ga_2O_3/GaN MRAs.

To assess the photoresponse characteristics of the planar Ga_2O_3/GaN film and Micro-architectonics of Ga_2O_3/GaN core-shell rod arrays, MSM PD based on a planar Ga_2O_3/GaN film (sample A) and Ga_2O_3/GaN MRAs (sample B) were fabricated. The device structure diagram of sample B is shown in inset of Figure 3b. The MSM structure, that is, the two electrodes constructed directly on the MRAs, had a large electrode area, which enabled the efficient collection of photogenerated carriers. Moreover, owing to the intrinsic high valence-band barrier between Ga_2O_3 and GaN (1.4 eV), the holes generated in GaN will be blocked at the interface of Ga_2O_3/GaN heterojunction underneath the electrode, resulting in a greater internal gain. The larger metal electrode area will cause more holes to be blocked at the interface of Ga_2O_3/GaN heterojunction, resulting in greater internal gain. The spectral response from 200 to 400 nm at applied bias of 7 V was investigated. The spectral responsivity values via wavelength are calculated using the relation [29]:

$$R = \frac{I_{ph} - I_{Dark}}{P_\lambda} \tag{1}$$

where I_{ph} and I_{Dark} are the currents measured under illumination and dark conditions, respectively, and P_λ is the incident power density at a specific wavelength of λ, which was calibrated using a standard silicon detector. The R dependences on wavelength under the applied bias of 7 V for samples A and B are shown in Figures 3a and 3b, respectively. Both PDs exhibited a wide-spectrum response in the BUV region. When the illumination wavelength exceeds 370 nm, the responsivity of sample A decreases rapidly, which corresponds to the spectral response cutoffs of GaN material. Sample B has an obvious spectral response at 370–400 nm, which is attributable to the damage of GaN crystal structure by etching process during the preparation of GaN rod arrays, forming many surface defects in GaN and inducing the defect response. Therefore, it is necessary to further optimize the preparation process in the future. As shown in Figure 3a, the peak responsivity was 6.3 A/W at 260 nm for sample A, and the responsivity at the range of 250–360 nm exceeded 2 A/W. For sample B, the peak responsivity was 2300 A/W at 280 nm, the peak responsivity is enhanced by approximately 4×10^2 times compared with sample A (6.3 A/W), and the responsivity at the 270–360 nm band exceeded 1500 A/W, as shown in Figure 3b. It is also superior to some reported values [34,35]. Both of the two samples have high responsivity, indicating that the two PDs have a large external quantum efficiency (EQE), which is defined as the number of electrons generated per incident photon and can be obtained using the following equation [27]:

$$EQE = \frac{Rhc}{q\lambda} \tag{2}$$

where R, h, c, q, and λ are the responsivity, Planck's constant, speed of light, electron charge, and wavelength of the incident optical light, respectively. It is observed that sample A exhibited an EQE of approximately 3×10^3% for 260 nm and sample B exhibited an ultrahigh EQE of approximately 1.02×10^6% for 280 nm. Compared with sample A, the internal gain of sample B is significantly increased. Obviously, it cannot be explained by the increase in light absorption efficiency owing to the large surface area of MRAs. This is assignable to the larger surface area of the MRA coupled with the Ga_2O_3/GaN heterostructure. This is discussed in detail later in this study.

Detectivity (D*), which usually describes the smallest detectable signal, is another key figure of merit of PDs. The detectivity (D*) can be expressed as [27]

$$D^* = \frac{R}{\sqrt{2eI_{dark}/S}} \tag{3}$$

where S (3.42×10^{-4} cm^2) is the effective area under illumination with the shot noise from I_{Dark} regarded as the major component of the total noise. The detectivity D* of sample A was calculated as 4.47×10^{12} Jones for 260 nm. The detectivity D* of sample B was calculated as 2.65×10^{13} Jones for 280 nm and shows that the ability to detect weak signals is improved.

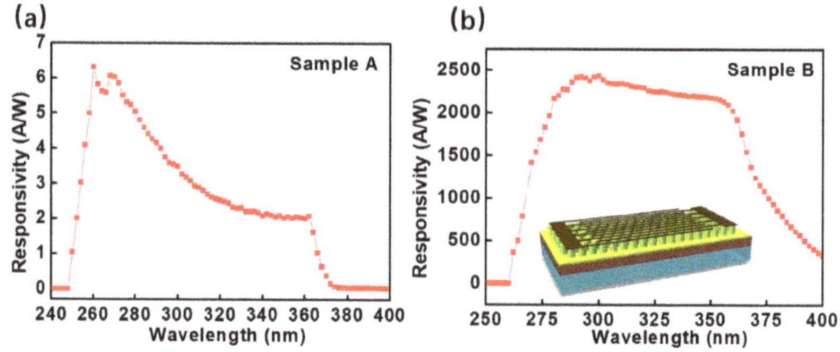

Figure 3. (**a**) Spectral responsivity values of sample A as a function of wavelength at 7 V bias, (**b**) spectral responsivity values of sample B as a function of wavelength at 7 V bias. The inset shows the device diagram of sample B.

Figure 4a,b shows the I–V characteristics of samples A and B under dark conditions and the illumination of 270 nm/360 nm light, respectively. For sample A, the dark current at 7 V bias is 1.14×10^{-9} A, and the photocurrents are 1.99×10^{-7} and 2.44×10^{-7} A at 270 nm and 360 nm light illuminations, respectively, while the photo-to-dark ratios for 360 nm and 270 nm/dark current both exceeded 10^2. Such low light-to-dark ratio is mainly due to the rather weak intensity of the 270 nm and 360 nm light illuminations, which is separated from the xenon lamp light passing through the monochromator. When the bias voltage is in the range of −3~3 V, the photocurrent at 360 nm light illumination is less than that at 270 nm light illumination, indicating that the depletion region is mainly in the Ga_2O_3 layer. As the bias voltage increases, the photocurrent for 360 nm increases rapidly, and exceeds the photocurrent for 270 nm, which indicates that the depletion layer of the device enters the GaN layer. The PD presents B-UV detection, which is consistent with the results of the response spectrum. For sample B, the dark current at 7 V bias is up to 10^{-6} A, which is three orders of magnitude higher than that of sample A. The significant increase in dark current is attributable to the following reasons. The larger electrode area of sample B not only increased the photocurrent, but also inevitably led to larger dark current. Meanwhile, the etching process of the GaN film also caused a deterioration of the GaN crystalline quality, resulting in a greater current leakage. Therefore, the photocurrent was evidently higher than that of dark current until the bias voltage exceeding 3.2 V. The photo-to-dark ratios for 360 nm and 270 nm/dark currents both exceeded 10. It is necessary to further optimize the process to improve the light/dark ratio.

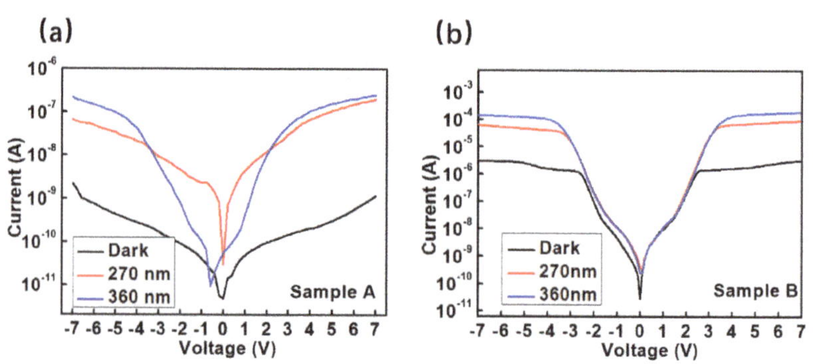

Figure 4. I−V characteristics of samples A (**a**) and B (**b**) in the dark and under 254 nm/365 nm UV illuminations.

To clearly explain the increased internal gain in sample B, the photodetection model and mechanism of sample B are schematically shown in Figure 5. As shown in Figure 5a,b, the microarchitectonics of Ga_2O_3/GaN core-shell rod arrays have a larger surface area, and more ambient oxygen molecules are adsorbed on the surface of the MRAs in the dark environment. The free electrons on the surface of the Ga_2O_3 film were trapped by oxygen molecules, forming a depletion layer and the energy band bending near the surface of the MRAs. When the device was illuminated under UV-C light, the photogenerated holes in Ga_2O_3 migrated to the surface of Ga_2O_3 due to the bending band. Compared with the photogenerated electrons in Ga_2O_3, the photogenerated holes in Ga_2O_3 that migrate to the surface of Ga_2O_3, were difficult to collect by the electrode, resulting in a large gain. When the device was illuminated under UV-A light, because negatively charged oxygen ions existed on the surface of the Ga_2O_3 film after the reaction with O_2, the photogenerated holes in the GaN film were aggregated at the interface of Ga_2O_3/GaN owing to the capacitance effect. These holes were also difficult for the electrode to collect, resulting in an enhanced gain. Therefore, sample B exhibited a greater internal gain than sample A in BUV band range.

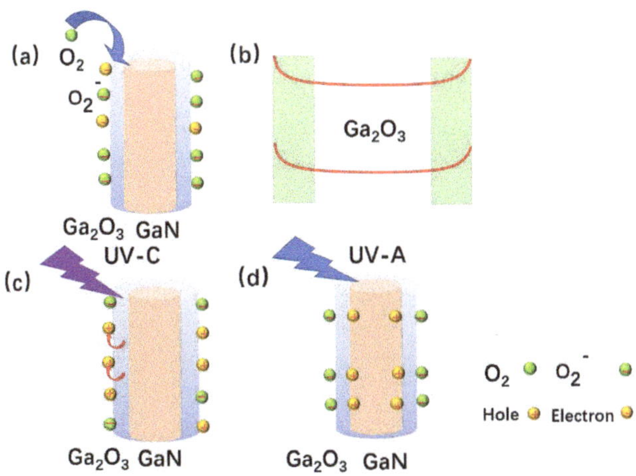

Figure 5. (**a**,**b**) Schematic of reaction of Ga_2O_3/GaN core-shell micron rod with oxygen and band of Ga_2O_3 films after reaction with oxygen respectively (**c**) Schematic of holes transferred to the surface of Ga_2O_3 film under UV-C illumination; (**d**) Schematic of photo-generated holes in GaN layer gathering at Ga_2O_3/GaN heterojunction interface owing to capacitance effect under UV-A illumination.

Moreover, the Ga_2O_3/GaN core-shell MRAs with higher surface-to-volume ratios effectively improved the light absorption efficiency and thus excited more photogenerated carriers. Because of the larger electrode area, these additional photogenerated electrons can be effectively collected and increase the photocurrent. In addition, more holes were blocked at the Ga_2O_3/GaN interface underneath the electrode, resulting in high responsivity. Since the Ga_2O_3/GaN heterojunction has the low conduction band barrier (0.1 eV) and high valence band barrier (1.4 eV), the photogenerated electrons in the GaN layer can pass through the Ga_2O_3/GaN heterojunction and be collected by the electrode. Meanwhile, the photogenerated holes in GaN are effectively blocked in the Ga_2O_3/GaN heterojunction, noting that these blocked photogenerated holes are mainly located at the Ga_2O_3/GaN heterojunction interface below the electrode. A larger metal electrode area can block more holes, thus leading to greater photoconductance gain. The designed structure effectively coupled the optical field enhancement effect with the internal gain of the heterojunction, resulting in ultra-high responsivity of the PDs.

Whether the increased gain caused by the oxygen adsorption, or the large electrode area coupled by the internal gain of the Ga_2O_3/GaN heterojunction, both are related to the surface area of Ga_2O_3/GaN core-shell rod arrays. To further confirm the above mechanism, Ga_2O_3/GaN NRAs with larger surface areas were fabricated using the same method as described above, except that the polystyrene microspheres with a diameter of 1 μm was applied as the mask template and the thermal oxidation time was reduced to 30 min. Figure 6a,b shows the surface and cross-sectional topography of the NRAs. As shown in Figure 6a,b, the highly ordered Ga_2O_3/GaN NRAs were obtained, the diameter of the nano-column is approximately 700 nm and the height is approximately 400–450 nm. A thin layer of Ga_2O_3 was identified in the outermost layer of the NRA. In order to demonstrate the formation of the Ga_2O_3/GaN core-shell structures, high-resolution HR-TEM inspections were performed. As shown in Figure 6c–e, the outmost shell with a thickness of 10–50 nm is confirmed, which corresponds to the β-Ga_2O_3 (201) and β-Ga_2O_3 (400), and the core area corresponds to GaN (0001). These results indicate that the GaN NRAs were oxidized at high temperature and form the polycrystalline Ga_2O_3 film on the GaN surface.

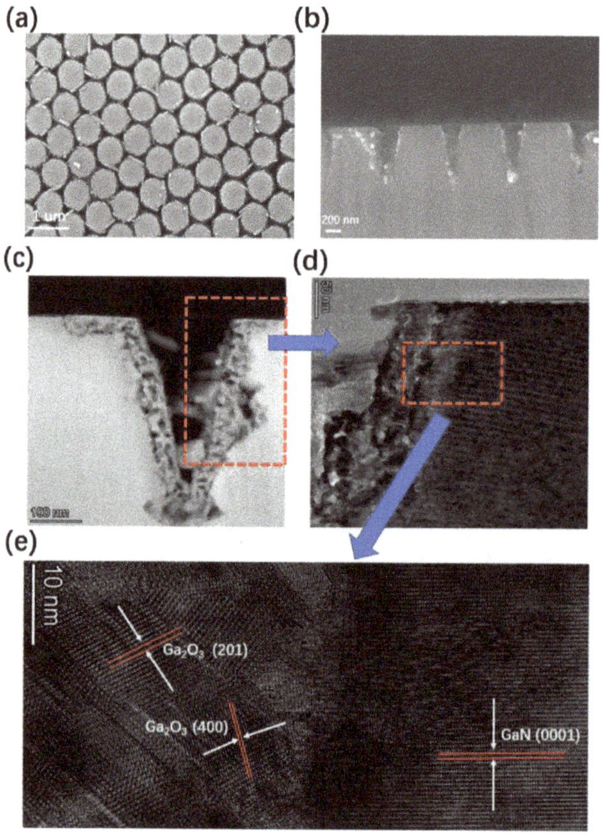

Figure 6. (**a**) Surface morphology of the oxidized Ga_2O_3/GaN NRAs. (**b**) Cross-section SEM image of Ga_2O_3/GaN NRAs. (**c**,**d**) Low-magnification TEM image of the core-shell NRA. (**e**) HR-TEM image of the core-shell NRA.

To further confirm the formation of the nanoarchitectonics of Ga_2O_3/GaN rod arrays with core-shell structure by thermal oxidation, energy dispersive X-ray (EDX) and element mapping characterizations were performed on the cross section of Ga_2O_3/GaN NRAs. As shown in Figure 7a–e, it can be intuitively seen that Ga and O elements are mainly distributed in the outermost layer of the nanorods, which represents the formed Ga_2O_3 thin film layer. The Ga and N elements are mainly distributed inside the nanorods, which corresponds to the GaN layer inside. These results further demonstrate the formation of the Ga_2O_3/GaN core-shell structures. To assess the photoresponse characteristics of the nano-architectonics of Ga_2O_3/GaN core-shell rod arrays, the PD (sample C) was fabricated in these structures with identical preparation method. Figure 8 shows the responsivity spectrum of sample C. The PDs exhibited a wide-spectrum response in the BUV region. An obvious optical response at 370–400 nm is observed in sample C, which is the same as sample B, indicating the existence of interband response. The peak responsivity of sample C is in the UV-A band range, while the case for sample B is in the UV-C band as described above. This is mainly because the thickness of Ga_2O_3 in sample C is ultrathin, only 10–50 nm, and the sample C mainly exhibits UV-A band response induced from GaN material. As shown in Figure 8, the peak responsivity was 2.65×10^4 A/W for 365 nm at 7 V applied bias, and the responsivity at the 270–360 nm band exceeded 1.5×10^4 A/W, which is a further order of magnitude enhancement than sample B. These results show that larger surface area leads to greater internal gain, which is consistent with the proposed mechanisms as aforementioned. Nevertheless, it is possible to further optimize the thickness of Ga_2O_3 film on GaN by adjusting the oxidation time, and realize the tunable detection in UV band. Overall, we proposed a simple method to fabricate the B-UV PD with high responsivity, which enables the detection under weak light illuminations.

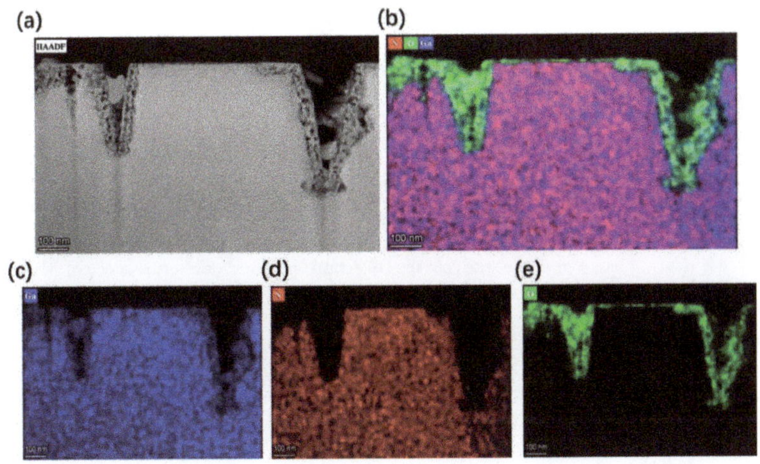

Figure 7. (**a**) Cross-section SEM image of Ga_2O_3/GaN NRA and (**b**) corresponding element mapping; (**c–e**) are distribution diagrams of Ga, N and O elements respectively.

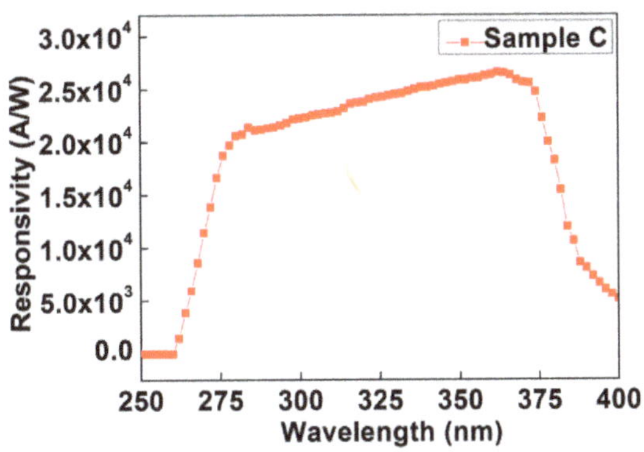

Figure 8. Spectral responsivity values of sample C as a function of wavelength at 7 V bias.

4. Conclusions

In this study, we designed and demonstrated a BUV PD based on Ga_2O_3/GaN core-shell MRAs and NRAs, which were fabricated using partially thermally oxidizing GaN MRAs and NRAs. The PD based on MRAs showed an ultrahigh responsivity of 2300 A/W for 280 nm and over 1500 A/W for the 270–360 nm band at 7 V. The detectivity and external quantum efficiency were approximately 2.65×10^{13} Jones and 1.02×10^{6}% at 280 nm, respectively. The responsivity was further increased to 2.65×10^{4} A/W for 280 nm and over 1.5×10^{4} A/W for 270–360 nm using the nanorod arrays. Such high performance can be attributed to the blocking of photo-generated holes. Specifically, the large surface area of Ga_2O_3/GaN MRAs and NRAs led to the migration of photogenerated holes to the surface of Ga_2O_3 film, which was caused by the band bending of Ga_2O_3 surface induced by oxygen absorption mechanism and resulted in the large internal gains. Moreover, the larger metal electrode area can block more holes at the heterojunction interface of Ga_2O_3/GaN, which was caused by the larger valence-band barrier (1.4 eV). Overall, the proposed metal–semiconductor–metal structure enables a large surface area and is suitable for mass production.

Author Contributions: R.T., J.L. and K.H. conceived the idea and designed the experiments. J.L. and K.H. supervised the project. R.T., G.L. and N.G. explored the oxidation process and the devices' fabrication. R.T., G.L. and J.L. performed the device characterization and the simulation. R.T., X.H., J.L., K.H., J.K. and R.Z. analyzed the data and contributed to the mechanism investigation. R.T., J.L. and K.H. wrote the manuscript. All authors participated in the discussion and commented on the manuscript. All authors have read and agreed to the published version of the manuscript.

Funding: This research was funded by Key Technologies Research and Development Program (2021YFB3600101); National Natural Science Foundation of China (62174141, 61974126, 51902273, and 61874090); Key scientific and technological Program of Xiamen (3502Z20191016); Natural Science Foundation of Fujian Province (No. 2021J01008).

Conflicts of Interest: The authors declare no conflict of interest. The funders had no role in the design of the study; in the collection, analyses, or interpretation of data; in the writing of the manuscript; or in the decision to publish the results.

References

1. Aslam, S.; Miko, L.; Stahle, C.; Franz, D.; Pugel, D.; Guan, B.; Zhang, J.; Gaska, R. Dual-band deep ultraviolet AlGaN photodetectors. *Electron. Lett.* **2007**, *43*, 1382–1384. [CrossRef]
2. Assefa, S.; Xia, F.; Vlasov, Y.A. Reinventing germanium avalanche photodetector for nanophotonic on-chip optical interconnects. *Nature* **2010**, *464*, 80–84. [CrossRef] [PubMed]
3. Gong, X.; Tong, M.; Xia, Y.; Cai, W.; Moon, J.S.; Cao, Y.; Yu, G.; Shieh, C.-L.; Nilsson, B.; Heeger, A.J. High-Detectivity Polymer Photodetectors with Spectral Response from 300 nm to 1450 nm. *Science* **2009**, *325*, 1665–1667. [CrossRef] [PubMed]
4. Konstantatos, G.; Sargent, E.H. Nanostructured materials for photon detection. *Nat. Nanotechnol.* **2010**, *5*, 391–400. [CrossRef] [PubMed]
5. Monroy, E.; Omn, F.; Calle, F. Wide-Bandgap Semiconductor Ultraviolet Photodetectors. *Semicond. Sci. Technol.* **2003**, *18*, R33–R51. [CrossRef]
6. Zheng, W.; Lin, R.; Zhang, D.; Jia, L.; Ji, X.; Huang, F. A new superconductor of cuprates with unique features. *Adv. Opt. Mater.* **2018**, *6*, 1800697. [CrossRef]
7. Li, J.Z.; Lin, J.Y.; Jiang, H.X.; Khan, M.A. Effects of persistent photoconductivity on the characteristic performance of an AlGaN/GaN heterostructure ultraviolet detector. *Appl. Phys. Lett.* **1998**, *72*, 2868–2870. [CrossRef]
8. Ji, L.W.; Peng, S.M.; Su, Y.K.; Young, S.-J.; Wu, C.Z.; Cheng, W.B. Ultraviolet photodetectors based on selectively grown ZnO nanorod arrays. *Appl. Phys. Lett.* **2009**, *94*, 203106. [CrossRef]
9. Lupan, O.T.; Braniste, M.; Deng, L.; Ghimpu, I.; Paulowicz, Y.K.; Mishra, L.; Kienle, R.; Adelung, I.; Tiginyanu, I. Rapid switching and ultra-responsive nanosensors based on individual shell–core Ga_2O_3/GaN:O_x@SnO_2 nanobelt with nanocrystalline shell in mixed phases. *Sens. Actuators B* **2015**, *221*, 544–555. [CrossRef]
10. Yu, J.; Shan, C.X.; Huang, X.M.; Zhang, X.W.; Wang, S.P.; Shen, D.Z. ZnO-based ultraviolet avalanche photodetectors. *J. Phys. D Appl. Phys.* **2013**, *46*, 305105. [CrossRef]
11. Yu, J.; Shan, C.X.; Liu, J.S.; Zhang, X.W.; Li, B.H.; Shen, D.Z. MgZnO avalanche photodetectors realized in Schottky structures. *Phys. Status Solidi. RRL—Rapid Res. Lett.* **2013**, *7*, 425–428. [CrossRef]

12. Shao, Z.G.; Chen, D.J.; Lu, H.; Zhang, R.; Cao, D.P.; Luo, W.J.; Zheng, Y.D.; Li, L.; Li, Z.H. High-Gain AlGaN Solar-Blind Avalanche Photodiodes. *IEEE Electron Device Lett.* **2014**, *35*, 372–374. [CrossRef]
13. Huang, Y.; Chen, D.J.; Lu, H.; Dong, K.X.; Zhang, R.; Zheng, Y.D.; Li, L.; Li, Z.H. Back-illuminated separate absorption and multiplication AlGaN solar-blind avalanche photodiodes. *Appl. Phys. Lett.* **2012**, *101*, 253516. [CrossRef]
14. Tut, T.; Gokkavas, M.; Inal, A.; Ozbay, E. AlxGa1-xN-based avalanche photodiodes with high reproducible avalanche gain. *Appl. Phys. Lett.* **2007**, *90*, 163506. [CrossRef]
15. Huang, Z.; Li, J.; Zhang, W.; Jiang, H. AlGaN Solar-Blind Avalanche Photodiodes with Enhanced Multiplication Gain Using Back-Illuminated Structure. *Appl. Phys. Express* **2013**, *6*, 054101. [CrossRef]
16. Tut, T.; Gokkavas, M.; Butun, B.; Butun, S.; Ulker, E.; Ozbay, E. Experimental evaluation of impact ionization coefficients in AlxGa1−xN based avalanche photodiodes. *Appl. Phys. Lett.* **2006**, *89*, 183524. [CrossRef]
17. Xie, X.H.; Zhang, Z.Z.; Shan, C.X.; Chen, H.Y.; Shen, D.Z. Dual-color ultraviolet photodetector based on mixed-phase-MgZnO/i-MgO/p-Si double heterojunction. *Appl. Phys. Lett* **2012**, *101*, 081104. [CrossRef]
18. Xuan, H.C.; Wen, X.M.; Yang, X.; Bo, F.; Zhi, T.J.; Jian, D.Y. Highly Narrow-Band Polarization-Sensitive Solar-Blind Photodetectors Based on β-Ga_2O_3 Single Crystals. *ACS Appl. Mater. Interfaces* **2019**, *11*, 7131–7137.
19. Kong, W.Y.; Wu, G.A.; Wang, K.Y.; Zhang, T.F.; Zou, Y.F.; Wang, D.D.; Luo, L.B. Graphene-β-Ga_2O_3 Heterojunction for Highly Sensitive Deep UV Photodetector Application. *Adv. Mater.* **2016**, *28*, 10725–10731. [CrossRef]
20. Pratiyush, A.; Krishnamoorthy, S.; Solanke, S.S.; Xia, V.Z.; Muralidharan, R.; Rajan, S.; Nath, D.N. High responsivity in molecular beam epitaxy grown β-Ga_2O_3 metal semiconductor metal solar blind deep-UV photodetector. *Appl. Phys. Lett.* **2017**, *110*, 221107. [CrossRef]
21. Qu, Y.; Wu, Z.; Ai, M.; Guo, D.; An, Y.; Yang, H.; Li, L.; Tang, W. Enhanced Ga_2O_3/SiC ultraviolet photodetector with graphene top electrodes. *J. Alloys Compd* **2016**, *680*, 247–251. [CrossRef]
22. Zhao, B.; Wang, F.; Chen, H.; Zheng, L.; Su, L.; Zhao, D.; Fang, X. An Ultrahigh Responsivity (9.7 mA W− ( ) Self-Powered Solar-Blind Photodetector Based on Individual ZnO-Ga_2O_3 Heterostructures. *Adv. Funct. Mater.* **2017**, *27*, 1700264. [CrossRef]
23. Guo, D.Y.; Liu, H.; Li, P.; Wu, Z.; Wang, S.; Cui, C.; Li, C.; Tang, W. Zero-Power-Consumption Solar-Blind Photodetector Based on β-Ga_2O_3/NSTO Heterojunction. *ACS Appl. Mater. Interfaces* **2017**, *9*, 1619–1628. [CrossRef] [PubMed]
24. Chen, X.; Liu, K.; Zhang, Z.; Wang, C.; Li, B.; Zhao, H.; Zhao, D.; Shen, D. Self-Powered Solar-Blind Photodetector with Fast Response Based on Au/β-Ga_2O_3 Nanowires Array Film Schottky Junction. *ACS Appl. Mater. Interfaces* **2016**, *8*, 4185–4191. [CrossRef]
25. Li, P.; Shi, H.; Chen, K.; Guo, D.; Cui, W.; Zhi, Y.; Wang, S.; Wu, Z.; Chen, Z.; Tang, W. Construction of GaN/Ga_2O_3 p–n junction for an extremely high responsivity self-powered UV photodetector. *J. Mater. Chem. C* **2017**, *5*, 10562–10570. [CrossRef]
26. Kanika, A.; Neeraj, G.; Mahesh, K.; Mukesh, K. Ultra-high-performance of Self-Powered β-Ga_2O_3 Thin Film Solar-blind Photodetector Grown on Cost-Effective Si Substrate using High-Temperature Seed Layer. *ACS Photonics* **2018**, *5*, 2391–2401.
27. Qian, L.-X.; Wu, Z.-H.; Zhang, Y.-Y.; Lai, P.T.; Liu, X.-Z.; Li, Y.-R. Ultrahigh-Responsivity, Rapid-Recovery, Solar-Blind Photodetector Based on Highly Nonstoichiometric Amorphous Gallium Oxide. *ACS Photon.* **2017**, *4*, 2203–2211. [CrossRef]
28. Weng, W.Y.; Hsueh, T.J.; Chang, S.J.; Huang, G.J. A β-Ga_2O_3/GaN Hetero-Structured Solar-Blind and Visible-Blind Dual-Band Photodetector. *IEEE Sens. J.* **2011**, *11*, 5. [CrossRef]
29. Guo, D.; Su, Y.; Shi, H.; Li, P.; Zhao, N.; Ye, J.; Wang, S.; Liu, A.; Chen, Z.; Li, C.; et al. Self-Powered Ultraviolet Photodetector with Superhigh Photoresponsivity (3.05 A/W) Based on the GaN/Sn:β-Ga_2O_3 pn Junction. *ACS Nano* **2018**, *12*, 12827–12835. [CrossRef]
30. Lin, W.; Zheng, D.; Zhang, Z.; Zhang, Q.; Liao, L.; Yang, F.; Huang, F. High-Performance Graphene/β-Ga_2O_3 Heterojunction Deep-Ultraviolet Photodetector with Hot-Electron Excited Carrier Multiplication. *ACS Appl. Mater. Interfaces* **2018**, *10*, 22419–22426. [CrossRef]
31. Kalra, A.; Vura, S.; Rathkanthiwar, S.; Muralidharan, R.; Raghavan, S.; Nath, D.N. Demonstration of High-Responsivity Epitaxial β-Ga_2O_3/GaN Metal–Heterojunction-Metal Broadband UV-A/UV-C Detector. *Appl. Phys. Express* **2018**, *11*, 064101. [CrossRef]
32. Nakagomi, S.; Sato, T.; Takahashi, Y.; Kokubun, Y. Deep Ultraviolet Photodiodes Based on the β-Ga_2O_3/GaN Heterojunction. *Sens. Actuators A Phys.* **2015**, *232*, 208–213. [CrossRef]
33. Zeng, C.H.; Ma, Y.J.; Zhang, B.S.; Xu, Y.M.; Kong, M. Broadband Ultraviolet Photodetector Based on Graphene/β-Ga_2O_3/GaN Heterojunction. *Mater. Sci. Forum* **2020**, *1014*, 131–136. [CrossRef]
34. He, T.; Zhang, X.; Ding, X.; Sun, C.; Zhao, Y.; Yu, Q.; Ning, J.; Wang, R.; Yu, G.; Lu, S.; et al. Broadband Ultraviolet Photodetector Based on Vertical Ga_2O_3/GaN Nanowire Array with High Responsivity. *Adv. Opt. Mater.* **2019**, *7*, 1801563. [CrossRef]
35. Wang, S.; Chen, K.; Zhao, H.; He, C.; Wu, C.; Guo, D.; Zhao, N.; Ungar, G.; Shen, J.; Chu, X.; et al. β-Ga_2O_3 Nanorod Arrays with High Light-to-Electron Conversion for Solar-Blind Deep Ultraviolet Photodetection. *RSC Adv.* **2019**, *9*, 6064–6069. [CrossRef] [PubMed]
36. Kalarickal, N.K.; Xia, Z.; McGlone, J.; Krishnamoorthy, S.; Moore, W.; Brenner, M.; Arehart, A.R.; Ringel, S.A.; Rajan, S. Mechanism of Si Doping in Plasma Assisted MBE Growth of β-Ga_2O_3. *Appl. Phys. Lett.* **2019**, *115*, 152106. [CrossRef]
37. Pratiyush, A.S.; Krishnamoorthy, S.; Kumar, S.; Xia, Z.; Muralidharan, R.; Rajan, S.; Nath, D.N. Demonstration of Zero Bias Responsivity in MBE Grown β-Ga_2O_3 Lateral Deep-UV Photodetector. *Jpn. J. Appl. Phys.* **2018**, *57*, 060313. [CrossRef]
38. Sasaki, K.; Higashiwaki, M.; Kuramata, A.; Masui, T.; Yamakoshi, S. MBE grown Ga_2O_3 and its power device applications. *J. Cryst. Growth* **2013**, *378*, 591–595. [CrossRef]

39. Sasaki, K.; Higashiwaki, M.; Kuramata, A.; Masui, T.; Yamakoshi, S. Growth temperature dependences of structural and electrical properties of Ga_2O_3 epitaxial films grown on β-Ga_2O_3 (010) substrates by molecular beam epitaxy. *J. Cryst. Growth* **2014**, *392*, 30–33. [CrossRef]
40. Alema, F.; Hertog, B.; Mukhopadhyay, P.; Zhang, Y.; Mauze, A.; Osinsky, A.; Schoenfeld, W.V.; Speck, J.S.; Vogt, T. Solar Blind Schottky Photodiode Based on an MOCVD-Grown Homoepitaxial β-Ga_2O_3 Thin Film. *APL Mater.* **2019**, *7*, 022527. [CrossRef]
41. Feng, Z.; Anhar Uddin Bhuiyan, A.F.M.; Karim, M.R.; Zhao, H. MOCVD Homoepitaxy of Si-Doped (010) β-Ga_2O_3 Thin Films with Superior Transport Properties. *Appl. Phys. Lett.* **2019**, *114*, 250601. [CrossRef]
42. Seryogin, G.; Alema, F.; Valente, N.; Fu, H.; Steinbrunner, E.; Neal, A.T.; Mou, S.; Fine, A.; Osinsky, A. MOCVD Growth of High Purity Ga_2O_3 Epitaxial Films Using Trimethylgallium Precursor. *Appl. Phys. Lett.* **2020**, *117*, 262101. [CrossRef]
43. Kneiß, M.; Hassa, A.; Splith, D.; Sturm, C.; von Wenckstern, H.; Schultz, T.; Koch, N.; Lorenz, M.; Grundmann, M. Tin-Assisted Heteroepitaxial PLD-Growth of κ-Ga_2O_3 Thin Films with High Crystalline Quality. *APL Mater.* **2019**, *7*, 022516. [CrossRef]
44. Wang, Q.; Chen, J.; Huang, P.; Li, M.; Lu, Y.; Homewood, K.P.; Chang, G.; Chen, H.; He, Y. Influence of growth temperature on the characteristics of β-Ga_2O_3 epitaxial films and related solar-blind photodetectors. *Appl. Surf. Sci.* **2019**, *489*, 101–109. [CrossRef]
45. Xu, C.; Liu, H.; Pan, X.; Ye, Z. Growth and characterization of Si-doped β-Ga_2O_3 films by pulsed laser deposition. *Opt. Mater.* **2020**, *108*, 110145. [CrossRef]
46. Altuntas, H.; Donmez, I.; Ozgit-Akgun, C.; Biyikli, N. Electrical characteristics of β-Ga_2O_3 thin films grown by PEALD. *J. Alloys Compd.* **2014**, *593*, 190–195. [CrossRef]
47. Choi, D.-W.; Chung, K.-B.; Park, J.-S. Low temperature Ga_2O_3 atomic layer deposition using gallium tri-isopropoxide and water. *Thin Solid Film.* **2013**, *546*, 31–34. [CrossRef]
48. Comstock, D.J.; Elam, J.W. Atomic Layer Deposition of Ga_2O_3 Films Using Trimethylgallium and Ozone. *Chem. Mater.* **2012**, *24*, 4011–4018. [CrossRef]
49. Zhou, H.; Alghamdi, S.; Si, M.; Qiu, G.; Ye, P.D. Al2O3/β-Ga_2O_3(-201) Interface Improvement Through Piranha Pretreatment and Postdeposition Annealing. *IEEE Electron Device Lett.* **2016**, *37*, 1411–1414. [CrossRef]
50. Choi, J.-H.; Ham, M.-H.; Lee, W.; Myoung, J.-M. Fabrication and characterization of GaN/amorphous Ga_2O_3 nanocables through thermal oxidation. *Solid State Commun.* **2007**, *142*, 437–440. [CrossRef]
51. Filippo, E.; Siciliano, M.; Genga, A.; Micocci, G.; Tepore, A.; Siciliano, T. Single crystalline β-Ga_2O_3 nanowires synthesized by thermal oxidation of GaSe layer. *Mater. Res. Bull.* **2013**, *48*, 1741–1744. [CrossRef]
52. Lin, L.; Luo, Y.; Lai, P.; Lau, K.M. Influence of oxidation and annealing temperatures on quality of Ga_2O_3 film grown on GaN. *Thin Solid Films* **2006**, *515*, 2111–2115. [CrossRef]
53. Oshima, T.; Kaminaga, K.; Mukai, A.; Sasaki, K.; Masui, T.; Kuramata, A.; Yamakoshi, S.; Fujita, S.; Ohtomo, A. Formation of Semi-Insulating Layers on Semiconducting β-Ga_2O_3 Single Crystals by Thermal Oxidation. *Jpn. J. Appl. Phys.* **2013**, *52*, 051101. [CrossRef]
54. Tang, R.; Li, G.; Li, C.; Li, J.; Zhang, Y.; Huang, K.; Ye, J.; Li, C.; Kang, J.; Zhang, R.; et al. Localized Surface Plasmon Enhanced Ga_2O_3 Solar Blind Photodetectors. *Opt. Express OE* **2020**, *28*, 5731–5740. [CrossRef] [PubMed]
55. Tang, R.F.; Li, G.Q.; Jiang, Y.; Gao, N.; Li, J.C.; Li, C.; Huang, K.; Kang, J.Y.; Wang, T.; Zhang, R. Ga_2O_3/GaN heterostructural ultraviolet photodetectors with exciton-dominated ultranarrow response. *ACS Appl. Electron. Mater.* **2021**, *4*, 188–196. [CrossRef]
56. Wang, S.; Li, Y.W.; Xiu, X.Q.; Zhang, L.Y.; Hua, X.M.; Xie, Z.L.; Tao, T.; Liu, B.; Chen, P.; Zhang, R. Synthesis and characterization of β-Ga_2O_3@ GaN nanowires. *Chin. Phys. B* **2019**, *28*, 028104. [CrossRef]
57. Zhang, L.Y.; Li, Y.W.; Xiu, X.Q.; Xin, G.Q.; Xie, Z.L.; Tao, T.; Liu, B.; Chen, P.; Zhang, R.; Zheng, Y.D. Preparation of vertically aligned GaN@Ga_2O_3 core-shell heterostructured nanowire arrays and their photocatalytic activity for degradation of Rhodamine B. *Superlattices Microstruct.* **2020**, *143*, 106556. [CrossRef]
58. Ariga, K. Nanoarchitectonics: What's coming next after nanotechnology? *Nanoscale Horiz.* **2021**, *6*, 364–378. [CrossRef]
59. Sudhir, G.S.; Peyrot, Y.; Krüger, J.; Kim, Y.; Klockenbrink, R.; Kisielowski, C.; Rubin, M.D.; Weber, E.R.; Kriegseis, W.; Meyer, B.K. Effect of Mg, Zn, Si, and O on the Lattice Constant of Gallium Nitride Thin Films. *MRS Online Proc. Libr.* **1997**, *482*, 569–574. [CrossRef]
60. Greco, G.; Fiorenza, P.; Giannazzo, F.; Alberti, A.; Roccaforte, F. Nanoscale electrical and structural modification induced by rapid thermal oxidation of AlGaN/GaN heterostructures. *Nanotechnology* **2013**, *25*, 025201. [CrossRef]
61. Hayes, J.M.; Kuball, M.; Bell, A.T.; Harrison, I.; Korakakis, D.; Foxon, C.T. High-temperature processing of GaN: The influence of the annealing ambient on strain in GaN. *Appl. Phys. Lett.* **1999**, *75*, 2097–2099. [CrossRef]

crystals

MDPI

Article

MOCVD-grown β-Ga_2O_3 as a Gate Dielectric on AlGaN/GaN-Based Heterojunction Field Effect Transistor

Samiul Hasan [1,*], Mohi Uddin Jewel [1], Scott R. Crittenden [2], Dongkyu Lee [3], Vitaliy Avrutin [4], Ümit Özgür [4], Hadis Morkoç [4] and Iftikhar Ahmad [1,*]

1 Department of Electrical Engineering, University of South Carolina, Columbia, SC 29208, USA
2 Department of Physics and Astronomy, University of South Carolina, Columbia, SC 29208, USA
3 Department of Mechanical Engineering, University of South Carolina, Columbia, SC 29208, USA
4 Department of Electrical and Computer Engineering, Virginia Commonwealth University, Richmond, VA 23284, USA
* Correspondence: shasan@email.sc.edu (S.H.); ahmad@cec.sc.edu (I.A.)

Abstract: We report the electrical properties of $Al_{0.3}Ga_{0.7}N/GaN$ heterojunction field effect transistor (HFET) structures with a Ga_2O_3 passivation layer grown by metal–organic chemical vapor deposition (MOCVD). In this study, three different thicknesses of β-Ga_2O_3 dielectric layers were grown on $Al_{0.3}Ga_{0.7}N/GaN$ structures leading to metal-oxide-semiconductor-HFET or MOSHFET structures. X-ray diffraction (XRD) showed the ($\bar{2}$01) orientation peaks of β-Ga_2O_3 in the device structure. The van der Pauw and Hall measurements yield the electron density of ~ 4×10^{18} cm^{-3} and mobility of ~770 $cm^2V^{-1}s^{-1}$ in the 2-dimensional electron gas (2DEG) channel at room temperature. Capacitance–voltage (*C-V*) measurement for the on-state 2DEG density for the MOSHFET structure was found to be of the order of ~1.5×10^{13} cm^{-2}. The thickness of the Ga_2O_3 layer was inversely related to the threshold voltage and the on-state capacitance. The interface charge density between the oxide and $Al_{0.3}Ga_{0.7}N$ barrier layer was found to be of the order of ~10^{12} cm^2eV^{-1}. A significant reduction in leakage current from ~10^{-4} A/cm^2 for HFET to ~10^{-6} A/cm^2 for MOSHFET was observed well beyond pinch-off in the off-stage at -20 V applied gate voltage. The annealing at 900 °C of the MOSHFET structures revealed that the Ga_2O_3 layer was thermally stable at high temperatures resulting in insignificant threshold voltage shifts for annealed samples with respect to as-deposited (unannealed) structures. Our results show that the MOCVD-gown Ga_2O_3 dielectric layers can be a strong candidate for stable high-power devices.

Keywords: GaN; MOSHFET; Ga_2O_3; MOCVD; gate dielectric

Citation: Hasan, S.; Jewel, M.U.; Crittenden, S.R.; Lee, D.; Avrutin, V.; Özgür, Ü.; Morkoç, H.; Ahmad, I. MOCVD-grown β-Ga_2O_3 as a Gate Dielectric on AlGaN/GaN-Based Heterojunction Field Effect Transistor. *Crystals* **2023**, *13*, 231. https://doi.org/10.3390/cryst13020231

Academic Editors: Peng Chen and Zhizhong Chen

Received: 6 January 2023
Revised: 21 January 2023
Accepted: 23 January 2023
Published: 28 January 2023

1. Introduction

GaN-based heterojunction field effect transistors (HFETs) have excellent properties such as high critical breakdown field, high current, and superior thermal/chemical stability, which are coveted for high power, both RF and high-frequency switching applications under harsh environments [1–4]. The high-frequency performance of HFETs is limited by a series of effects associated with charge leakage, trapping/de-trapping, and conduction characteristics at different locations of the devices [5]. One of the most significant performance-limiting phenomena is the injection of electrons from the gate electrode to the surface region of the semiconductor on the drain side of the gate electrode, which results in reliability issues as well as limitations on the input drive in power applications [6,7]. For high drain and gate bias, the magnitude of the electric field under the gate region can cause tunneling/leakage of electrons from the gate metal to the semiconductor. Thus, the tunneling/gate leakage effect becomes critical for radio frequency (RF) applications because the electric field oscillates with the total terminal (dc+RF) voltage [8]. During the high voltage portion of the RF cycle, a pulse of electrons will tunnel from the gate by acquiring sufficient energy and can cause gate breakdown [9,10]. The electron transport on the surface is sluggish due to high effective mass, and dispersion is introduced due to the charging/discharging time constant [11]. The electrons escaping from the gate metal gather on the surface, creating a "virtual gate" effect that functions as an effective increase in gate length on the drain side can result in leakage current [7]. As a result, the conducting channel depletes of free electrons, and the device dc current and RF power decreases [7]. To overcome this problem, a dielectric material, often an oxide layer, is introduced between the metal gate and the semiconductor, creating metal-oxide-semiconductor-HFET (MOSHET) for high-frequency applications. Thus, the gate tunneling is reduced or eliminated with improved surface charge modulation by an insulating oxide layer under the gate [7,12].

The incorporation of the oxide layer improves the GaN-based HFET performance by minimizing the gate leakage current, maximizing the output power (as the input can be driven harder), improving the breakdown voltage, etc [13]. The interface between the semiconductor and oxide layer, however, plays an important role in transistor performance [14]. Chemically and thermally stable oxides with low density of interface states between the insulator (oxide) and semiconductor are required [15]. Several groups have reported the use of different dielectrics, such as aluminum oxide (Al_2O_3) [16], hafnium dioxide (HfO_2) [17], zirconium dioxide (ZrO_2) [18], silicon dioxide (SiO_2) [19], silicon nitride (Si_3N_4), and hexagonal boron nitride (h-BN) [20], fabricated by several deposition methods, including atomic layer deposition (ALD) [14], pulsed laser deposition (PLD) [21], and plasma enhanced chemical vapor deposition (PECVD) [21]. SiO_2/hBN has also been reported to be used as a substrate material for field effect transistors [22].

Ga_2O_3 is a promising material with potential dielectric applications for high-power devices because of its wide bandgap (4.4–5.3 eV) and high breakdown voltage (8 $MVcm^{-1}$) [23,24]. It has a moderate dielectric constant (k = 10.6), which is higher than those of traditional dielectrics such as SiO_2 (k = 3.9), and Si_3N_4 (k = 7.4) [16]. This dielectric-constant value allows gate scaling and a smaller voltage for the same charge. The feasibility of using Ga_2O_3 as a gate dielectric was demonstrated by employing ALD with compounds such as trimethylgallium, and triethylgallium as a Ga precursor and ozone or oxygen (O_2) plasma as an oxygen precursor [23,25,26]. The Ga_2O_3 layers grown by ALD for MOSHFETs are amorphous and prone to change their properties, especially threshold voltage, with annealing due to crystallization [27]. The use of crystalline metal–organic chemical vapor deposition (MOCVD) grown β-Ga_2O_3 as a gate dielectric has so far not been reported. A comprehensive study of the electrical properties is necessary to determine the feasibility of MOCVD Ga_2O_3 as a gate dielectric.

In this work, we study the electrical properties of $Al_{0.3}Ga_{0.7}N$/GaN HFET with MOCVD-grown β-Ga_2O_3 as a passivation/dielectric layer. β-Ga_2O_3 layers of three different thicknesses of 10 nm, 20 nm, and 30 nm were grown on top of $Al_{0.3}Ga_{0.7}N$/GaN creating MOSHFET structures. Furthermore, the variations in electrical properties, particularly the shift in threshold voltage, are studied for annealed and as-deposited MOSHFET structures with MOCVD-grown β-Ga_2O_3.

2. Experimental Methods

The epilayers used for devices studied in this paper were deposited in a vertical cold wall metal–organic chemical vapor deposition (MOCVD) system using nitrogen (N_2) as a carrier gas on a 2-inch diameter c-axis sapphire substrate with 0.2° offcut. Trimethylaluminum (TMAl), triethylgallium (TEG), ammonia (NH_3), and ultra-high purity oxygen were used as aluminum (Al), gallium (Ga), N_2, and oxygen (O_2) precursors, respectively. First, a thin 150 nm aluminum nitride (AlN) buffer layer was grown on the sapphire substrate using the process described in [28,29]. Then, on top of the AlN layer, a 500 nm thick gallium nitride (GaN) channel layer with a V/III ratio of 8000 at a temperature of 960 °C, a 2 nm AlN spacer, and a 25 nm thick barrier aluminum gallium nitride ($Al_{0.3}Ga_{0.7}N$) layers with V/III ratio of 5000 at a temperature of 1020 °C were grown at 100 Torr chamber pressure, creating heterojunction field effect transistor (HFET) structure [30–32]. The van der Pauw and Hall effect measurements show that the GaN layers used for our device structures were highly resistive. Figure 1 shows the schematic of the FET epilayer structures used in this paper where, Figure 1a exhibits the schematic of the HFET structure. For this study, the β-Ga_2O_3 layers were grown on top of the HFET structure using the MOCVD process at 700 °C, 50 Torr chamber pressure, and a VI/III ratio of ~900. The details of the Ga_2O_3 growth can be found elsewhere [33]. This creates a MOSHFET structure with the dielectric layer thickness varying from 10 nm to 30 nm, as shown in Figure 1b. The MOSHFET structure with 30 nm oxide thickness was annealed to gauge its electrical and thermal stability. Annealing was performed at 900 °C for 30 min in a 50/50 O_2/N_2 (nitrogen was used as carrier gas) environment to avoid desorption from the Ga_2O_3 layer.

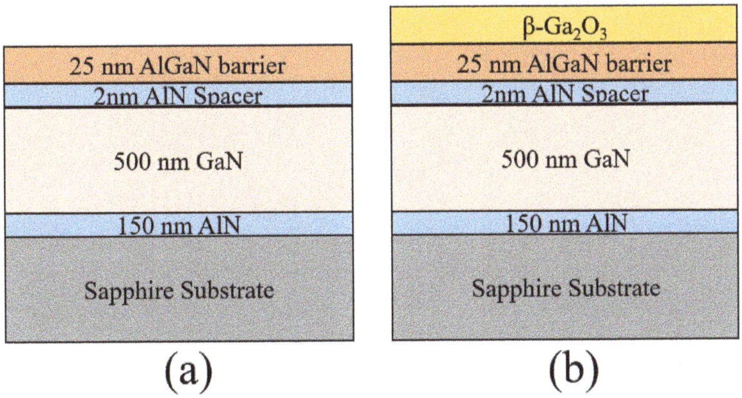

Figure 1. The epilayer structure of the (**a**) $Al_{0.3}Ga_{0.7}N$/GaN HFET, (**b**) Ga_2O_3 MOSHFET.

We characterized the structural quality and electrical properties of all the device structures obtained in this work. A Rigaku Miniflex II Desktop X-ray diffractometer with Cu-K$\alpha 1$ x-ray source (λ = 1.5406 Å) operated at 30 mA current and 15 kV voltage was used to evaluate the structural properties of the epilayers. The capacitance–voltage *(C-V)* measurements were performed using a mercury probe controller model 802B connected with a HP 4284A Precision LCR Meter capable of measuring the impedance as a function of frequency. Gate leakage current was measured in the same mercury probe set up with Keysight B2910 Precision Source/Measure Units (SMU). The gate diameter of the mercury probe was 797 µm with 0.1 pF stray capacitance. The van der Pauw/Hall effect measurements were performed on the samples using the MMR Technologies Inc. H-50 controller and MPS-50 programmable power supply with indium contacts.

3. Results and Discussions

Figure 2 shows the X-ray diffraction (XRD) 2θ scan of the MOSHFET structure (before and after annealing). The peaks at 18.8° and 38.2° are related to the ($\bar{2}$01) and ($\bar{4}$02) Ga_2O_3 of the β phase [34]. The peak at 34.5° and the adjacent higher angle shoulder are consistent with the (002) and (002) reflection from the GaN channel and AlGaN barrier layers, respectively [35]. Note that the GaN channel layer was grown on 0.15 µm AlN. The peak at 36.1° is due to the (002) AlN reflection. The peaks at 20.4° and 41.6° correspond to the (003) and (006) sapphire reflections [34]. The most common method of oxide dielectric deposition is ALD, which is mostly used for depositing amorphous materials [36]. The problem with amorphous layers is that, during the rapid thermal annealing (RTA) step required to form ohmic contacts to MOSHFETs, there occurs a phase transformation from amorphous to crystalline structure [27]. This transformation results in the formation of a microcrystalline structure with multiple grain boundaries, which creates leakage paths, rendering it unsuitable for device applications [27]. The process used for the MOCVD oxide deposition favors the growth of single-crystal β-Ga_2O_3 dielectric layers, as confirmed by the XRD data. Therefore, it is expected that, as Ga_2O_3 is already in the crystalline form, the thermal treatment will have a minimal impact on its electrical properties, as we demonstrate by the annealing experiments in the later description.

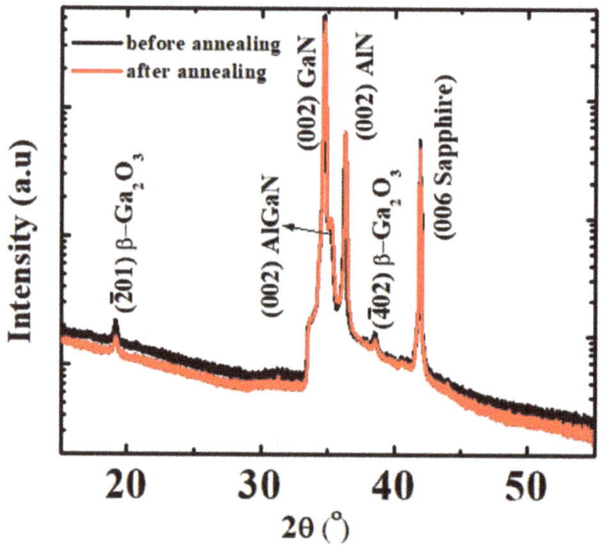

Figure 2. XRD 2θ scan of the MOSHFET structure (before and after annealing) confirming the presence of the peaks consistent with crystalline Ga_2O_3 (β phase), GaN, AlGaN, AlN, and sapphire substrate.

The origin of the highly conductive quantum confined two-dimensional electron gas (2DEG) at the AlGaN/GaN interface is due to the lack of inversion symmetry along the [0001] axis of GaN coupled with AlN being relatively more electronegative [2]. The difference between spontaneous and piezoelectric polarization and band offset at the interface introduces a fixed polarization-induced sheet of carrier charges, indicated by the shift of Fermi level in the conduction band. Schrödinger and Poisson's equation-based charge distribution and band diagram can be calculated using different available tools [37]. However, a simpler approach using the Langer and Heinrich rule helps to estimate different parameters and understand the band structure at the heterojunction interfaces [38,39]. Figure 3 shows the schematical band diagram for the Ga_2O_3 MOSHFET. The bandgaps for GaN, $Al_{0.3}Ga_{0.7}N$, and Ga_2O_3 are 3.4 eV, 4.03 eV (using Vegard's law), and 4.9 eV (β phase), respectively [40]. The position of the Fermi level in GaN near the channel-barrier interface is in the conduction band; where n_s is the sheet carrier density, C_b is the $Al_{0.3}Ga_{0.7}N$ barrier layer capacitance, C_{ox} is the Ga_2O_3 layer capacitance. We can express the total gate capacitance (C_G) using the series capacitance formula as below:

$$\frac{1}{C_G} = \frac{1}{C_b} + \frac{1}{C_{ox}} \quad (1)$$

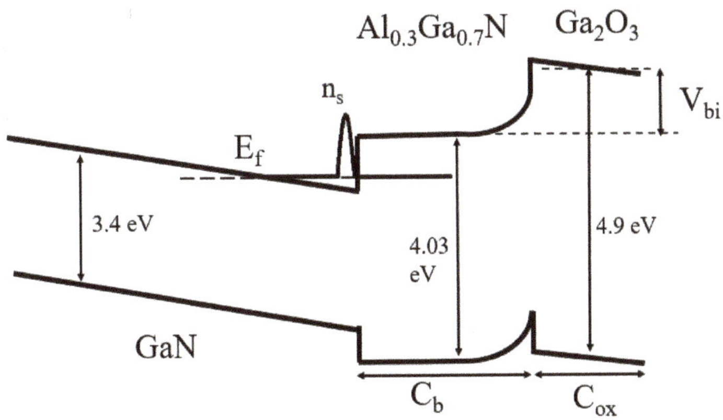

Figure 3. The representative band diagram of GaN/AlGaN MOSHFET with Ga_2O_3 as a gate dielectric.

The capacitance–voltage (*C-V*) measurements of the MOSHFET structure were performed using a mercury probe gate contact to extract pertinent electrical parameters of the device structures, such as threshold voltage (V_{th}), zero gate voltage capacitance, and 2DEG electron density. The total gate capacitance is given by Equation (1), i.e., the addition of the oxide dielectric layer capacitance in series with the barrier layer capacitance. The benefit of an oxide dielectric is to increase the gate breakdown voltage and/or reduce the gate leakage current by suppressing the surface states, sometimes superseded by the impact of different device electrical parameters such as threshold voltage and gate leakage current. The threshold voltage of the MOSHFET structure is given by the Equation [41]:

$$V_{th,MOSHFET} = V_{th,HFET}\left(1 + \frac{d_i \varepsilon_{s,b}}{\varepsilon_i d_{s,b}}\right) \tag{2}$$

where $V_{th,MOSHFET}$ and $V_{th,HFET}$ are the threshold voltage of the MOSHFET and HFET (that without the oxide layer- otherwise identical), respectively. d_i, $d_{s,b}$, ε_i and $\varepsilon_{s,b}$ are the thicknesses and dielectric constants of the insulator/oxide (indexed by *i*) and semiconductor barrier (indexed by *s,b*) layers, respectively. From equation (2), it is clear that if d_i increases, $V_{th,MOSHFET}$ increases, whereas $V_{th,MOSHFET}$ decreases with the increase in ε_i. Thus, a higher dielectric constant and lower dielectric thickness are desirable for minimum threshold voltage shifts.

Figure 4 shows the capacitance voltage characteteristics, threshold voltage and zero capacitance dispersion of our samples. Figure 4a shows the *C-V* data for HFET and MOSHFET structures with oxide thicknesses of 10 nm, 20 nm, and 30 nm. The addition of oxide on the HFET barrier adds capacitance in series with the existing barrier layer capacitance, which would lower the overall gate capacitance. The relative dielectric constants for the AlGaN barrier layer (9.2) and that for the Ga_2O_3 (10.6) are very close [42], and the AlGaN barrier layers is undoped. Due to these reasons, we did not observe any change in the *C-V* curve shape near the zero-gate voltage position when increasing the negative gate voltage before depleting the channel. From equation (2), we observe that as the thickness of the oxide layer increases, the V_{th} should increase. The *C-V* measurements on the MOSHFET structure with MOCVD-grown Ga_2O_3 gate dielectric confirm the increase in the V_{th} value. As the thickness of the Ga_2O_3 layer increased, the V_{th} exhibited a negative shift. Thus, with the increase in dielectric layer thickness, a higher voltage is required to deplete the 2DEG. During the *C-V* measurement, when 2DEG starts to deplete, the capacitance decreases drastically, ultimately leading to pinching off of the channel. By further increasing the gate voltage beyond pinch-off, the depletion layer extends to the GaN channel. Figure 4b summarizes the above discussion.

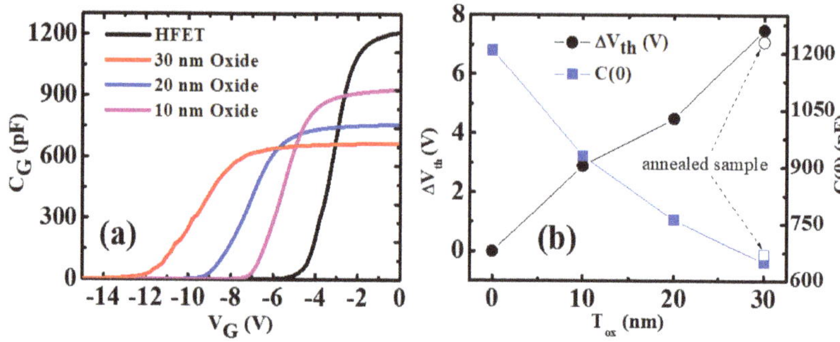

Figure 4. *C-V* characteristics at 1MHz frequency of (a) HFET and Ga_2O_3-based MOSHFETs with different oxide thicknesses, (b) Threshold voltage and zero gate-voltage capacitance dispersion for HFET and Ga_2O_3 MOSHFETs.

Figure 5 compares the *C-V* data for annealed and as-deposited MOSHFETs with a 30 nm thick β-Ga_2O_3 as a gate oxide layer. The 30-min annealing at 900 °C was performed in the MOCVD reactor, used for the Ga_2O_3 growth, as described in the experimental method section. As shown in Figure 2, the pre and post-annealed XRD was similar. As can be concluded from the *C-V* data, the annealing experiment did not result in any discernable change in the V_{th} and zero capacitance values, demonstrating the excellent thermal stability of the crystalline β-Ga_2O_3. The change in threshold voltage (open circle) and capacitance (open square) for the annealed MOSHFET structure are also shown in Figure 4b.

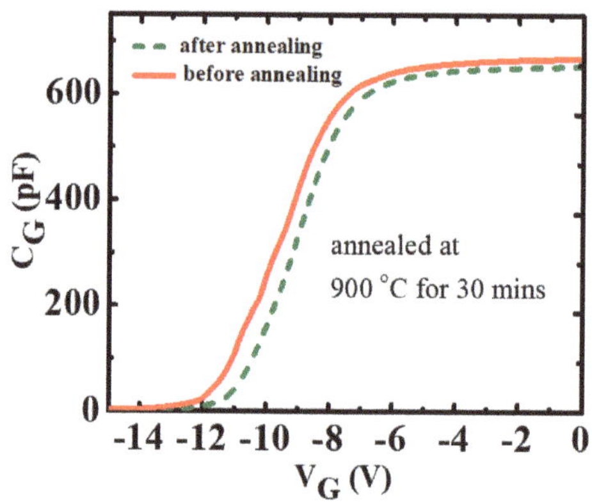

Figure 5. *C-V* data of at 1MHz frequency Ga_2O_3-based MOSHFET with 30 nm oxide thickness annealed (dashed line) and unannealed (continuous line).

The carrier density N_d was calculated using the Hall effect data and found to be on the order of 10^{18} cm^{-3}, the exact values of the carrier density can be found in Table 1. The built-in voltage V_{bi} (as shown in the band diagram: Figure 2) can be measured from $1/C^2$ intercept with the x-axis and expressed by equation (3) [43]:

$$V_{bi} = \frac{qN_d x_d^2}{2\epsilon_s} + \frac{qN_d x_d}{\epsilon_{ox}} t_{ox} + \frac{Q_{ox}}{\epsilon_{ox}} t_{ox} \quad (3)$$

Table 1. The summary of the key electrical parameters measured/calculated from C-V and Hall measurements.

	V_{th} (V)	μ ($cm^2V^{-}s^{-1}$) (Hall)	R_{sh} (Ωcm^{-2}) (Hall)	N_d (cm^{-3}) (Hall)	n_s (cm^{-2}) (C-V)	n_s (cm^{-2}) (Hall)	Q_{ox} (Ccm^{-2}) (C-V)	D_{it} ($cm^{-2}eV^{-1}$) (C-V)	Leakage Current at −20 V (A/cm^{-2})
HFET	−5	750	537	6.8×10^{18}	1.25×10^{13}	1.55×10^{13}	None	None	1.66×10^{-4}
10 nm	−7.9	772	630	5.0×10^{18}	1.28×10^{13}	1.4×10^{13}	-6.68×10^{12}	7.47×10^{12}	1.3×10^{-6}
20 nm	−9.5	770	650	4.2×10^{18}	1.24×10^{13}	1.42×10^{13}	-1.64×10^{13}	7.57×10^{12}	1.12×10^{-6}
30 nm	−12.5	776	685	4.4×10^{18}	1.4×10^{13}	1.4×10^{13}	-2×10^{13}	4.98×10^{12}	9.54×10^{-7}
annealed	−12.1	760	680	4.5×10^{18}	1.4×10^{13}	1.4×10^{13}	-3.06×10^{12}	3×10^{12}	8.33×10^{-7}

Here, t_{ox} and ϵ_{ox} are the thickness and permittivity of the oxide layer, N_d is the carrier concentration, x_d and ϵ_s are the depletion width and permittivity of the AlGaN barrier layer. If we set $t_{ox} = 0$, then the 2nd and 3rd term of Equation (3) becomes zero, and the equation represents the built-in voltage for conventional HFET structure. Using Equation (1) and the parallel plate capacitance formula for each series capacitor component, we can calculate the value of x_d at zero gate voltage. Inserting the determined value of x_d, and previously calculated/measured N_d, t_{ox}, and known ϵ_{ox}, ϵ_s into Equation (3), we can obtain the oxide charge Q_{ox}. The calculated values of Q_{ox} for all the samples is shown in Table 1. We observe a trend in the value of Q_{ox}; as the oxide thickness increases, the value of Q_{ox}, become more negative which increases V_{th} shift supporting Equation (2). There may be impact of stress on the charge, but we did not isolate that in our calculation. Due to low thickness, we did not observe any signature peak of the dielectric Ga_2O_3 or barrier AlGaN layer in Raman measurement (Figure S2 in Supplementary Section).

Figure 6 shows the 2DEG carrier density (n_s) versus the gate voltage (V_G) for 10, 20, and 30 nm gate oxide thicknesses calculated using the following equation:

$$qn_s = (V_G - V_{th})C_G \tag{4}$$

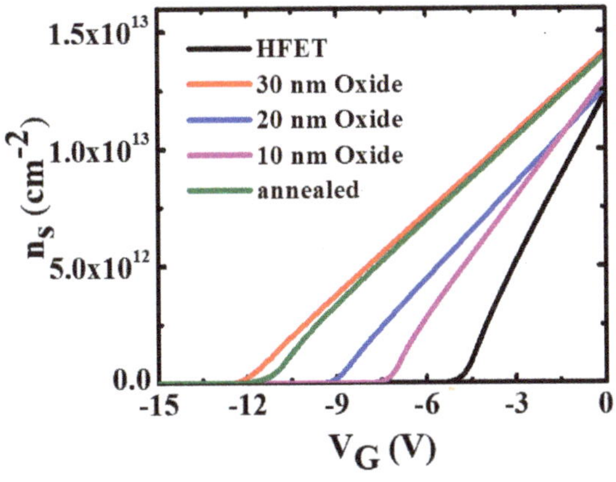

Figure 6. 2DEG carrier density in FETs vs. applied gate voltage for different gate oxide thicknesses.

In all three cases, the zero-gate voltage value of sheet carrier density is very close in the range of (1.25–1.5) × 10^{13} cm^{-2}, a slightly higher value of n_s could be due to sample-to-sample variations. The sheet carrier concentrations measured using van der Pauw/Hall effect method are also in the range of (1–1.5) × 10^{13} cm^{-2}, which validates our calculations using the value from the C-V measurement by employing equation (4). The sheet carrier concentration does not change with oxide layer thickness or annealing of the oxide layer. In previous studies for the ALD-grown amorphous oxide dielectrics-based MOSHFETs, it has been demonstrated that the annealing of the oxide layers drastically shifts V_{th}. As we can see in Figure 5, annealing of the crystalline Ga_2O_3-based MOSHFET, there is no noticeable change in the threshold voltage in contrast to the case for previously reported ALD-grown oxides [44]. To further validate the MOCVD oxide material property we can use Equation (1), to calculate the experimental dielectric constant for β-Ga_2O_3. The values of the gate capacitance, C_G, before and after annealing are 1.64×10^{-7} F/cm^{-2} and 1.67×10^{-7} F/cm^{-2} (for 1 MHz frequency measurement). C_b is the barrier capacitance of the MOSHFET, and the value is 3.3×10^{-7} F/cm^{-2}. Based on $C_{ox} = \frac{\varepsilon_r \varepsilon_0}{t_{ox}}$, we get $\varepsilon_r = 10.9$ (using $C_G = 1.67 \times 10^{-7}$ F/cm^{-2}), whereas the reference value is 10.6. Thus, it can be inferred that the MOCVD-grown oxide has a dielectric property that is very close to the previously reported literature value [42].

The interface trapped charge or interface traps stem(s) from dangling bonds at the semiconductor–insulator interface. The frequency-dependent High–Low method is commonly used to determine interface charge density (D_{it}). Figure 7 shows the frequency-dependent CV measurement used to calculate the interface charge densities for MOSHFET with 10 nm oxide thickness (the frequency-dependent C-V measurements of MOSHFET with 20 nm, 30 nm oxide thickness along with the annealed sample are shown in the Supplementary Section Figure S1).

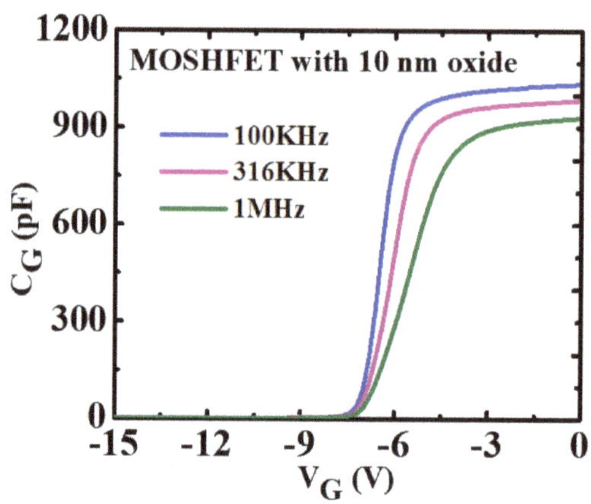

Figure 7. Frequency-dependent C-V characteristic of a MOSHFET with 10 nm thick gate oxide at frequencies of 100 kHz, 316 kHz, and 1 MHz.

The High–Low frequency *CV* method compares a low-frequency *C-V* curve with one that is free of interface traps. The latter is usually referred to as a high-frequency *C-V* measurement, where interface traps with relatively long-time constants cannot respond, leading to decreased measured capacitance. At low frequencies, the interface traps can respond, if not deep, thus resulting in higher capacitance; 100 kHz and 1 MHz frequencies are the typical values can be used for CV-based calculations of the density of interface states (D_{it}) of Nitride systems [45]. Consequently, from the difference between high- and low-frequency CV measurements, the D_{it} can be obtained based on Equation (5) at a specific applied gate voltage [46]:

$$D_{it}(V_G) = \frac{C_{ox}}{q}\left(\frac{C_{LF}}{C_{ox} - C_{LF}} - \frac{C_{HF}}{C_{ox} - C_{HF}}\right) \tag{5}$$

where C_{ox} is the capacitance of the oxide dielectric layer calculated using the parallel plate capacitor formula, q is the unit elementary charge, C_{LF} is the MOSHFET low-frequency capacitance value and C_{HF} is the MOSHFET high-frequency capacitance value. The total trap densities for all the samples are tabulated in Table 1. For all samples, the calculated interface trap densities are in the range (3–7.57) × 10^{12} $cm^{-2}eV^{-1}$; these values are close to the typically calculated values of MOSHFETs (typical values are in the order of ~10^{11} $cm^{-2}eV^{-1}$–10^{13} $cm^{-2}eV^{-1}$) [27]. The D_{it} value is expected to be lower for the processed devices due to the mesa isolation [47]. Our data revealed no specific correlation between the oxide thickness and interface trap densities. Ideally, this is the case, but the total number of bulk defects in the oxide under the gate would depend on the thickness, more data are needed to find any correlation. It is observed that the annealed sample showed a slightly smaller trap density, which can correlate to the higher n_s, the origin of which is not yet explored.

To further investigate the β-Ga_2O_3 viability as a gate dielectric leakage current measurements were performed for all the samples. Figure 8 shows the gate leakage current in the HFET and different thicknesses β-Ga_2O_3 MOSHFET structure. There is a significant reduction in leakage current for the MOSHFET structure compared to the HFET structure in the off-stage. The leakage current at -20 V for HFET is ~10^{-4} A/cm^2, and it reduces to ~10^{-6} A/cm^2 for MOSHFET. This remarkable improvement in the gate leakage current shows that β-Ga_2O_3 can be used as an effective dielectric layer for GaN/AlGaN MOSHFETs. Table 1 summarizes the key electrical parameters of the GaN/AlGaN-based HFET and GaN/AlGaN/β-Ga_2O_3-based MOSHFET determined in this work.

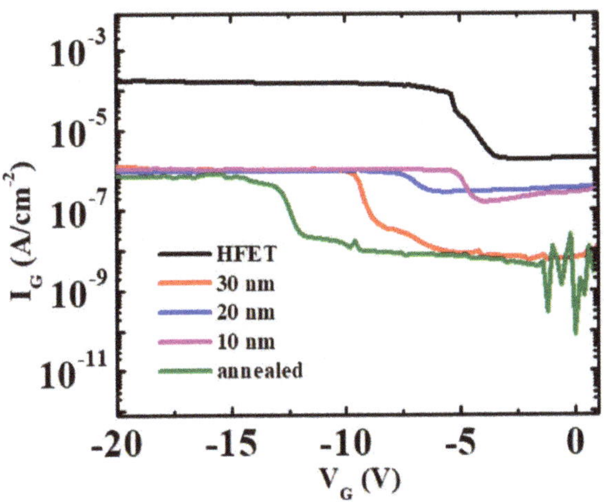

Figure 8. Gate leakage current characteristics for HFET and for different gate oxide thicknesses MOSHFET.

4. Conclusions

We have demonstrated MOCVD-grown single-crystal Ga_2O_3 thin films as a gate dielectric on AlGaN/GaN HFETs. We have found that an increase in the thickness of the dielectric layer has an impact on threshold voltage V_{th}, shifting it to more negative values and reducing the zero capacitance as additional C_{ox} is added in series. The sheet carrier densities for HFET and MOSHFETs were determined to be ~10^{13} cm^{-2}, well within the typical range- of 10^{12} cm^{-2}–10^{13} cm^{-2} for AlGaN/GaN-based devices. The leakage current was reduced by approximately 2 order from ~10^{-4} A/cm^2 for HFET to ~10^{-6} A/cm^2 for MOSHFET at −20 V. Moreover, the addition of the oxide layer did not change the sheet carrier concentration but had an impact on the calculated value of oxide charge Q_{ox}. The calculated Q_{ox} value was found to be negative and mainly responsible for depleting the 2DEG. As the thickness of the Ga_2O_3 layer increases, the Q_{ox} becomes more negative, following a trend similar to the change in V_{th} with increasing gate oxide thickness. The charge density in the oxide–AlGaN barrier interface was found to be of the order of ~10^{12} cm^2eV^{-1}. The thermal stability, as confirmed by the annealing experiment, suggests that the MOCVD-grown single-crystal Ga_2O_3 layer could be more suitable for the gate dielectric application compared to the ALD-grown oxide due to threshold voltage stability. The moderate interface trap density and good thermal stability indicate that MOCVD-grown β-Ga_2O_3 is an excellent candidate for gate dielectric as well as a passivation layer for III-Nitride-based high-power RF MOSHFET devices.

Supplementary Materials: The following supporting information can be downloaded at: https://www.mdpi.com/article/10.3390/cryst13020231/s1. Figure S1. Frequency-dependent CV measurements of MOSHFET with (a) 20 nm thick gate oxide (b) 30 nm thick gate oxide, and (c) annealed 30 nm thick gate oxide. Figure S2. (a) Raman spectra of HFET and MOSHFET (30 nm oxide), (b) Raman spectra of Ga_2O_3 on sapphire to identify the Ga_2O_3 signature peak positions.

Author Contributions: S.H. Conceptualization of this study, Experimentation, Data curation, Methodology, Writing—original draft, M.U.J.: Revision, Experimentation, S.R.C.: Revision, Experimentation, Editing, D.L.: Revision, Experimentation, Editing, V.A.: Revision, Editing, Ü.Ö.: Revision, Editing, H.M.: Revision, Editing, I.A.: Conceptualization of this study, Visualization, Editing, Revision, Principal Investigator of this study. All authors have read and agreed to the published version of the manuscript.

Funding: This research was funded by National Science Foundation (NSF), grant number 2124624.

Data Availability Statement: The data generated during and/or analyzed during the current study are available from the corresponding authors on reasonable request.

Acknowledgments: This research was supported by National Science Foundation (NSF) award No. 2124624 managed by Dominique M. Dagenais.

Conflicts of Interest: The authors declare that they have no conflict of interest.

References

1. Morkoç, H. *Handbook of Nitride Semiconductors and Devices, Materials Properties, Physics and Growth;* Wiley-VCH: Weinhem, Germany, 2009; ISBN 9783527408375.
2. Mohammad, S.N.; Salvador, A.A.; Morkoc, H. Emerging gallium nitride-based devices. *Proc. IEEE.* **1995**, *83*, 1306–1355. [CrossRef]
3. Morkoç, H. *Nitride Semiconductors and Devices;* Springer: Berlin/Heidelberg, Germany; New York, NY, USA, 1999; ISBN 3-540-64038.
4. Zeng, F.; An, J.X.; Zhou, G.; Li, W.; Wang, H.; Duan, T.; Jiang, L.; Yu, H. A Comprehensive Review of Recent Progress on GaN High Electron Mobility Transistors: Devices, Fabrication and Reliability. *Electronics* **2018**, *7*, 377. [CrossRef]
5. Islam, N.; Mohamed, M.F.P.; Khan, M.F.A.J.; Falina, S.; Kawarada, H.; Syamsul, M. Reliability, Applications and Challenges of GaN HEMT Technology for Modern Power Devices: A Review. *Crystals* **2022**, *12*, 1581. [CrossRef]
6. Kim, H.; Lee, J.; Liu, D.; Lu, W. Gate Current Leakage and Breakdown Mechanism in Unpassivated AlGaNGaN High Electron Mobility Transistors by Post-Gate Annealing. *Appl. Phys. Lett.* **2005**, *86*, 143505. [CrossRef]
7. Leach, J.H.; Morkoç, H. Status of Reliability of GaN-Based Heterojunction Field Effect Transistors. *Proc. IEEE* **2010**, *98*, 1127–1139. [CrossRef]
8. Rao, G.P.; Singh, R.; Lenka, T.R. Performance Analysis of AlGaN/GaN HEMT for RF and Microwave Nanoelectronics Applications. In *HEMT Technology and Applications;* Springer Nature: Singapore, 2023; pp. 139–153. [CrossRef]
9. Trew, R.J. Wide bandgap semiconductor transistors for microwave power amplifiers. *IEEE Microw. Mag.* **2000**, *1*, 46–54. [CrossRef]
10. Brar, B.; Boutros, K.; DeWarnes, R.E.; Tilak, V.; Shealy, R.; Eastman, L. Impact ionization in high performance AlGaN/GaN HEMTs. In Proceedings of the IEEE Lester Eastman Conference on High Performance Devices, Newark, DE, USA, 8 August 2002; pp. 487–491.
11. Trew, R.J.; Green, D.S.; Shealy, J.B. AlGaN/GaN HFET reliability. *IEEE Microw. Mag.* **2009**, *10*, 116–127. [CrossRef]
12. Zhou, H.; Lou, X.; Sutherlin, K.; Summers, J.; Kim, S.B.; Chabak, K.D.; Gordon, R.G.; Ye, P.D. DC and RF Performance of AlGaN/GaN/SiC MOSHEMTs with Deep Sub-Micron T-Gates and Atomic Layer Epitaxy MgCaO as Gate Dielectric. *IEEE Electron Device Lett.* **2017**, *38*, 1409–1412. [CrossRef]
13. Liu, C.; Chor, E.F.; Tan, L.S. Investigations of HfO_2/AlGaN/GaN metal-oxide-semiconductor high electron mobility transistors. *Appl. Phys. Lett.* **2006**, *88*, 173504. [CrossRef]
14. Asubar, J.T.; Yatabe, Z.; Gregusova, D.; Hashizume, T. Controlling surface/interface states in GaN-based transistors: Surface model, insulated gate, and surface passivation. *J. Appl. Physics.* **2021**, *129*, 121102. [CrossRef]
15. Ramanan, N.; Lee, B.; Misra, V. Accurate characterization and understanding of interface trap density trends between atomic layer deposited dielectrics and AlGaN/GaN with bonding constraint theory. *Appl. Phys. Lett.* **2015**, *106*, 243503. [CrossRef]

16. Gusev, E.P.; Copel, M.; Cartier, E.; Baumvol, I.J.; Krug, C.; Gribelyuk, M.A. High-resolution depth profiling in ultrathin Al_2O_3 films on Si. *Appl. Phys. Letters.* **2000**, *76*, 176–178. [CrossRef]
17. Lee, B.H.; Kang, L.; Nieh, R.; Qi, W.J.; Lee, J.C. Thermal stability and electrical characteristics of ultrathin hafnium oxide gate dielectric reoxidized with rapid thermal annealing. *Appl. Phys. Letters.* **2000**, *76*, 1926–1928. [CrossRef]
18. Copel, M.; Gribelyuk, M.; Gusev, E. Structure and stability of ultrathin zirconium oxide layers on Si (001). *Appl. Phys. Letters.* **2000**, *76*, 436–438. [CrossRef]
19. Husna, F.; Lachab, M.; Sultana, M.; Adivarahan, V.; Fareed, Q.; Khan, A. High-Temperature Performance of AlGaN/GaN MOSHEMT with SiO_2 Gate Insulator Fabricated on Si (111) Substrate. *IEEE Trans. Electron Devices* **2012**, *59*, 2424–2429. [CrossRef]
20. Illarionov, Y.Y.; Rzepa, G.; Waltl, M.; Knobloch, T.; Grill, A.; Marco, M.; Furchi, M.M.; Mueller, T.; Grasser, T. The role of charge trapping in MoS_2/SiO_2 and MoS2/hBN field-effect transistors. *2d Mater.* **2016**, *3*, 035004. [CrossRef]
21. Balakrishnan, G.; Babu, R.V.; Shin, K.S.; Song, J.I. Growth of highly oriented γ-and α-Al_2O_3 thin films by pulsed laser deposition. *Opt. Laser Technol.* **2014**, *56*, 317–321. [CrossRef]
22. Khan, M.F.; Rehman, S.; Akhtar, I.; Aftab, S.; Ajmal, H.M.S.; Khan, W.; Kim, D.-k.; Eom, J. High mobility $ReSe_2$ field effect transistors: Schottky-barrier-height-dependent photoresponsivity and broadband light detection with Co decoration. *2d Mater.* **2019**, *7*, 015010. [CrossRef]
23. Passlack, M.; Schubert, E.F.; Hobson, W.S.; Hong, M.; Moriya, N.; Chu, S.N.; Konstadinidis, K.; Mannaerts, J.P.; Schnoes, M.L.; Zydzik, G.J. Ga_2O_3 films for electronic and optoelectronic applications. *J. Appl. Phys.* **1995**, *77*, 686–693. [CrossRef]
24. Green, A.J.; Speck, J.; Xing, G.; Moens, P.; Allerstam, F.; Gumaelius, K.; Neyer, T.; Arias-Purdue, A.; Mehrotra, V.; Kuramata, A.; et al. β-Gallium Oxide Power Electronics. *APL Mater* **2022**, *10*, 029201. [CrossRef]
25. Choi, D.W.; Chung, K.B.; Park, J.S. Low temperature Ga_2O_3 atomic layer deposition using gallium tri-isopropoxide and water. *Thin Solid Films* **2013**, *546*, 31–34. [CrossRef]
26. Donmez, I.; Ozgit-Akgun, C.; Biyikli, N. Low temperature deposition of Ga_2O_3 thin films using trimethylgallium and oxygen plasma. *J. Vac. Sci. Technol. A Vac. Surf. Films* **2013**, *31*, 01A110. [CrossRef]
27. Hashizume, T.; Nishiguchi, K.; Kaneki, S.; Kuzmik, J.; Yatabe, Z. State of the Art on Gate Insulation and Surface Passivation for GaN-Based Power HEMTs. *Mater. Sci. Semicond. Process.* **2018**, *78*, 85–95. [CrossRef]
28. Hasan, S.; Mamun, A.; Hussain, K.; Gaevski, M.; Ahmad, I.; Khan, A. Growth Evolution of High-Quality MOCVD Aluminum Nitride Using Nitrogen as Carrier Gas on the Sapphire Substrate. *J. Mater. Res.* **2021**, *36*, 4360–4369. [CrossRef]
29. Hasan, S.; Mamun, A.; Hussain, K.; Patel, D.; Gaevski, M.; Ahmad, I.; Khan, A. Investigation of MOCVD Grown Crack-Free 4 Mm Thick Aluminum Nitride Using Nitrogen as a Carrier Gas. *MRS Adv.* **2021**, *6*, 456–460. [CrossRef]
30. Shrestha, N.; Shrestha, N.M.; Wang, Y.Y.; Li, Y.; Chang, E.Y. Effect of AlN Spacer Layer on AlGaN/GaN HEMTs. In Proceedings of the 16th International Workshop on Computational Electronics, Nara, Japan, 4–7 June 2013; ISBN 978-3-901578-26-7.
31. Hubáček, T.; Hospodková, A.; Kuldová, K.; Zíková, M.S.; Pangrác, J.; Čížek, J.; Liedke, M.O.; Butterilng, M.; Wagner, A.; Hubík, P.; et al. Improvement of Luminescence Properties of N-GaN Using TEGa Precursor. *J. Cryst. Growth* **2020**, *531*, 125383. [CrossRef]
32. Roy, P.; Jawanpuria, S.; Vismita; Prasad, S.; Islam, A. Characterization of AlGaN and GaN Based HEMT with AlN Interfacial Spacer. In Proceedings of the IEEE 2015 5th International Conference on Communication Systems and Network Technologies, CSNT, Gwalior, India, 4–6 April 2015; pp. 786–788.
33. Jewel, M.U.; Hasan, S.; Crittenden, S.; Avrutin, V.; Özgür, Ü.; Morkoç, H.; Ahmad, I. Growth of Thick Phase-Pure β-Ga2O3 on a c-Plane Sapphire Substrate Using MOCVD. *SPIE Conf. Proceeding*, 2023; *submitted*.
34. Ghose, S.; Rahman, S.; Hong, L.; Rojas-Ramirez, J.S.; Jin, H.; Park, K.; Klie, R.; Droopad, R. Growth and Characterization of β-Ga_2O_3 Thin Films by Molecular Beam Epitaxy for Deep-UV Photodetectors. *J. Appl. Phys.* **2017**, *122*, 095302. [CrossRef]
35. Moram, M.A.; Vickers, M.E. X-ray diffraction of III-nitrides. *Rep. Prog. Phys.* **2009**, *72*, 036502. [CrossRef]
36. George, S.M. Atomic Layer Deposition: An Overview. *Chem. Rev.* **2010**, *110*, 111–131. [CrossRef]
37. Grundmann, M. BandEng. 2005. Available online: http://my.ece.ucsb.edu/mgrundmann/ (accessed on 6 January 2023).
38. Langer, J.M.; Heinrich, H. Deep-level impurities: A possible guide to prediction of band-edge discontinuities in semiconductor heterojunctions. *Phys. Rev. Letters.* **1985**, *55*, 1414. [CrossRef]
39. Jena, D. Polarization Induced Electron Populations in III–V Nitride Semiconductors Transport, Growth, and Device Applications. Ph.D. Thesis, University of California Santa Barbara, Santa Barbara, CA, USA, 2003.
40. Langer, J.M.; Delerue, C.; Lannoo, M.; Heinrich, H. Transition-metal impurities in semiconductors and heterojunction band lineups. *Phys. Rev. B* **1988**, *38*, 7723. [CrossRef] [PubMed]
41. Rai, S.; Adivarahan, V.; Tipirneni, N.; Koudymov, A.; Yang, J.; Simin, G.; Khan, M.A. Low threshold-14 W/mm ZrO_2/AlGaN/GaN metal–oxide–semiconductor heterostructure field effect transistors. *Jpn. J. Appl. Phys.* **2006**, *45*, 4985. [CrossRef]
42. Shih, H.Y.; Chu, F.C.; Das, A.; Lee, C.Y.; Chen, M.J.; Lin, R.M. Atomic Layer Deposition of Gallium Oxide Films as Gate Dielectrics in AlGaN/GaN Metal–Oxide–Semiconductor High-Electron-Mobility Transistors. *Nanoscale Res. Lett.* **2016**, *11*, 235. [CrossRef]
43. Sze, S.; Li, Y.; Ng, K. *Physics of Semiconductor Devices*; John and Wiley and Sons: Hoboken, NJ, USA, 2021.
44. Mollah, S.; Hussain, K.; Mamun, A.; Alam, M.D.; Chandrashekhar, M.V.; Simin, G.; Khan, A. Threshold voltage control with high-temperature gate-oxide annealing in ultrawide bandgap AlGaN-channel MOSHFETs. *Appl. Phys. Express* **2022**, *15*, 104001. [CrossRef]

45. Osvald, J. Interface traps contribution to capacitance of Al2O3/(GaN) AlGaN/GaN heterostructures at low frequencies. *Phys. E Low-Dimens. Syst. Nanostructures* **2017**, *93*, 238–242. [CrossRef]
46. Schroder, D.K. Frontmatter. In *Semiconductor Material and Device Characterization*; John and Wiley and Sons: Hoboken, NJ, USA, 2005.
47. Remesh, N.; Kumar, S.; Guiney, I.; Humphreys, C.J.; Raghavan, S.; Muralidharan, R.; Nath, D.N. A Novel Technique to Investigate the Role of Traps in the Off-State Performance of AlGaN/GaN High Electron Mobility Transistor on Si Using Substrate Bias. *Phys. Status Solidi* **2020**, *217*, 1900794. [CrossRef]

crystals

MDPI

Article

Effect of Amorphous Photonic Structure Surface Mounted on Luminous Performances of White LED

Fei Huang [1], Yiyong Chen [2], Jingxin Nie [2], Chunsheng Shen [1], Jiulong Yuan [1], Yukun Guo [1], Boyan Dong [2], Lu Liu [3], Weihua Chen [2,*], Zhizhong Chen [1,2] and Bo Shen [2]

1 Jiangsu Nuanyang Semiconductor Technology Co., Ltd., Sheyang 224300, China
2 State Key Laboratory for Artificial Microstructure and Mesoscopic Physics, School of Physics, Peking University, Beijing 100871, China
3 State Key Laboratory of Advanced Optical Communication Systems and Networks, School of Electrical Engineering and Computer Science, Peking University, Beijing 100871, China
* Correspondence: chenwh@pku.edu.cn

Abstract: We fabricated amorphous photonic structures (APSs) with different periods and hole diameters. The GaN-based white light emitting diodes (LEDs) at nominal correlated color temperatures (CCTs) of 5000 and 6000 K were surface mounted by these APSs. The electroluminescence (EL) measurements showed less luminous efficiency (LE) and higher CCT than the ones of the virginal white LEDs. However, the LEs of many APS-mounted white LEDs increased compared to white the LEDs without APSs at the same CCTs. A finite-difference time-domain (FDTD) simulation was carried out on the ASPs surface-mounted white LEDs and bidirectional scattering distribution functions (BSDFs) of different emissions were transferred to a Monte Carlo ray tracing simulation. The simulated LEs and CCTs conformed well to the experimental ones. The effects of the blue emission transmission and phosphor concentration were simulated to predict the absolute LE enhancement methods for white LEDs. At last, the hopeful APSs for high Les' general lighting were discussed.

Keywords: white light emitting diode (WLED); amorphous photonic structure (APS); luminous efficiency (LE); correlated color temperature (CCT); Fresnel loss; finite-difference time-domain (FDTD)

Citation: Huang, F.; Chen, Y.; Nie, J.; Shen, C.; Yuan, J.; Guo, Y.; Dong, B.; Liu, L.; Chen, W.; Chen, Z.; et al. Effect of Amorphous Photonic Structure Surface Mounted on Luminous Performances of White LED. *Crystals* **2023**, *13*, 6. https://doi.org/10.3390/cryst13010006

Academic Editor: Robert F. Klie

Received: 31 October 2022
Revised: 25 November 2022
Accepted: 17 December 2022
Published: 20 December 2022

1. Introduction

Phosphor-based white light-emitting diodes (LEDs) are low-cost and popular lighting devices because of their high luminous efficiency, long lifetime, and environmental protection. Although the luminous efficiency of white LEDs with an InGaN active layer in the laboratory have been achieved more than 300 lm/W [1], the commercial ones usually showed a lower efficiency below 200 lm/W. In a phosphor-coated white LED, the interplay between multiple scatterings, emissions, and absorptions of light would be concerned carefully [2]. To reduce the back-scattering and inelastic scattering of blue and yellow lights, some techniques, such as scattered photon extraction (SPE), remote phosphor, multi-layered phosphor, and enhanced light extraction by internal reflection (ELiXIR) are developed [3–6]. The phosphor layer features include the encapsulant and phosphor materials, volume or mass concentration and size distribution of particles, matrix blending, and so on, which can adjust the global performances of the phosphor layer [7–9]. The efficient matching of refractive indexes among the encapsulant, package structure, chip, and phosphor is the key issue in light extraction [10,11]. The light extraction efficiency (LEE) nearly approaches 100% when the resin with a refractive index above 1.8 encapsulated the GaN-based LED [11]. Such high index resin is not available due to its reliability. To improve the LEE of white LEDs is still important and significant in the fields of general lighting and backlight of liquid crystal display. With the maturation of the material and processes of the white LED encapsulation, even several percent enhancement of LEE becomes very difficult [12].

Recently, some nanostructures were applied on the white LED packaging to improve the color conversion efficiency, directionality, and spontaneous emission rate (SER) [13–22]. Quantum dot (QD) material is a type of good phosphor for white LEDs because of its high quantum yield, narrow spectral width, and high SER [13–15]. However, the agglomeration, self-absorption of luminescence, and surface state of the QDs lead to low efficiency in the encapsulant. Dang et al. presented a nonporous (NP) GaN by a simple electrochemical (EC) etching technique, where the QDs do not agglomerate and Förster resonance energy transfer may occur by being immediately proximal to the active layer [13]. The metallic nanostructures cause the surface plasmon coupling to both QD and quantum wells to enhance color conversion [15,22]. The ZnO nanorod is fabricated on the AZO seed layer using the aqueous solution method, which increases the luminous flux of white LED by 60% [16]. The directionality of light emission can be determined by dielectric Vogel spiral arrays or rare-earth phosphor-patterned films [17,18]. The directional emissions are beneficial to imaging and projecting devices. Due to the scattering of the phosphor, the total internal reflection (TIR) at the light extraction surface is not paid much attention. Many people find that flat devices show similar LEEs compared to the structured devices [2,10,23,24]. However, the TIR and Fresnel losses still remain on the light extraction surface [11,16,25–27], which should be alleviated to enhance the LEE of white LEDs.

There are some reports on mounting photonic structures onto the light extraction surface of monochromatic LED packages [28–30]. It is found that the LEE of blue LEDs can be improved even if it has reached over 90% [29]. The moth-eye technique can eliminate the Fresnel reflection effectively [31]. Some moth-eye structures are amorphous photonic structures (APSs) that show short-range order only. So, it is expected that the LEE will be improved when the white LED package surface is mounted with APSs. Moreover, APSs show a strong light extraction ability for a broad spectrum because of its antireflective structures and coherent scattering [32–34]. However, it has not been used in the white LED yet, which consists of a narrow blue emission and a broad yellow emission. In APS optimization, the finite-difference time-domain (FDTD) is used [35]. However, the FDTD simulation is limited to be operated on several microns by the computation capacity. It should combine with other simulators, such as Monte Carol ray tracing to complete the large-size white LED package simulation [36]. In this work, we proposed an easy and feasible method that nanoscale intermediate polymer stamp (IPS) APSs were transferred to the silicone surface of white the LED package by the surface-mounting method, which was well used in the white LED package by Lumileds Co. [37]. Combined with the nanoimprint technique, it will open a door to apply the nanostructures in the white LED structures. The different periods and hole diameters of APSs were used on the different correlated color temperatures (CCTs). Although all the APS-mounted white LEDs showed less luminous efficiency (LE) than the virginal white LEDs, the LEs of some APS-mounted white LEDs were enhanced when compared to the same CCT white LEDs without APSs. Moreover, some light output powers (LOPs) were enhanced by surface-mounted APSs. At last, the FDTD combined with ray tracing simulations were carried out, which explained the LE reduction with the APSs and predicted the LE enhancement with appropriate APS structures.

2. Material and Methods

APS structures were fabricated from commercial nanoporous anodic aluminum oxide (AAO) templates, typically as shown in Figure 1a. The templates were the silicon replicas of the original AAO structures on aluminum sheets. In this work, the quasi-periods of AAO templates were about 450 and 800 nm. The hole diameters were 400, 300, and 200 nm for the 450 nm-period AAO templates and the depths were about 500 nm, which corresponded to APS-1, APS-2, and APS-3. The hole diameters were 600 and 700 nm for the 800 nm-period AAO templates and the depths were about 600 and 700 nm, respectively, which corresponded to APS-4 and APS-5. Figure 1b shows the schematic diagram of APS transferring and surface-mounting process. First, the APS was transferred to the intermediate polymer stamp (IPS) using nanoimprint technology. The detailed processes of nanoimprint and demolding would be found in [29]. Subsequently, PDMS (Sylgard184, Dow Corning 10:1 ratio with the curing agent) acting as the adhesive coating was spin-coated onto the commercial white LED packages (length = 5.6 mm, width = 3.0 mm, height = 0.77 mm, model: HV56301ZMQ20DC from Hualian Electronics Corp., Ltd., Xiamen, China). The schematic diagram and feature size of the white LED package is shown in Figure 1c. The IPS and PDMS covered the surface of white LED and the thickness of IPS and PDMS was about 0.19 mm. The nominal CCTs of the white LEDs were 5000 and 6000 K, respectively. Finally, the IPS was bonded onto the white LED using PDMS adhesive. The whole device was then heated for 30 min at 100 °C to cure the PDMS. The top-view photo of the completed device is shown in Figure 1d.

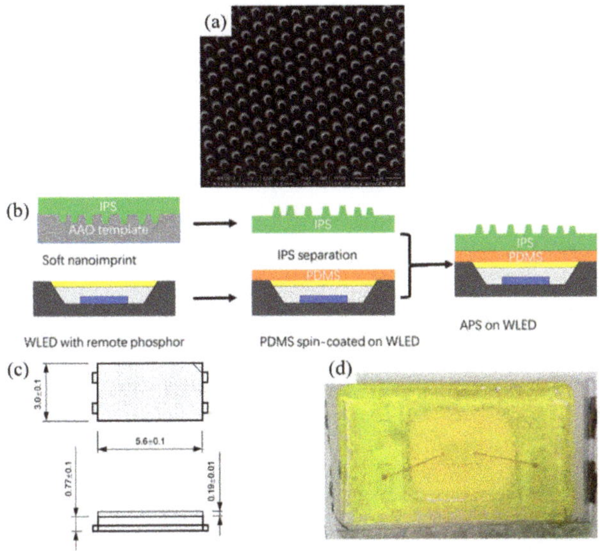

Figure 1. (**a**) SEM images of typical APS imprinted using AAO template. (**b**) The schematic diagram of APS transferring and surface-mounting process. (**c**) The schematic diagram and feature size of the white LED with and without package. (**d**) The top-view photo of the completed device.

The surface morphology of the APSs was examined using a scanning electron microscope (SEM, Nova Nano SEM 430, FEI, New York, USA). The electroluminescence (EL) spectra of the LEDs with and without APSs were measured using an Everfine HAAS-2000 spectroradiometer under the constant mode by using an integrating sphere (Everfine Co., Huangzhou, China). The integral duration is set as the constant of 100 ms. The white LED was located on the copper support. In the measurement, the increase in the junction temperature was less than 2 °C and could be ignored. Three-dimensional FDTD simulations were carried out to study the effect of APSs on the light extraction using Lumerical software (FDTD Solutions v8.21, Vancouver, BC, Canada) [35]. However, FDTD simulation dealt with light propagation in a small area due to the computation capacity and cannot work well in the phosphor layer. So, the Monte Carlo ray-tracing method was combined with an asymmetric bidirectional scattering distribution function (BSDF) [36]. After obtaining a BSDF for certain APSs using FDTD simulation, the corresponding APS surface was defined by the BSDF in a ray-tracing simulation.

3. Results and Discussion

Figure 2 shows the EL results of white LEDs at 5000 K with and without APSs mounting. The nominal CCTs of the virginal white LEDs are 5000 K. The APSs include APS-1,

APS-2, and APS-3, which correspond to the patterns of AAO400, AAO 300, and AAO200, respectively, in [29]. The periods of the APSs are about 450 nm. Figure 2a shows that all the integral intensities of blue emissions increase, while the ones of yellow emissions decrease with the APSs mounted. The white LEDs with APS-1 show more yellow emission reductions and more blue emission increase than those with APS-2 and APS-3. It is found that the spectral shapes are also modified by APSs. The EL intensities at the short-wavelength side of the yellow emissions are higher than the virginal ones. Figure 2b,c show the LOPs and LEs ratios to the virgins with different currents. It can be seen that the LOPs of the white LED with APS-2 are enhanced about 0.5% and the ones with APS-3 are similar to the virginal ones. However, the white LED with APS-1 shows that its LOP reduces significantly. Additionally, the LOP ratios increase with current monotonously. In Figure 2c, all the LEs of white LEDs decrease when the APSs are mounted. The white LEDs with APS-2 and APS-3 show about a 1% reduction in LEs, while APS-1 shows a more than 3% reduction.

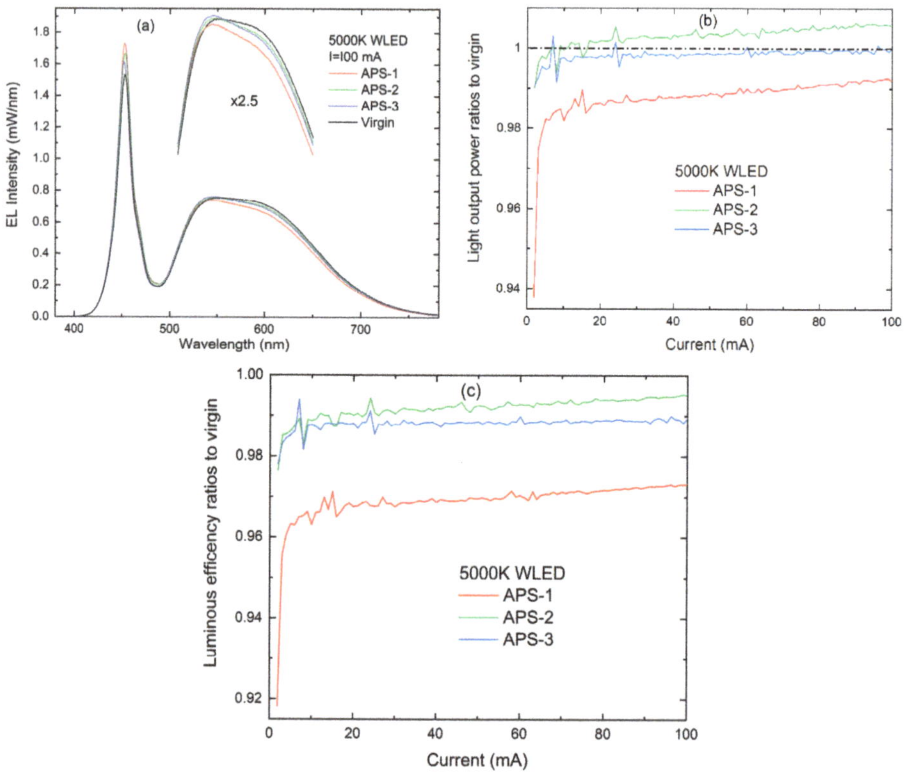

Figure 2. EL results of white LEDs of 5000 K mounted with and without APSs. (**a**) EL spectra at current of 100 mA. Current dependencies of (**b**) light output power ratios to virgin and (**c**) luminous efficiency ratios to virgin for different APSs. The periods of the APSs are about 450 nm. In the inset of (**a**), four magnified segments are shown as 2.5 times of those in the original ones.

The blue emissions are enhanced by APSs mounting, which is due to the Fourier power spectra of APSs within a certain spatial frequency, as described in our previous work [29]. As for the decreasing yellow emissions, it may be due to the quenching of yellow emission by APSs or the less blue emissions to excite the yellow phosphor. Using the multistage FDTD simulations [29], the LEE enhancements of the monochromatic LEDs with wavelengths of 450, 525, and 580 nm were calculated as 3.7%, 1.2%, and 1.6% by APS-1, respectively. So, the yellow emissions decreasing is caused by less blue emission excitation. It also indicates APS-1 may be a candidate nanostructure for broadband LEE enhancement. The LEE enhancements of blue LEDs mounted with APS-1, APS-2, and APS-3 are 3.0%, 2.3%, and 1.3%, respectively [29], which demonstrates the reduction in yellow emission. As for the increasing yellow emission at the short-wavelength side, it may be due to the smaller hole diameters of APS-2 and APS-3 than those of APS-1. Although the blue emission excitation for yellow phosphor reduces with APS-2 and APS-3, the more yellow emission at short-wavelength side is extracted from the package. The enhancement of blue emission and part yellow emission lead to the LOPs with APS-2 and APS-3 not decreasing, as shown in Figure 2b. However, because the photopic vision function of blue light is much lower than that of the yellow one, all the LEs in Figure 2c show the reduction of 1–3%. With the current increasing, the CCT increases monotonously (not shown here), which means an increase in the blue emission content in the spectrum of white LED. For the virginal LED and LEDs with APSs, the increasing current would result in the droop effects on LOPs and LEs. However, considering the ratios to the virginal LEDs, the ratios of LOPs and LEs increase monotonously because there are more blue emissions with the increasing current and the APSs would further enhance the blue emission.

The EL measurements were also carried out on the white LEDs at 5000 K with APS-4 and APS-5 and 6000 K with APS-1. The LEs, LOPs, and CCTs of white LEDs with and without different APSs at the current of 100 mA are obtained and calculated, as listed in Table 1. All the LEs of white LEDs decrease when the APSs are mounted. The low CCTs of white LEDs with suitable APS seemly correspond to small LE variations. The LOPs of white LEDs with APSs change little compared to the flat ones. With APS-2 and APS-3, they show the LOP enhanced or unchanged compared to the virginal LEDs. The CCT increases when the APSs are mounted. APS-1 causes more CCT increase than APS-2 and APS-3. It is also seen that the nominal 6000 K white LED with APS-1 shows more CCT increase than the 5000 K one with the same APS. Combined with LEs and LOPs results, the blue emission increase seems to overlap the yellow emission decrease in LOPs, which agrees with the early report when the color conversion efficiency is high [38]. It is reasonable that the CCT increases with APSs because the blue emission contents increase in the white LEDs' spectra. Similarly, the smaller loss of yellow emission LOP will cause the smaller CCT increase. Lower nominal CCTs correlate to the smaller blue emission content in the white LED spectrum, which means a smaller effect of APS-1 on LE and CCT. It can be supposed that the smaller blue emission contents and appropriate APSs would enhance the LEs and decrease the CCT increment after APS is mounted on the white LED packages.

Table 1. LEs, LOPs, and CCTs of white LEDs with and without different APSs at the current of 100 mA.

Structures		LE (lm/W)	LOP (mW)	CCT (K)
APS-1	without	153.66	154.1	6304
	with	148.06	153.0	7354
	without	164.74	158.5	4940
	with	160.08	157.2	5615
APS-2	without	165.47	159.6	4823
	with	165.07	160.5	5240
APS-3	without	167.84	158.4	4779
	with	166.12	158.4	5198
APS-4	without	166.40	158.7	4712
	with	163.55	155.9	4942
APS-5	without	164.54	157.4	4820
	with	160.62	154.4	5119

Figure 3 shows the EL spectra of white LEDs with and without APSs at CCTs of 5000 K. The periods of the APSs are 800 nm. The hole diameters are 600 and 700 nm for APS-4 and APS-5, respectively. The operating current is 100 mA. Compared to the EL spectra in Figure 2, the peak intensities of blue emissions increase insignificantly, while the ones of yellow emissions decrease similarly to with APS-2 and APS-3 after the APSs are mounted. So, both LOPs and LEs decrease for the white LEDs with APS-4 and APS-5. The CCTs also increase, whose changes are lower than 300 K. It indicates that the APSs with 800 nm period cannot enhance the blue light extraction significantly. The small blue light extraction enhancement would not cause significant yellow emission reduction. The yellow emission reduction may be due to the destructive interference. The non-iridescent structural colors in an APS come as a result of coherent scattering [34], while the other wavelength lights may not be extracted from the semiconductors due to the destructive interference. As for the small CCT changes of white LEDs by APS-4 and APS-5, it is due to the very small blue emission increase. If the APSs cause the yellow emission coherent scattering and the blue emission decreasing, the LOPs and LEs will be enhanced.

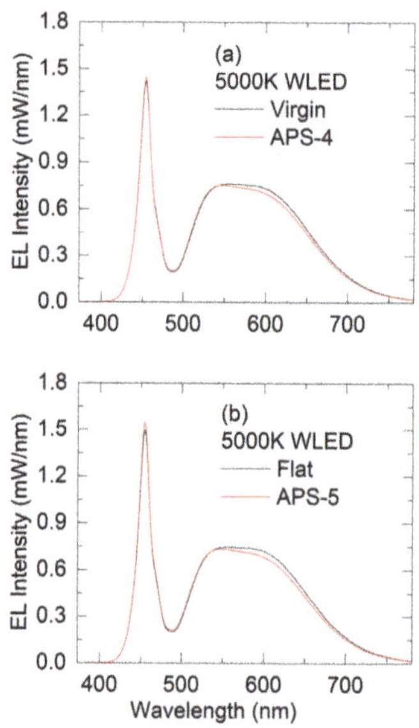

Figure 3. EL spectra of white LEDs mounted with and without (**a**) APS-4 and (**b**) APS-5 at CCTs of 5000 K. The periods of the APSs are about 800 nm.

The commercial white LEDs at the nominal CCTs of 5000 and 6000 K show a standard deviation of 200 K. So, ten typical white LEDs with APSs and their virgins are chosen to inspect the effects of the APSs on luminous performances. Their LEs and CCTs are plotted in Figure 4. It is obvious that all the white LEDs with APSs show lower LEs and higher CCTs than their virgins. With the CCT increase from 5000 K, the LEs seems to decrease monotonously. The record LE of white LEDs is 303 lm/W at about 5000 K [1]. More yellow emissions show higher LEs of white LEDs at the same LOPs. However, the red phosphor is less efficient than the yellow one, so the LEs will decrease when the CCT is lower than about 4500 K. Figure 4 shows that, at the same CCTs, many white LEDs show higher LEs with APSs than those without APSs, such as red points 2 and 4. The red points 9 and 10 also show higher LEs if the CCTs of the virginal white LEDs extrapolate to 8500 K. According to the linear fitting results, when the CCT is higher than 6500 K, the white LEDs with these APSs will show LE enhancements to the virgins at the same CCT. It indicates that the blue light emission enhancement by APSs can benefit the LE increases in white LEDs at the same CCTs. However, the absolute LE increases in white LEDs are impossible with the above APSs because the CCT increases compared to the virgins. The experimental data points fluctuate along the fitting lines, which indicates the different effects of APSs and phosphor contents on the blue and yellow emissions. More blue and yellow emissions than the virginal ones originate from the emission enhancement by APSs. The yellow emission coherent scattering APSs may be concerned in light extraction.

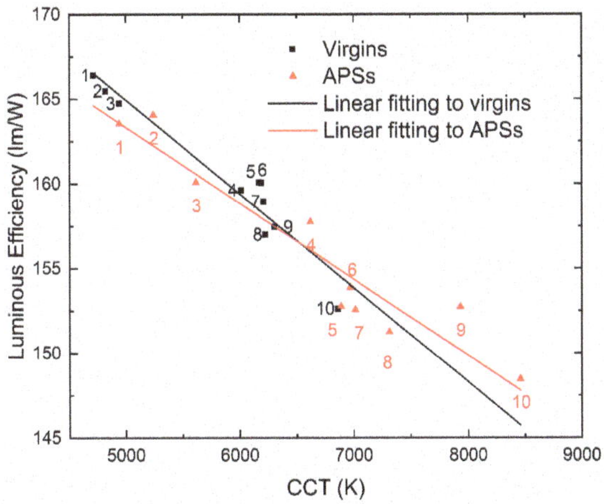

Figure 4. Relationship between luminous efficiencies and CCTs for white LEDs with and without APSs. The data are chosen with APS-1, APS-2, APS-3, APS-4, and APS-5 at the nominal CCTs of 5000 and 6000 K. The same numbers of black and red data points correspond to the same white LEDs with and without APSs. The fitting lines show same color as the experimental data.

In general, the LEs can be improved with appropriate APSs and corresponding virginal white LED spectra. At the same CCTs, the white LEDs with APSs can be easily achieved with higher LEs than those without APSs. However, the absolute enhancement of LE does not compare the same white LEDs with and without APSs. Here, we carried out the FDTD and ray tracing simulations to explore the LEs enhancement method. Because FDTD cannot simulate the phosphor effects and ray tracing cannot deal with the nanostructures [35,36], the BSDF of APS-1 was calculated by FDTD, which was then used in a ray tracing simulation. Figure 5a shows the schematic diagram of FDTD model according to our previous study [29]. The light source emitted the plane wave and it propagated from IPS through APS and then into the air. The refractive index values of IPS and air were set as 1.41 and 1.00, respectively. The lateral dimensions of the computational domain were set as 6 μm × 6 μm, considering the limitation of computer memory and computation capacity. The boundary condition in the top simulation area was set as a perfectly matched layer boundary condition and the four lateral boundaries were set as periodic boundary conditions. Figure 5b presents the LED model in the ray tracing simulator. The blue rectangular represents the LED chip and the yellow layer above represents the phosphor. In the ray tracing simulator, we figured out the emission of phosphor from the SPD of white LED without APS. In Figure 5c, the EL spectra of white LEDs are simulated, where the virginal and APS-1 white LEDs are simulated according to the experimental ones. The percentages of 100%, 90%, 80%, and 60% correspond to the blue emission transmission ratios to the real one with APS-1 by FDTD. The yellow emission transmissions remain unchanged. Then, the modified BSDFs are transferred to a ray tracing simulator. The blue emission reductions using nanostructures cause significantly yellow emission enhancements. It is observed that the simulated EL spectra of virginal and APS-1 (100% transmission) white LEDs are similar to the experimental ones. With the whole effects of APS-1, the LE decreases about 4% and the CCT increases from 6336 to 7589 K compared to the virginal ones. These results also agree with the experiment ones, whose LE decreases about 3.6% and the CCT increases from 6304 to 7354 K, as listed in Table 1. It means the FDTD combined with ray tracing with the BSDFs is suitable for the white LEDs with and without APSs.

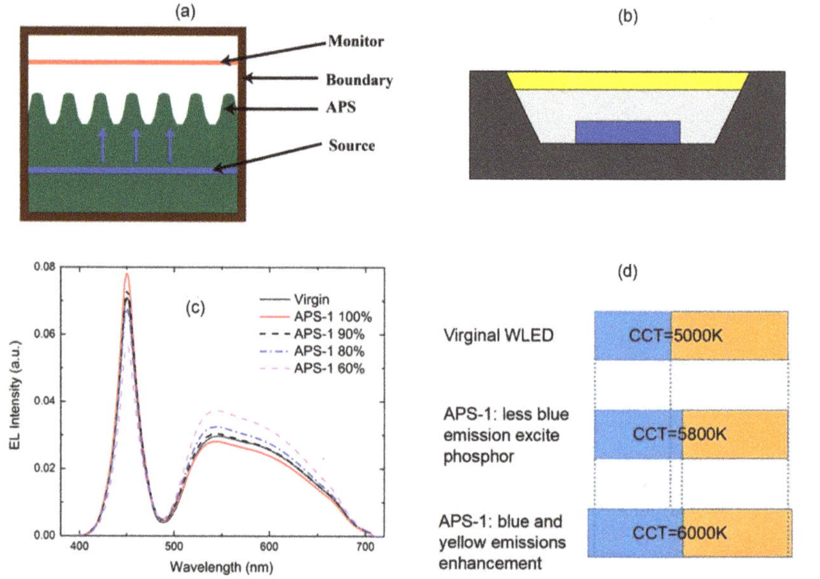

Figure 5. (**a**) The schematic diagram of white LED with APS in the FDTD simulation. (**b**) The white LED model in the ray tracing simulator. (**c**) Simulated EL spectra of white LEDs with APS with different blue emission transmission ratios. (**d**) The diagrammatic sketch of LE enhancement of white LED by APS-1 mounted.

APS-1 90% means a 10% reduction in blue emissions transmission of APS-1, which leads to a small increase in blue and yellow emission LOPs compared to the flat ones, as shown in Figure 5c. With a further decrease in blue emissions transmission for APS-1 80%, blue emission LOP reduces a bit and the yellow one increases significantly. As for APS-1 60%, both blue emission reduction and yellow emission enhancement are significant. The LEs of APS-1 80% and APS-1 60% samples are 10% and 25% enhancements compared to the virginal ones. Moreover, the CCTs decrease to 5868 and 4978 K, respectively. The absolute LEs enhancement requires more blue emissions to excite the yellow phosphor instead of light extraction from APS-1 directly. It is notable that the blue emission transmission reductions are set artificially, which is also difficult to find such APSs in practice. However, it can be found that the new APSs should change its period and size to decrease the blue emission transmission.

At last, the LE enhancement with APSs at the same CCTs can be discussed by a diagrammatic sketch, as shown in Figure 5d. It is supposed that the blue emission is 40% of LOP for 6000 K white LED, while 30% for 5000 K white LED. If their virginal LOPs are same, the virginal LE for 5000 K white LED is higher than that for 6000 K when APS-1 is mounted to the 5000 K LED package, more blue emission will be extracted directly. Less emission blue excitation led to yellow emission reduction. In the semiconductor, APS-1 reassigns the blue emission to extraction and excitation for phosphor. The total LOP does not change in the semiconductor with APS-1. Because the blue emission increases, the CCT will increase, which is supposed as 5800 K. Furthermore, the blue and yellow extracted emissions to air will be enhanced by APS-1. As mentioned above, the blue emission enhanced more than the yellow one as the CCT increased to 6000 K. Both the LOP and LE increase compared to those of the virginal white LEDs at 6000 K.

With the phosphor concentration increasing further, the scattering in the phosphor is more significant. So, the blue emissions reaching the surfaces of light extraction become less. The luminous flux will be enhanced absolutely compared to the virginal one, which is similar to the small transmission ratios of the blue emissions through APS-1 in Figure 5c. However, the quenching effect will appear when the phosphor concentration becomes high, because the self-absorption of phosphor and back scattering to chip will be significant. In the further research, the higher CCT white LEDs will be tried by sophisticated phosphor assembly. On the other hand, the yellow emission coherent scattering APSs may be hopeful to enhance the LEs for white LEDs. The broadband transmission of most yellow emissions and some blue emissions will be better for the absolute LE enhancements. The effective APSs need more design and experiments.

4. Conclusions

In summary, the APSs with different periods and hole diameters were fabricated and surface mounted on the white LEDs at nominal CCTs of 5000 and 6000 K. The EL measurements on these white LEDs were performed. All the white LEDs with APSs show less LE and higher CCT than the ones of the virginal white LEDs. However, the LEs of many APS-mounted white LEDs enhanced compared to the white LEDs without APSs at the same CCTs. Some LOPs of the white LEDs are enhanced by APSs surface mounted. At last, FDTD combined with Monte Carlo ray tracing simulations were carried out on the ASPs surface-mounted white LEDs using BSDF transferring. The simulation results of EL spectra, LEs, and CCTs conform well to the experimental ones. The simulations on the effects of the blue emission transmission and phosphor concentration predict that the absolute LE enhancement for white LEDs requires high phosphor concentrations for the existing APSs beneficial to blue light extraction or the optimal APSs for yellow emission coherent scattering. The APSs with the structural color are more hopeful for high LE general lighting.

Author Contributions: Conceptualization, Z.C.; Data curation, F.H., Y.C., J.N., C.S., J.Y., Y.G., B.D. and L.L.; Funding acquisition, Z.C.; Investigation, C.S., J.Y. and Y.G.; Methodology, Y.C.; Project administration, B.S.; Resources, J.N., L.L. and W.C.; Supervision, W.C. and B.S.; Writing—original draft, F.H. and Y.C.; Writing—review and editing, Z.C. All authors have read and agreed to the published version of the manuscript.

Funding: This work was supported by National Key Research and Development Program [grant number 2021YFB3600100]; National Natural Science Foundation of China [grant numbers 62174004, 61927806]; and Guangdong Basic and Applied Basic Research Foundation [grant number 2020B1515120020].

Institutional Review Board Statement: Not applicable.

Informed Consent Statement: Not applicable.

Data Availability Statement: The data presented in this study are available on request from the corresponding author.

Conflicts of Interest: The authors declare no conflict of interest.

References

1. Cho, J.; Park, J.H.; Kim, J.K.; Schubert, E.F. White light-emitting diodes: History, progress, and future. *Laser Photonics Rev.* **2017**, *11*, 1600147. [CrossRef]
2. Leung, V.Y.F.; Lagendijk, A.; Tukker, T.W.; Mosk, A.P.; Jzerman, W.L.I.; Vos, W.L. Interplay between multiple scattering, emission, and absorption of light in the phosphor of a white light-emitting diode. *Opt. Express* **2014**, *22*, 8190. [CrossRef] [PubMed]
3. Narendran, N.; Gu, Y.; Freyssinier-Nova, J.P.; Zhu, Y. Extracting phosphor-scattered photons to improve white LED efficiency. *Phys. Status Solidi (a)* **2005**, *202*, R60–R62. [CrossRef]
4. Kim, J.K.; Luo, H.; Schubert, E.F.; Cho, J.; Sone, C.; Park, Y. Strongly Enhanced Phosphor Efficiency in GaInN White Light-Emitting Diodes Using Remote Phosphor Configuration and Diffuse Reflector Cup. *Jpn. J. Appl. Phys.* **2005**, *44*, L649–L651. [CrossRef]
5. Allen, S.C.; Steckl, A.J. ELiXIR—Solid-state luminaire with enhanced light extraction by internal reflection. *J. Disp. Technol.* **2007**, *3*, 155–159. [CrossRef]

6. You, J.P.; Tran, N.T.; Shi, F.G. Light extraction enhanced white light-emitting diodes with multi-layered phosphor configuration. *Opt. Express* **2010**, *18*, 5055–5060. [CrossRef]
7. Kang, D.-Y.; Wu, E.; Wang, D.-M. Modeling white light-emitting diodes with phosphor layers. *Appl. Phys. Lett.* **2006**, *89*, 231102. [CrossRef]
8. Huang, S.C.; Wu, J.K.; Hsu, W.-J.; Chang, H.H.; Hung, H.Y.; Lin, C.L.; Su, H.-Y.; Bagkar, N.; Ke, W.-C.; Kuo, H.T.; et al. Particle Size Effect on the Packaging Performance of YAG:Ce Phosphors in White LEDs. *Int. J. Appl. Ceram. Technol.* **2009**, *6*, 465–469. [CrossRef]
9. Liu, Z.; Liu, S.; Wang, K.; Luo, X. Measurement and numerical studies of optical properties of YAG:Ce phosphor for white light-emitting diode packaging. *Appl. Opt.* **2010**, *49*, 247–257. [CrossRef]
10. Kim, J.P.; Jeon, S.-W. Investigation of Light Extraction by Refractive Index of an Encapsulant, a Package Structure, and Phosphor. *IEEE Trans. Compon. Packag. Manuf. Technol.* **2016**, *6*, 1815–1819. [CrossRef]
11. Ma, M.; Mont, F.W.; Yan, X.; Cho, J.; Schubert, E.F.; Kim, G.B.; Sone, C. Effects of the refractive index of the encapsulant on the light-extraction efficiency of light-emitting diodes. *Opt. Express* **2011**, *19*, A1135–A1140. [CrossRef] [PubMed]
12. Kuritzky, L.Y.; Espenlaub, A.C.; Yonkee, B.P.; Pynn, C.D.; DenBaars, S.P.; Nakamura, S.; Weisbuch, C.; Speck, J.S. High wall-plug efficiency blue III-nitride LEDs designed for low current density operation. *Opt. Express* **2017**, *25*, 30696–30707. [CrossRef] [PubMed]
13. Dang, C.; Lee, J.; Zhang, Y.; Han, J.; Breen, C.; Steckel, J.S.; Coe-Sullivan, S.; Nurmikko, A. A Wafer-Level Integrated White-Light-Emitting Diode Incorporating Colloidal Quantum Dots as a Nanocomposite Luminescent Material. *Adv. Mater.* **2012**, *24*, 5915–5918. [CrossRef] [PubMed]
14. Kang, J.-H.; Li, B.; Zhao, T.; Johar, M.A.; Lin, C.-C.; Fang, Y.-H.; Kuo, W.-H.; Liang, K.-L.; Hu, S.; Ryu, S.-W.; et al. RGB Arrays for Micro-Light-Emitting Diode Applications Using Nanoporous GaN Embedded with Quantum Dots. *ACS Appl. Mater. Interfaces* **2020**, *12*, 30890–30895. [CrossRef] [PubMed]
15. Chen, Y.-P.; Ni, C.-C.; Wu, R.-N.; Kuo, S.-Y.; Su, Y.-C.; Huang, Y.-Y.; Chen, J.-W.; Hsu, Y.-C.; Wu, S.-H.; Chen, C.-Y.; et al. Combined effects of surface plasmon coupling and Förster resonance energy transfer on the light color conversion behaviors of colloidal quantum dots on an InGaN/GaN quantum-well nanodisk structure. *Nanotechnology* **2020**, *32*, 135206. [CrossRef] [PubMed]
16. Lee, C.-T.; Wu, T.-J. Light distribution and light extraction improvement mechanisms of remote GaN-based white light-emitting-diodes using ZnO nanorod array. *J. Lumin.* **2013**, *137*, 143–147. [CrossRef]
17. Gorsky, S.; Zhang, R.; Gok, A.; Wang, R.; Kebede, K.; Lenef, A.; Raukas, M.; Negro, L.D. Directional light emission enhancement from LED-phosphor converters using dielectric Vogel spiral arrays. *APL Photonics* **2018**, *3*, 126103. [CrossRef]
18. Cabello-Olmo, E.; Molet, P.; Mihi, A.; Lozano, G.; Míguez, H. Enhanced Directional Light Extraction from Patterned Rare-Earth Phosphor Films. *Adv. Opt. Mater.* **2021**, *9*, 2001611. [CrossRef]
19. Ji, J.; Zhang, G.; Yang, S.; Feng, X.; Zhang, X.; Yang, C.C. Theoretical analysis of a white-light LED array based on a GaN nanorod structure. *Appl. Opt.* **2020**, *59*, 2345–2351. [CrossRef]
20. Su, V.-C.; Gao, C.-C. Remote GaN metalens applied to white light-emitting diodes. *Opt. Express* **2020**, *28*, 38883. [CrossRef]
21. Elahi, A.M.N.; Xu, J. Modeling of Nanophosphor-Coupled Porous Layers for Color Conversion in III-Nitride Micro-LED Arrays. *J. Electron. Mater.* **2022**, *51*, 3023–3032. [CrossRef]
22. Shin, D.-Y.; Kima, T.; Akyuz, O.; Demir, H.V.; Lee, I.-H. Ag@SiO_2-embedded InGaN/GaN nanorod array white light-emitting diode with perovskite nanocrystal films. *J. Alloys Compd.* **2022**, *898*, 162974. [CrossRef]
23. Zheng, H.; Zhao, Z.; Wang, Y.; Li, L.; Liu, S.; Luo, X. Effect of patterned substrate on light extraction efficiency of chip-on-board packaging LEDs. In Proceedings of the 2014 IEEE 64th Electronic Components and Technology Conference, Orlando, FL, USA, 27–30 May 2014; pp. 1876–1879.
24. Kim, K.H.; Kim, W.H.; Jeon, S.-W.; Choi, M.; Bin Song, S.; Kim, J.P. Effects of the optical absorption of a LED chip on the LED package. *Solid-State Electron.* **2015**, *111*, 166–170. [CrossRef]
25. Allen, S.C.; Steckl, A.J. A nearly ideal phosphor-converted white light-emitting diode. *Appl. Phys. Lett.* **2008**, *92*, 143309. [CrossRef]
26. Choi, H.-S.; Park, J.-S.; Moon, C.-H. Improvement of Light Extraction Efficiency of LED Packages Using an Enhanced Encapsulant Design. *J. Opt. Soc. Korea* **2014**, *18*, 370–376. [CrossRef]
27. Li, J.-S.; Tang, Y.; Li, Z.-T.; Chen, J.-X.; Ding, X.-R.; Yu, B.-H. Precise Optical Modeling of Phosphor-Converted LEDs With Arbitrary Concentration and Thickness Using Bidirectional Scattering Distribution Function. *IEEE Photonics J.* **2018**, *10*, 8201117. [CrossRef]
28. Zhu, Z.; Liu, B.; Cheng, C.; Chen, H.; Gu, M.; Chen, L.; Liu, J.; Ouyang, X.; Xu, J.; Zhang, C. Convenient method for improving the light output of scintillators by using buffer layers coated with photonic crystals. *Opt. Express* **2020**, *28*, 11301–11308. [CrossRef]
29. Chen, Y.; Chen, Z.; Jiao, F.; Zhan, J.; Pan, Z.; Deng, C.; Xi, X.; Kang, X.; Chen, W.; Wang, Q.; et al. Enhanced light extraction efficiency of an LED package by a surface-mounted amorphous photonic structure. *Opt. Express* **2021**, *29*, 31594. [CrossRef]
30. Shin, J.-H.; Choi, H.-J.; Han, K.-S.; Ra, S.; Choi, K.-W.; Lee, H. Effect of anti-reflective nano-patterns on LED package. *Curr. Appl. Phys.* **2013**, *13*, S93–S97. [CrossRef]
31. Clapham, P.B.; Hutley, C. Reduction of Lens Reflexion by the "Moth Eye" Principle. *Nature* **1973**, *244*, 281–282. [CrossRef]
32. Kim, J.-J.; Lee, Y.; Kim, H.G.; Choi, K.-J.; Kweon, H.-S.; Park, S.; Jeong, K.-H. Biologically inspired LED lens from cuticular nanostructures of firefly lantern. *Proc. Natl. Acad. Sci. USA* **2012**, *109*, 18674–18678. [CrossRef] [PubMed]

33. Chan, L.W.; Morse, D.E.; Gordon, M.J. Moth eye-inspired anti-reflective surfaces for improved IR optical systems & visible LEDs fabricated with colloidal lithography and etching. *Bioinspir. Biomim.* **2018**, *13*, 41001.
34. Goerlitzer, E.; Taylor, R.N.K.; Vogel, N. Bioinspired Photonic Pigments from Colloidal Self-Assembly. *Adv. Mater.* **2018**, *30*, e1706654. [CrossRef] [PubMed]
35. Lumerical Inc. FDTD Solutions. Available online: https://www.lumerical.com (accessed on 1 June 2021).
36. Ding, X.; Li, J.; Chen, Q.; Tang, Y.; Li, Z.; Yu, B. Improving LED CCT uniformity using micropatterned films optimized by combining ray tracing and FDTD methods. *Opt. Express* **2015**, *23*, A180–A191. [CrossRef] [PubMed]
37. Bechtel, H.; Schmidt, P.; Busselt, W.; Schreinemacher, B.S. Lumiramic: A new phosphor technology for high performance solid state light sources. *Proc. SPIE* **2008**, *7058*, 70580E.
38. Chen, Z.-Z.; Qin, Z.-X.; Hu, X.-D.; Yu, T.-J.; Yang, Z.-J.; Zhang, B.; Yao, G.-Q.; Qiu, X.-M.; Zhang, G.-Y. Fabrication and Characterization of High-power White LED. *Chin. J. Liq. Cryst. Disp.* **2004**, *19*, 83.

crystals

MDPI

Article

Investigation of the Effect of ITO Size and Mesa Shape on the Optoelectronic Properties of GaN-Based Micro LEDs

Aoqi Fang, Hao Xu, Weiling Guo *, Jixin Liu, Jiaxin Chen and Mengmei Li

Key Laboratory of Optoelectronics Technology, Ministry of Education, Beijing University of Technology, Beijing 100124, China

* Correspondence: guoweiling@bjut.edu.cn

Abstract: In this paper, in order to explore the influence of indium tin oxide (ITO) size and mesa shape on the performance of GaN-based micro light emitting diodes (Micro LEDs) on sapphire substrates, Micro LEDs of different sizes with ITO area smaller than or equal to the light-emitting area were designed and fabricated. The experiment results show that when the ITO area of the Micro LED is equal to the area of the light-emitting area, its optoelectronic performance is significantly better than that of the Micro LEDs whose ITO area is smaller than the area of the light-emitting area. When the light-emitting area size is 40 μm, the wall-plug efficiency (WPE) of the two structures of Micro LEDs can differ by more than 50%. Based on above experiment results, this paper designed and fabricated Micro LEDs with different sizes of square and circular mesa with the same ITO area as the light-emitting area. The experimental results show that the WPE of the circular mesa Micro LED is slightly higher than that of the square mesa Micro LED at low current density. However, as the current density and chip size increase, the performance of the Micro LED with a square mesa is better.

Keywords: GaN; Micro LED; ITO; wall-plug efficiency; micro-electronics devices

Citation: Fang, A.; Xu, H.; Guo, W.; Liu, J.; Chen, J.; Li, M. Investigation of the Effect of ITO Size and Mesa Shape on the Optoelectronic Properties of GaN-Based Micro LEDs. *Crystals* **2022**, *12*, 1593. https://doi.org/10.3390/cryst12111593

Academic Editor: Ikai Lo

Received: 6 October 2022
Accepted: 2 November 2022
Published: 9 November 2022

1. Introduction

In recent years, with the maturity of III-V compound semiconductor devices and integrated circuit technology [1–3], miniature portable electronic products have begun to emerge. Among them, Micro LEDs are widely used in smart phones, visible light communication, implantable devices and other fields due to their unique advantages [4,5]. However, the conventional LED structure has problems such as the difficulty of achieving good ohmic contact on the P-GaN surface, and the dense current distribution under the P electrode, which will lead to uneven current and heat distribution of the LED, which will degrade the optoelectronic performance of the LED and affect the reliability of the device [6]. Therefore, ITO films with high light transmittance, good electrical conductivity and easy-to-form good ohmic contact with P-GaN [7,8] are widely used as extended electrodes to improve the above problems. In recent years, scholars have continued to improve the structure and preparation methods of ITO films to further improve the optoelectronic properties of LEDs: in 2010, Cheng, L. et al. used photoresist as a mask for inductively coupled plasma etching (ICP) to transfer the twisted shape of the photoresist to the ITO surface; the study found that more scattering occurred on ITO with a rougher surface, and moreover, the performance of LEDs increased [9]. In 2018, J. Xu et al. deposited a thin layer of Al after sputtering ITO, and alloyed Al atoms with ITO atoms after thermal annealing; when the overall thickness of the film is 395nm, the light transmittance reaches 93.2%, and the optical output power of the device is increased by 13% compared with the pure ITO film [10]. For Micro LEDs, the mesa structure is an important factor affecting the optoelectronic performance of the device. The improvement of the light-emitting mesa mainly focuses on the treatment of the side wall [11]. In 2010, Chen P.H. et al. proposed to engrave nano-textures on the sidewall of the LED mesa to improve the light extraction efficiency, which increased the output power of the LED by more than 45% [12]; In 2010,

P.H. Chen et al. proposed an alternative approach to investigate these defects directly after MESA formation, by coupling optical characterization techniques together with time-of-flight secondary ion mass spectrometry (TOF-SIMS) on AlGaInP square shaped pixels of different sizes formed by BCl_3-based reactive ion etching (RIE) [13], which provides convenience for the processing of mesa sidewall defects.

Although the research on ITO and mesa is relatively advanced, most of the studies focus on the improvement of the preparation process and the influence of the surface mechanism [14–16], and there are few studies on the specific size of ITO and the shape of the mesa in Micro LEDs. Therefore, this paper proposes Micro LEDs with an ITO area less than or equal to the area of the light-emitting area and square and circular mesa, to explore the influence of the size of the ITO relative to the area of the light-emitting area and the shape of the mesa on the optoelectronic property of Micro LEDs.

2. Materials and Methods

A GaN-based LED commercial epitaxial layer grown on sapphire was used, from bottom to top comprised of: 3.4 μm U-GaN; 2.3 μm N-GaN; 150 nm InGaN/GaN multiple quantum wells (MQW) layer; 43 nm P-GaN. For the Micro LED with ITO area equal to the mesa area, in order to prevent the shrinkage of ITO from affecting the final area, we firstly deposited 110 nm ITO and 1.5 nm Al on the epitaxial layer by electron beam evaporation. Secondly, we annealed at 580 °C for 5 min to achieve good ohmic contact between ITO and P-GaN. Then, we etched the epitaxial layer 1.2 μm to N-GaN layer by inductively coupled plasma (ICP) etching to form light-emitting mesa. For the Micro LED with ITO area smaller than the light-emitting area, the above two steps were done in reverse. Then, for both kinds of Micro LED, 1000 nm-thick SiO_2 was grown by PECVD as the passivation layer (PV). Because SiO_2 is hydrophilic and photoresist is hydrophobic, HDMS treatment was used to remove the -OH on the surface of SiO_2, so that the surface of SiO_2 could form hydrophobicity to adhere to the photoresist well. The PV pattern was obtained by photolithography and etching. Cr/Al/Ti/Pt/Au multilayer metal as P/N electrode with a thickness of 2.4 μm was deposited by electron beam evaporation. Finally, the wafer was thinned to about 200 μm by mechanical grinding method. A 39-layer distributed Bragg reflection (DBR) layer was fabricated by alternating SiO_2 and TiO_2 on the back of the wafer to improve the optical extraction efficiency.

Figure 1a is the Micro LED with the size of ITO smaller than the size of the light-emitting area. The ITO of this structure is 2 μm away from the boundary of the light-emitting area, and the ITO shrank inward by 2–3 μm during the process preparation. Therefore, the distance between the ITO boundary and the sidewall was maintained at 4–5 μm, so the overall area of the ITO was smaller than the mesa area, which had some impacts on the light output of the device. To investigate the effect of ITO size change on the performance of Micro LED, a novel Micro LED sample with the ITO size equal to the light-emitting area size was designed for this paper, as shown in Figure 1b.

In order to explore the influence of different mesa shapes on the optoelectronic characteristics of Micro LEDs, square and circular mesa Micro LEDs with the same light-emitting area were designed and fabricated. The schematic diagram of the structure of the circular mesa Micro LED is shown in Figure 1c.

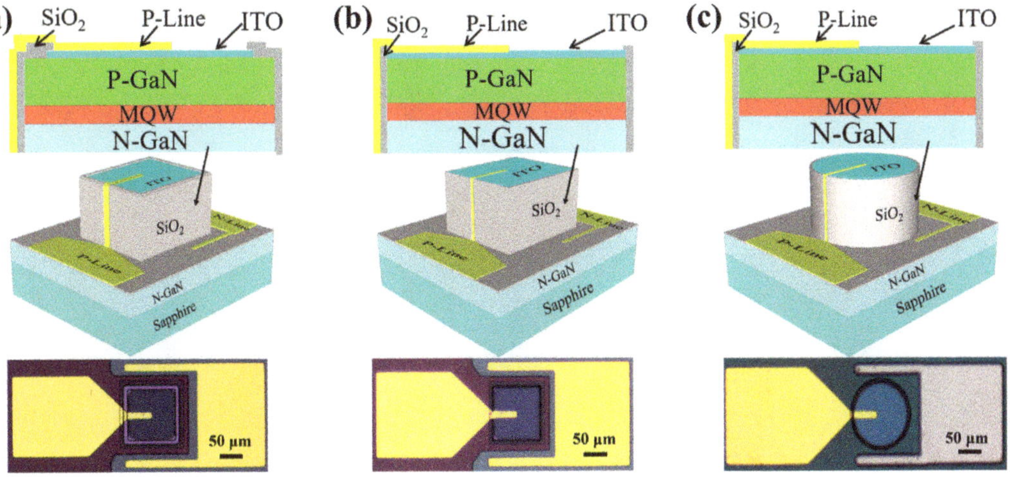

Figure 1. Micro LED (**a**) with ITO smaller than light emitting area (**b**) with ITO equal to light emitting area (**c**) with circular mesa (ITO equal to light emitting area).

3. Results and Discussion

3.1. Influence of Different ITO Sizes on Optoelectronic Properties

Structure A and B represent Micro LEDs whose ITO size is smaller than and equal to the area of the light-emitting area, respectively. Figure 2 shows the I-V characteristic curves of the two structures under different light-emitting area sizes; it can be seen that at the same light-emitting area size, the series resistance of A is smaller than that of B, which is because of the increase of ohmic contact area between the P-GaN and ITO, which reduces the voltage of the device. As the size of the light-emitting area of the Micro LED decreases, the series resistance increases; as the contact area between the P-GaN and the ITO decreases with the decrease, this results in an increase in the contact resistance.

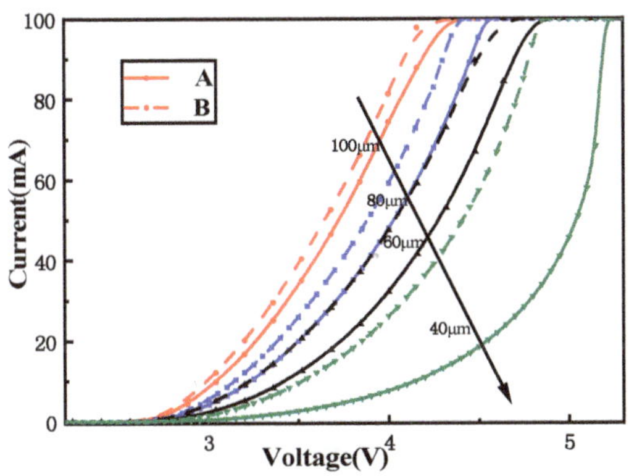

Figure 2. I–V curves of Micro LEDs with ITO size smaller than (A) and equal to (B) the light-emitting area size.

Figure 3 shows the variation curves of the WPE of the two structures with the current density under different light-emitting sizes. Because under the dual action of Auger recombination and carrier leakage, the rise in luminous power will be limited, while the current and voltage can be kept in a high range, so the WPE shows a downward trend. The WPE of B is higher than that of A, because the larger ITO area of B not only means larger light-emitting area and more photons emitted compared with the conventional structure A, but also means larger ohmic contact area and lower voltage. In addition, under the same light-emitting size, the larger the ITO is, the smaller the series resistance is (Figure 2). According to Formula (3), the value of voltage-temperature coefficient S decreases with the decrease of series resistance, so the thermal stability of B is better, which makes the current spreading and the optoelectronic performance of structure B better.

$$S = \left|\frac{dV_F}{dT}\right| = \frac{nk}{q}\ln\left(\frac{I_F}{I_S}\right) + \frac{dR_s}{dT}I_F \tag{1}$$

where, I_S is the reverse saturation current, q is the electronic quantity, k is the Boltzmann constant, T is the absolute temperature and n is the ideal factor. I_S is a function of temperature. Under the condition that all impurities in the semiconductor material are ionized and the intrinsic excitation can be ignored, I_S can be summarized as:

$$I_S = Ae\left(\sqrt{\frac{D_n}{\tau_n}}\frac{n_i^2}{N_A} + \sqrt{\frac{D_P}{\tau_P}}\frac{n_i^2}{N_D}\right) = CT^3\exp\left(-\frac{eV_{g0}}{KT}\right) \tag{2}$$

A is the knot area, C is a constant related to junction area and impurity concentration and V_{g0} is the potential difference between the conduction band bottom and the valence band top of the PN junction material at absolute zero. Substituting Formula (2) into Formula (1), S can be summarized as:

$$S = \left|\frac{dV_F}{dT}\right| = \frac{nk}{q}\ln\left(\frac{C}{I_F}\right) + \frac{3nk}{q}\ln T + \frac{3nk}{q} + \frac{1.5kT + E_a}{2kT^2}I_F R_s \tag{3}$$

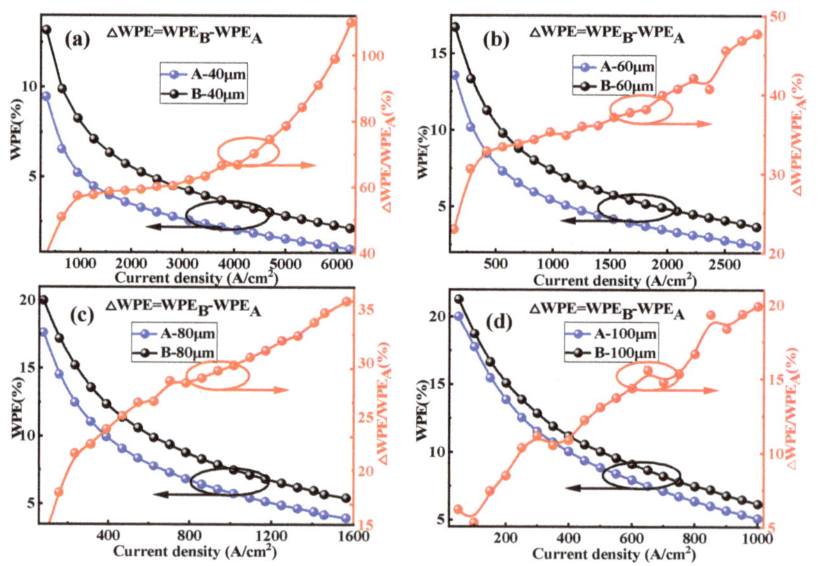

Figure 3. WPE vs. current densities of Micro LED A and B (**a**) 40 μm (**b**) 60 μm (**c**) 80 μm (**d**) 100 μm.

Figure 3 also shows the variation of the novel structure (B) compared with the conventional structure (A) under four kinds of size. It can be seen that as the chip size decreases, the ITO area has a greater impact on the optoelectronic performance of the Micro LED. Where the size of the light-emitting area is 40 μm and 100 μm, the WPE of the novel structure improves by up to 110% and 18% as shown in Figures 3a and 3d, respectively. A mathematical model can be used to describe this phenomenon. If the size of the light-emitting area is "L" and the distance between the ITO boundary and the boundary of the light-emitting area is "a" (approximately equal to 2 μm), then the area ratio of the ITO to the light-emitting area is "$(L - a)^2/L^2$". Thus, the larger the size "L", the closer the ratio to "1", and the smaller the effect of the shrinkage of ITO on the current spreading is. Therefore, the performance of small size Micro LED can be significantly improved on account of the growth of the ITO area.

3.2. Influence of Different Mesa Shapes on Optoelectronic Properties

In this paper, C and D represent the square mesa Micro LED and circular mesa Micro LED in this comparative experiment, respectively. The specific dimensions of Micro LED with two structures are shown in Table 1.

Table 1. Specific dimensions of C and D.

Group	Length of C (L)	Area of C	Diameter of D (R)	Area of D
I	30 μm	900 μm^2	34 μm	908 μm^2
II	40 μm	1600 μm^2	45 μm	1590 μm^2
III	80 μm	6400 μm^2	90 μm	6362 μm^2
IV	100 μm	10,000 μm^2	113 μm	10,028 μm^2

Figure 4 shows the variation curves of WPE and luminous power of Micro LEDs with different mesa shapes under different current densities. As a whole, the WPE is almost the same for C and D; the luminous power of C is a little bit higher than that of D at high current density. However, with the increase in device size and current density, the square Micro LED has better optoelectronic performance. The WPE of the square mesa Micro LED is larger than that of the circular one, because the side surface area of the square light-emitting area is about 13% larger than that of the circular one, which leads to more light-emitting from the sidewall of C, thereby increasing the luminous power. As the current density increases further, more light emits from the sidewall, and the optoelectronic performance gap between the two structures will be widened, and as the size of the light-emitting area increases, the Micro LED with a square mesa will have better stability than the Micro LED with a circular mesa.

To further explore the performance differences of Micro LEDs with different mesa shapes in different sizes, the thermal characteristics of devices with two structures were tested under the current of 20 mA. Figure 5 shows the thermal resistance of Micro LEDs with square and circular mesa of different sizes. The thermal resistance of the square mesa Micro LED is slightly higher than that of the circular one when the size is small, which is the reason why the WPE of the D-I is slightly higher than that of C-I. With the increase in chip size, the thermal resistance of the two structures achieves the same level, that is, the thermal characteristics of the two structures are not the main factor affecting WPE in the case of large mesa size; the difference of luminous power caused by the light-emitting from the sidewall is the main reason for the difference in the optoelectronic performance of devices with two kinds of structures. Therefore, in Micro LEDs of small size, thermal factors are the cause why the square mesa cannot achieve a better optoelectronic performance than the circular one. However, with the increase in chip size, the Micro LED of the square mesa has a better optoelectronic performance due to the better luminous power.

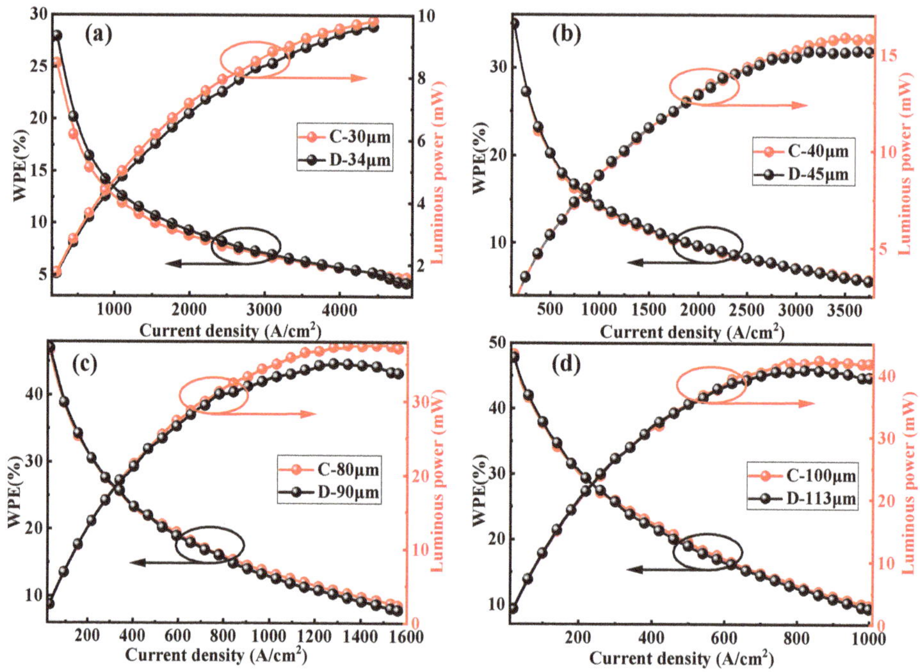

Figure 4. WPE and luminous power of C and D at different current densities, (**a**)–(**d**) correspond to experimental group I–IV (The size of the mesa increases gradually).

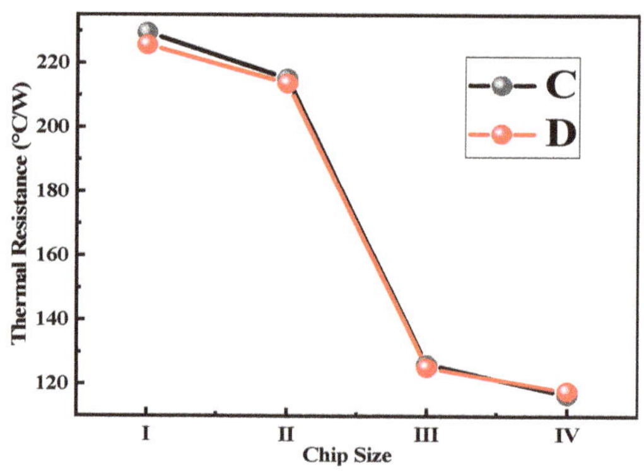

Figure 5. Thermal resistance of C and D with different mesa sizes at 20 mA current.

4. Conclusions

In this paper, Micro LEDs with a light-emitting area size of 40–100 μm and an ITO size smaller than or equal to the light-emitting area size were designed and fabricated respectively. After testing and analyzing these Micro LEDs with different structures, it is found that when the area of ITO is equal to the area of the light-emitting area, the optoelectronic performance of the Micro LED is obviously better than that of conventional Micro LEDs with the area of ITO smaller than the area of the light-emitting area. When the area of the light-emitting area is 40 μm, the improvement of WPE by the new structure exceeds 50%, and with the increase in the current density, there is still a trend to improve WPE. According to this experimental result, we can improve the overall performance of the device by increasing the contact area between ITO and P-GaN, especially in small size

Micro LEDs. Furthermore, Micro LEDs with square and circular mesa of different sizes were designed and fabricated, which has the structure that the area of ITO is equal to the area of the light-emitting area. It is found that the WPE of the Micro LED with a square mesa is slightly higher than that of the Micro LED with a circular mesa at a high current density on account that the light emits from sidewall, which is instructive for us to choose the mesa shape of Micro LED as Micro LED usually works under high current density. And as the size of the light-emitting area increases, the Micro LED with a square mesa will have better stability than the Micro LED with a circular mesa. In the future display array fabrication, we can extract the light from the sidewall of Micro LED on the square mesa to the pixel position, which will be of great help to the future research of Micro LED display array. In general, the square mesa Micro LED with the ITO size equal to the area of the light emitting area exhibits better optoelectronic properties, which provides a new idea for further reducing the size of the Micro LED and improving the performance of the Micro LEDs in the future.

Author Contributions: Conceptualization, W.G. and A.F.; methodology, A.F.; formal analysis, A.F. and H.X.; investigation, A.F., J.C. and M.L.; writing—original draft preparation, A.F. and H.X.; writing—review and editing, A.F., H.X. and J.L.; supervision, W.G. All authors have read and agreed to the published version of the manuscript.

Funding: This research was funded by National key R&D program of China (No. 2017YFB0402803).

Institutional Review Board Statement: Not applicable.

Informed Consent Statement: Not applicable.

Data Availability Statement: Not applicable.

Conflicts of Interest: The authors declare no conflict of interest. The funders had no role in the design of the study; in the collection, analyses, or interpretation of data; in the writing of the manuscript; or in the decision to publish the results.

References

1. Zhou, F.; Xu, W.; Ren, F.; Zhou, D.; Chen, D.; Zhang, R.; Zheng, Y.; Zhu, T.; Lu, H. High-Voltage Quasi-Vertical GaN Junction Barrier Schottky Diode with Fast Switching Characteristics. *IEEE Electron Device Lett.* **2021**, *42*, 974–977. [CrossRef]
2. Hwangbo, S.; Hu, L.; Hoang, A.T.; Choi, J.Y.; Ahn, J.H. Wafer-scale monolithic integration of full-colour micro-LED display using MoS_2 transistor. *Nat. Nanotechnol.* **2022**, *17*, 500–506. [CrossRef] [PubMed]
3. Du, Z.; Li, D.; Guo, W.; Xiong, F.; Tang, P.; Zhou, X.; Zhang, Y.; Guo, T.; Yan, Q.; Sun, J. Quantum Dot Color Conversion Efficiency Enhancement in Micro-Light-Emitting Diodes by Non-radiative Energy Transfer. *IEEE Electron Device Lett.* **2021**, *42*, 1184–1187. [CrossRef]
4. Huang, C.; Cheng, Y.; Tsao, Y.; Liu, X.; Kuo, H. Micro-LED backlight module by deep reinforcement learning and micro-macro-hybrid environment control agent. *Photon. Res.* **2022**, *10*, 269–279. [CrossRef]
5. Shan, X.; Zhu, S.; Qiu, P.; Qian, Z.; Lin, R.; Wang, Z.; Cui, X.; Liu, R. Multifunctional Ultraviolet-C Micro-LED With Monolithically Integrated Photodetector for Optical Wireless Communication. *J. Lightwave Technol.* **2022**, *40*, 490–498. [CrossRef]
6. Kim, T.; Yoon, Y.; Oh, S.; Cha, Y.; Hong, I.; Cho, M.; Hong, C.; Choi, H.; Kwak, J. Opposite Behavior of Multilayer Graphene/Indium-Tin-Oxide p-Electrode for Gallium Nitride Based-Light Emitting Diodes Depending on Thickness of Indium-Tin-Oxide Layer. *J. Nanosci. Nanotechnol.* **2018**, *18*, 6106–6111. [CrossRef] [PubMed]
7. Min, J.; Kwak, H.; Kim, K.; Jeong, W.; Lee, D. Very thin ITO/metal mesh hybrid films for a high-performance transparent conductive layer in GaN-based light-emitting diodes. *Nanotechnology.* **2017**, *28*, 045201. [CrossRef] [PubMed]
8. Lu, W.; Aplin, D.; Clawson, A.R.; Yu, P. Effects of the gas ambient in thermal activation of Mg-doped p-GaN on Hall effect and photoluminescence. *J. Vac. Sci. Technol. A* **2013**, *31*, 011502. [CrossRef]
9. Cheng, L.; Wu, Y. InGaN-GaN Light Emitting Diode Performance Improved by Roughening Indium Tin Oxide Window Layer via Natural Lithography. Electrochem. *Solid-State Lett.* **2010**, *13*, J8–J10. [CrossRef]
10. Xu, J.; Zhang, W.; Peng, M.; Dai, J.; Chen, C. Light-extraction enhancement of GaN-based 395 nm flip-chip light-emitting diodes by an Al-doped ITO transparent conductive electrode. *Opt. Lett.* **2018**, *43*, 2684–2687. [CrossRef] [PubMed]
11. Yang, F.; Xu, Y.; Li, L.; Cai, X.; Li, J.; Tao, J.; Zheng, S.; Cao, B.; Xu, K. Optical and microstructural characterization of Micro-LED with sidewall treatment. *J. Phys. D-Appl. Phys.* **2022**, *55*, 43. [CrossRef]
12. Chen, P.; Chang, L.C.; Tsai, C.H.; Lee, Y.C.; Lai, W.C.; Wu, M.L.; Kuo, C.H.; Sheu, J.k. GaN-Based Light-Emitting Diodes with Pillar Structures Around the Mesa Region. *IEEE J. Quantum Electron.* **2010**, *46*, 1066–1071. [CrossRef]

13. Boussadi, Y.; Rochat, N.; Barnes, J.P.; Bakir, B.; Ferrands, P.; Masenelli, B.; Licitra, C. Investigation of sidewall damage induced by reactive ion etching on AlGaInP MESA for micro-LED application. *J. Lumines.* **2021**, *234*, 117937. [CrossRef]
14. Wang, P.; Gan, Z.; Liu, S. Improved light extraction of GaN-based light-emitting diodes with surface-patterned ITO. *Opt. Laser Technol.* **2009**, *41*, 823–826. [CrossRef]
15. Lee, Y.; Reddy, M.; Kim, B.; Park, C. Surface morphological, structural, electrical and optical properties of GaN-based light-emitting diodes using submicron-scaled Ag islands and ITO thin films. *Opt. Mater.* **2018**, *81*, 109–114. [CrossRef]
16. Park, J.; Kim, J.; Kim, J.; Kim, D.; Na, J.; Kim, S. Formation of an indium tin oxide nanodot/Ag nanowire electrode as a current spreader for near ultraviolet AlGaN-based light-emitting diodes. *Nanotechnology* **2017**, *28*, 045205. [CrossRef] [PubMed]

crystals

Article

Nickel-Assisted Transfer-Free Technology of Graphene Chemical Vapor Deposition on GaN for Improving the Electrical Performance of Light-Emitting Diodes

Penghao Tang [1], Fangzhu Xiong [1], Zaifa Du [1], Kai Li [1], Yu Mei [1], Weiling Guo [1,*] and Jie Sun [2,3,*]

[1] Key Laboratory of Optoelectronics Technology, College of Microelectronics, Beijing University of Technology, Beijing 100124, China
[2] Fujian Science & Technology Innovation Laboratory for Optoelectronic Information of China, Fuzhou 350100, China
[3] National and Local United Engineering Laboratory of Flat Panel Display Technology, College of Physics and Information Engineering, Fuzhou University, Fuzhou 350100, China
* Correspondence: guoweiling@bjut.edu.cn (W.G.); jie.sun@fzu.edu.cn (J.S.)

Abstract: With the rapid development of graphene technology, today graphene performs well in the application of light-emitting diode (LED) transparent electrodes. Naturally, high-quality contact between the graphene and the GaN underneath is very important. This paper reports a process for nickel-assisted transfer-free technology of graphene chemical vapor deposition on GaN. The nickel film plays the dual role of etching mask and growth catalyst, and is removed by the subsequent "penetration etching" process, achieving good direct contact between the graphene and GaN. The results show that the graphene effectively improves the current spreading of GaN-based LEDs and enhances their electrical performance. This scheme avoids the wrinkles and cracks of graphene from the transfer process, and is not only suitable for the combination of graphene and GaN-based LEDs, but also provides a solution for the integration of graphene and other materials.

Keywords: GaN; LEDs; graphene; transparent electrode; CVD; transfer-free

Citation: Tang, P.; Xiong, F.; Du, Z.; Li, K.; Mei, Y.; Guo, W.; Sun, J. Nickel-Assisted Transfer-Free Technology of Graphene Chemical Vapor Deposition on GaN for Improving the Electrical Performance of Light-Emitting Diodes. *Crystals* **2022**, *12*, 1497. https://doi.org/10.3390/cryst12101497

Academic Editors: Peng Chen and Zhizhong Chen

Received: 7 October 2022
Accepted: 19 October 2022
Published: 21 October 2022

1. Introduction

With the development of science and technology, light-emitting diodes (LEDs) play an increasingly important role in our daily life. Since the birth of gallium nitride (GaN)-based LEDs, they have greatly promoted the development of lighting and display fields. However, there are still problems in the development, such as the difficulty of heavy doping of p-type GaN. This means that the conductivity of p-type GaN is relatively poor, so that the current is difficult to spread to achieve sufficient current injection. For this reason, transparent electrodes came into being. Indium tin oxide (ITO) is the most representative transparent electrode material, which has excellent electrical conductivity and high transparency in the visible band [1–4]. Nevertheless, from the perspective of sustainable development, ITO is inferior to graphene because of the scarcity of indium [5,6]. In addition, graphene performs better than ITO on flexible substrates. These factors make graphene one of the most viable potential transparent electrode materials [7].

In the growth of graphene, metals play a crucial catalytic role. The catalytic mechanism is mainly divided into two types [8]: one is the carbon segregation and precipitation mechanism represented by nickel, under which more layers of graphene are grown; the other is the surface adsorption mechanism represented by copper, under which the number of graphene layers grown is lesser. Both of these mechanisms are widely used in the growth of graphene.

However, the vital role played by metal catalysts in graphene growth means that the growth of high-quality graphene is often inseparable from metals. As a result, as-grown graphene is always attached to the surface of metal substrates [9–11], which is difficult to use in electronic devices directly. There are some solutions to this problem: for example, graphene can be transferred from the metal substrates to the target substrates, but in this process greater or lesser contamination and defects will inevitably be introduced. In addition, graphene can be directly grown on the target substrates, but without metal catalysis the growth of graphene is often more difficult [12,13].

To solve this problem, in this paper we propose a nickel-assisted transfer-free technology of CVD graphene on GaN. In this method, nickel film is used to catalyze the growth of high-quality graphene, and it is removed by the penetration etching process to realize the contact of graphene with the target substrates, so as to avoid the transfer process. In addition, nickel is not only used as a catalyst, but also as a mask to etch the LED mesas. The measurement results show that the as-grown graphene film effectively improves the electrical performance of LEDs. The effects of different growth time on the performance of LEDs are compared as well. The process is simple, efficient and highly repeatable, which directly realizes good, tight contact between graphene and GaN, providing a solution for the integration of graphene and GaN devices.

2. Materials and Methods

Since lift-off photolithography of metal electrodes after graphene growth may cause some damage to graphene, in order to pursue more stable and excellent electrical performance nickel is selected as the catalyst to grow graphene with more layers, reducing the impact of subsequent processes on the continuity of graphene.

Figure 1 shows a schematic diagram of the process flow of the LEDs in this paper. The epitaxial wafers used were provided by Xiangneng Hualei Optoelectronics Company. First, a 300 nm nickel film was photolithographically sputtered on the epitaxial wafer as the subsequent etching mask and growth catalyst. The $260 \times 515\ \mu m^2$ LED mesas were etched with inductively coupled plasma (ICP) at an etching ratio of approximately 1:80 for nickel and GaN, as shown in Figure 1b. The samples were subsequently grown in cold-wall plasma-enhanced chemical vapor deposition (PECVD) at 600 °C and 6 mbar, under an atmosphere of $CH_4/H_2/Ar$ (5/20/960 sccm) and at a plasma power of 40 W. As shown in Figure 1d, after the growth of graphene, PMMA (poly(methyl-methacrylate)) was spin-coated on the sample (3000 rpm for 30 s) and baked at 150 °C for 10 min. It is approximately 150 nm thick. Then the sample was put into the etching solution for more than four hours ($CuSO_4$:HCl:H_2O = 10 g:50 mL:50 mL). The etching solution can pass through the PMMA and graphene to slowly etch Ni film. Since PMMA is in close contact with graphene and n-GaN, after the metal is etched, PMMA and graphene will fall onto the surface of the sample instead of floating away in the solution. Then PMMA was removed by acetone and isopropyl alcohol (Figure 1f). Due to the weak van der Waals force on the surface of graphene, it is necessary to remove the graphene on the electrode area with oxygen plasma etching to prevent metal electrodes from falling off, as shown in Figure 1g. Finally, 15 nm Ti and 300 nm Au were sputtered as electrodes and annealed in a 450 °C vacuum for five minutes (Figure 1h).

Figure 2a shows the structure of the LEDs. Figure 2b shows the scanning electron microscope (SEM) image of the LED arrays (SE2 mode). An area of intentionally broken graphene has a higher contrast for better observation of graphene film, which is characterized by SEM (in-lens mode), as shown in Figure 2c. Since the in-lens mode has higher resolution, which is more suitable for observing the microscopic morphology, the graphene layer can be clearly seen in Figure 2c. The white boundary line in the image is the edge of the LED mesa, below which is the p-GaN and graphene on the mesa, and above which is the n-GaN beside the mesa. The dark film is graphene, and p-GaN is exposed in the broken area of graphene, corresponding to the white area in the image. It can be clearly seen that the graphene film has been broken and folded in this area, and even some graphene falls on n-GaN.

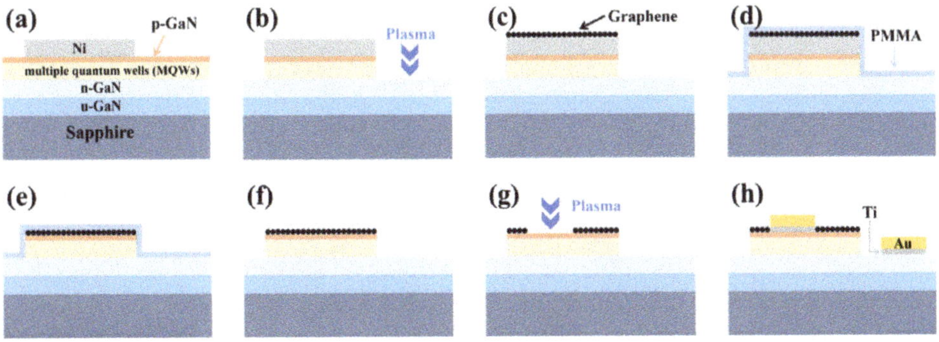

Figure 1. Process flow diagram of growth of graphene and fabrication of LEDs. (**a**) Sputter Ni film. (**b**) Etch the mesas. (**c**) Grow graphene. (**d**) Spin-coat PMMA. (**e**) Etch Ni film. (**f**) Remove the PMMA. (**g**) Remove some graphene. (**h**) Sputter metal electrodes.

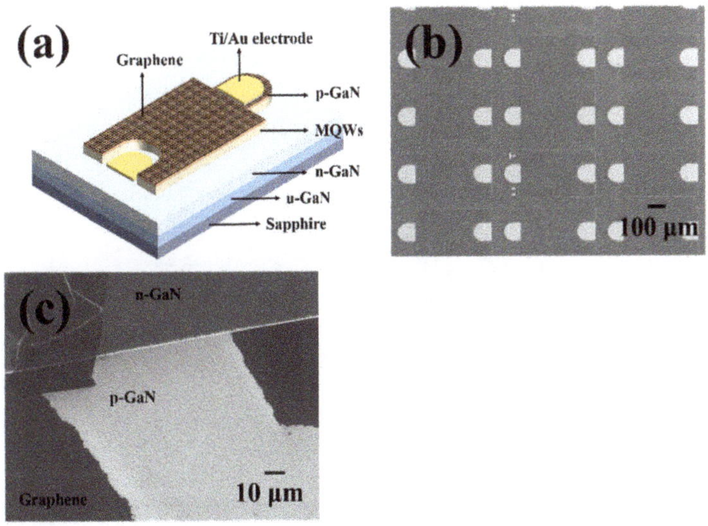

Figure 2. (**a**) Structure diagram of LEDs. (**b**) SEM image of LED arrays in SE2 mode. (**c**) SEM image of graphene on the mesa in in-lens mode.

3. Results and Discussion

In order to study the effect of graphene of different qualities on LED performance, LEDs with different graphene growth time were fabricated. Figure 3 shows the Raman characterization of graphene with different growth time after the fabrication of LEDs. The number of graphene layers is usually estimated in terms of the 2D/G ratio and the shape of the peaks [14]. It can be seen that with the increase of growth time, the 2D/G ratio gradually decreases, indicating that the number of graphene layers gradually increases. The graphene is estimated to be about 5–10 layers thick, based on both the Raman spectra [15] and our previous atomic force microscopy (AFM) measurement [12]. In addition, the Raman characterization was performed after the LEDs were fabricated, indicating that the graphene still maintains good quality after the lift-off photolithography of the metal electrodes.

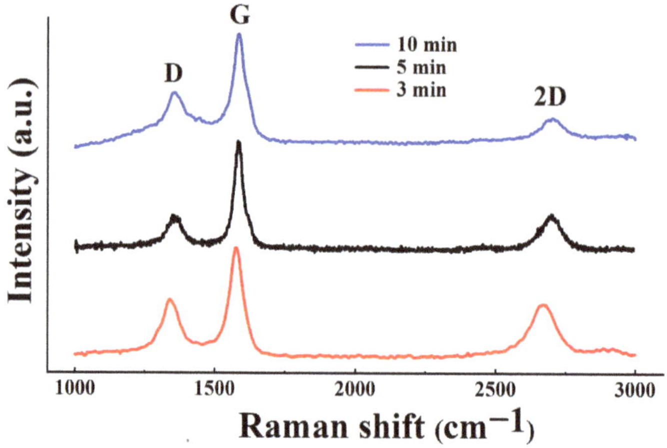

Figure 3. Raman characterization of graphene grown for 3 min, 5 min and 10 min, respectively.

Figure 4 is the energy dispersive spectroscopy (EDS) surface scanning result, inset is the characterized area and the characterized elements are Ni, Ga and Au. Among them, the molecular ratio of Ni is 0.01%, and the weight ratio is 0.02%. This indicates that the Ni film has been substantially etched away.

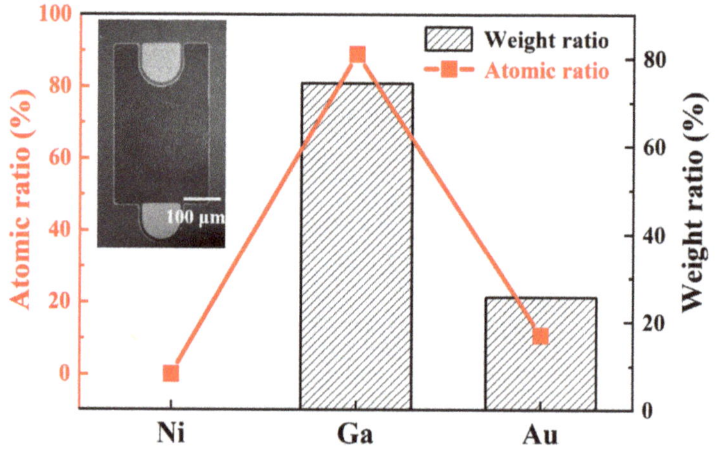

Figure 4. EDS characterization of the LED. Characterized elements are Ni, Ga and Au. The inset is the characterized area.

In addition to the fabrication of LEDs with graphene, graphene-free LEDs were also fabricated for comparison using basically the same process, and their current–voltage characteristics are shown in Figure 5. The inset is the current–voltage characteristics in semi-log scale. The dotted auxiliary lines are the reverse extension lines of the nearly linear part of the current–voltage characteristic curves, and their intersections with the abscissa represent the turn-on voltages of LEDs. The turn-on voltages of LEDs with growth time of 10 min, 5 min, 3 min and without graphene are about 3.2 V, 3.5 V, 3.7 V and 4 V, respectively; the operating voltages at 20 mA are about 3.5 V, 4.2 V, 4.4 V and 4.7 V, respectively. The slopes (k_1–k_4) of these auxiliary lines, which are mainly related to the LED series resistance, are also shown in Figure 5. Compared with graphene-free LEDs, the current spreading effect of graphene will reduce the series resistance of graphene-coated LEDs. Therefore, with the increase of growth time, the slope increases and the series resistance decreases. It can be inferred that graphene improves electrical performance of LEDs, and the thicker the graphene the greater the improvement, which may be attributed to its better continuity. From the inset, it is clear that these LEDs have a certain degree of leakage current before reaching their respective turn-on voltage, which may be related to the recombination current in the barrier area. Compared with graphene-free LEDs, the leakage current of LEDs with a growth time of 3 min and 5 min is not much different. However, the LEDs with graphene grown for 10 min have serious leakage current. This may be due to the fact that many carbon atoms are adsorbed on the sidewalls of the mesas because the growth time is too long, even though graphene is not formed because there is no metal catalyst on the sidewalls. Therefore, properly increasing the growth time can effectively improve the electrical performance of LEDs.

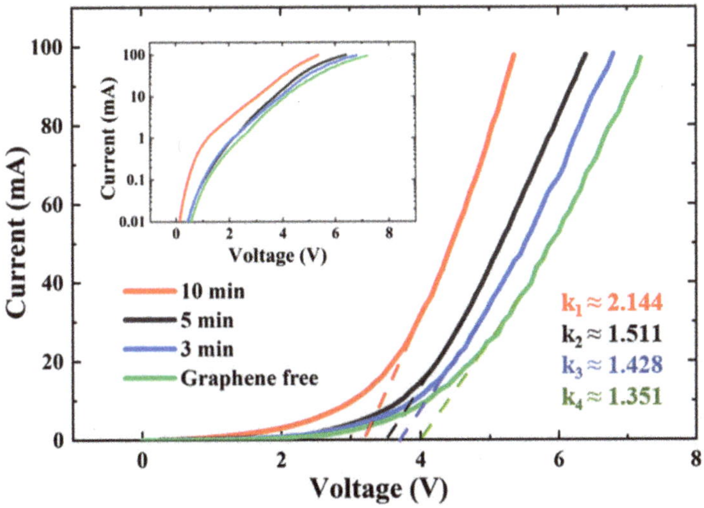

Figure 5. Current–voltage characteristics of LEDs with and without graphene. The dotted auxiliary lines are the reverse extension lines of the nearly linear part of the current–voltage characteristic curves, and k1–k4 represent their slopes. The inset is the current–voltage characteristics in semi-log scale.

Figure 6 is the optical microscope luminescence images of LEDs with and without graphene at 20 mA. Even though thicker graphene theoretically blocks more light [10], the difference in luminescence of LEDs coated with different graphene is still small when observed under an optical microscope, so only a representative graphene-coated LED is shown. It can be seen that the graphene-free LED only emits light near the metal electrodes, while the graphene-coated LED emits light evenly on the entire mesa. This shows that graphene plays a good role in current expansion, which indicates that graphene plays a good role in current spreading.

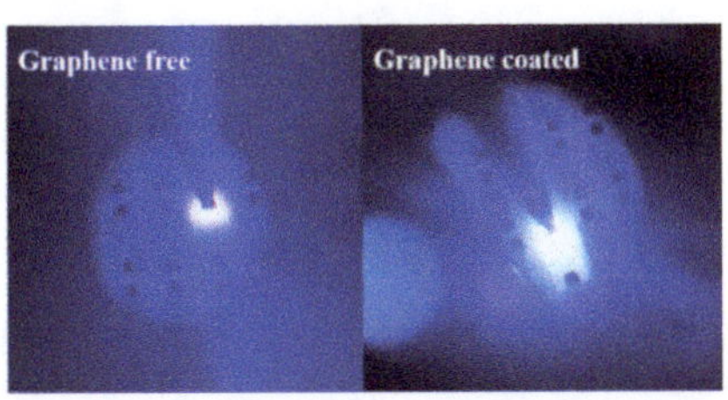

Figure 6. Images of LEDs without and with graphene at 20 mA.

4. Conclusions

In this paper, a method of growing transfer-free graphene on GaN by PECVD is introduced. LEDs with different thickness of graphene are fabricated by controlling the growth time. In the whole process, the Ni film is not only used as the mask for mesa etching but also as the catalyst for graphene growth, which is subsequently removed to achieve direct contact between graphene and GaN. When the growth time is less than ten minutes, the thickness of graphene increases with the growth time. The measurement results show that graphene realizes current spreading and effectively improves the electrical performance of LEDs, although excessive growth time may lead to current leakage of LEDs. This method makes CVD graphene achieve good contact with GaN substrate, avoiding the defects and impurities introduced in the transfer process.

Author Contributions: Conceptualization, P.T. and J.S.; methodology, P.T., F.X. and J.S.; investigation, K.L.; data curation, P.T. and Y.M.; writing—original draft preparation, P.T.; writing—review and editing, P.T., Z.D. and J.S.; supervision, W.G. and J.S. All authors have read and agreed to the published version of the manuscript.

Funding: This research was funded by the Fujian Provincial Projects (2021HZ0114, 2021J01583, 2021L3004), the Fujian Science & Technology Innovation Laboratory for Optoelectronic Information of China (2021ZZ122, 2020ZZ110), the National Key R&D Program of China (2018YFA0209004) and the Beijing Municipal Commission of Education (KM201810005029).

Data Availability Statement: Not applicable.

Conflicts of Interest: The authors declare no conflict of interest.

References

1. Amano, H.; Kito, M.; Hiramatsu, K.; Akasaki, I. P-type conduction in Mg-doped GaN treated with low-energy electron-beam irradiation (LEEBI). *Jpn. J. Appl. Phys.* **1989**, *28*, 2112–2114. [CrossRef]
2. Nakamura, S.; Mukai, T.; Senoh, M.; Lwasa, N. Thermal annealing effects on p-type Mg-doped GaN films. *Jpn. J. Appl. Phys.* **1992**, *31*, 139–142. [CrossRef]
3. Neugebauer, J.; Vandewalle, C.G. Role of hydrogen in doping of GaN. *Appl. Phys. Lett.* **1996**, *68*, 1829–1831. [CrossRef]
4. Hamberg, I.; Granqvist, C.G. Evaporated Sn-doped In2O3 films: Basic optical properties and applications to energy-efficient windows. *J. Appl. Phys.* **1986**, *60*, 123–160. [CrossRef]
5. Li, F.; Lin, Z.; Zhang, B.; Zhang, Y.; Wu, C.; Guo, T. Fabrication of flexible conductive graphene/Ag/Al-doped zinc oxide multilayer films for application in flexible organic light-emitting diodes. *Org. Electron.* **2013**, *14*, 2139–2143. [CrossRef]
6. Jiang, G.; Tian, H.; Wang, X.-F.; Hirtz, T.; Wu, F.; Qiao, Y.-C.; Gou, G.-Y.; Wei, Y.-H.; Yang, J.-M.; Yang, S.; et al. An efficient flexible graphene-based light-emitting device. *Nanoscale Adv.* **2019**, *1*, 4745–4754. [CrossRef]
7. Bolotin, K.; Sikes, K.; Jiang, Z.; Klima, M.; Fudenberg, G.; Hone, J.; Kim, P.; Stormer, H. Ultrahigh electron mobility in suspended graphene. *Solid State Commun.* **2008**, *146*, 351–355. [CrossRef]
8. Li, X.; Cai, W.; Colombo, L.; Ruoff, R.S. Evolution of graphene growth on Ni and Cu by carbon isotope labeling. *Nano Lett.* **2009**, *9*, 4268–4272. [CrossRef] [PubMed]
9. Sun, J.; Lindvall, N.; Cole, M.T.; Angel, K.T.T.; Wang, T.; Teo, K.B.K.; Chua, D.H.C.; Liu, J.; Yurgens, A. Low partial pressure chemical vapor deposition of graphene on copper. *IEEE Trans. Nanotechnol.* **2012**, *11*, 255–260. [CrossRef]
10. Kim, B.-J.; Lee, C.; Jung, Y.; Baik, K.H.; Mastro, M.A.; Hite, J.K.; Eddy, C.R., Jr.; Kim, J. Large-area transparent conductive few-layer graphene electrode in GaN-based ultra-violet light-emitting diodes. *Appl. Phys. Lett.* **2011**, *99*, 143101. [CrossRef]
11. Sutter, P.; Sadowski, J.T.; Sutter, E. Graphene on Pt (111): Growth and substrate interaction. *Phys. Rev. B* **2009**, *80*, 245411. [CrossRef]
12. Xiong, F.; Sun, J.; Cole, M.T.; Guo, W.; Yan, C.; Dong, Y.; Wang, L.; Du, Z.; Feng, S.; Li, X.; et al. GaN LEDs with in situ synthesized transparent graphene heat-spreading electrodes fabricated by PECVD and penetration etching. *J. Mater. Chem. C* **2022**, *10*, 6794–6804. [CrossRef]
13. Giannazzo, F.; Fisichella, G.; Greco, G.; La Magna, A.; Roccaforte, F.; Pecz, B.; Yakimova, R.; Dagher, R.; Michon, A.; Cordier, Y. Graphene integration with nitride semiconductors for high power and high frequency electronics. *Phys. Status Solidi A* **2017**, *214*, 1600460. [CrossRef]
14. Ferrari, A.C.; Meyer, J.C.; Scardaci, V.; Casiraghi, C.; Lazzeri, M.; Mauri, F.; Piscanec, S.; Jiang, D.; Novoselov, K.S.; Roth, S.; et al. Raman spectrum of graphene and graphene layers. *Phys. Rev. Lett.* **2006**, *97*, 187401. [CrossRef] [PubMed]
15. Bi, H.; Sun, S.; Huang, F.; Xie, X.; Jiang, M. Direct growth of few-layer graphene films on SiO2 substrates and their photovoltaic applications. *J. Mater. Chem.* **2012**, *22*, 411–416. [CrossRef]

crystals MDPI

Article

Effect of Source Field Plate Cracks on the Electrical Performance of AlGaN/GaN HEMT Devices

Ye-Nan Bie [1,2], Cheng-Lin Du [1,2,*], Xiao-Long Cai [1,2,*], Ran Ye [1,2], Hai-Jun Liu [2], Yu Zhang [1,2], Xiang-Yang Duan [1,2] and Jie-Jie Zhu [3]

1 State Key Laboratory of Mobile Network and Mobile Multimedia Technology, Shenzhen 518055, China
2 Wireless Product Planning Department, ZTE Corporation, Shenzhen 518057, China
3 School of Microelectronics, Xidian University, Xi'an 710071, China
* Correspondence: du.chenglin@zte.com.cn (C.-L.D.); cai.xiaolong@zte.com.cn (X.-L.C.)

Abstract: In the current study, the effects of cracks in source field plates (SFPs) on the electrical performance of AlGaN/GaN high electron mobility transistors (HEMTs) are investigated systematically using numerical simulation. In detail, the influence of crack width and junction angle in SFPs on device performance is studied. The results indicate that the SFP structure increases the breakdown voltage of a device, but the occurrence of cracks causes premature breakdown, which is confirmed experimentally by the structural analysis of these devices after breakdown. With an increase in crack width, the electrical performance becomes worse. A beveled SFP architecture is proposed by increasing the angle at the SFP junction to reduce the probability of cracking and enhance the reliability of the device. However, with an increase in bevel angle, the modulation effect of the SFP on the channel electric field is gradually weakened. Therefore, it is necessary to balance the relationship between electrical performance and bevel angle according to the actual demands. This work provides potential support for SFP structural optimization design for AlGaN/GaN HEMTs.

Keywords: AlGaN/GaN HEMTs; source field plate cracks; GaN TCAD modeling; electrical performance

Citation: Bie, Y.-N.; Du, C.-L.; Cai, X.-L.; Ye, R.; Liu, H.-J.; Zhang, Y.; Duan, X.-Y.; Zhu, J.-J. Effect of Source Field Plate Cracks on the Electrical Performance of AlGaN/GaN HEMT Devices. *Crystals* **2022**, *12*, 1195. https://doi.org/10.3390/cryst12091195

Academic Editors: Peng Chen and Zhizhong Chen

Received: 29 July 2022
Accepted: 23 August 2022
Published: 25 August 2022

1. Introduction

Compared with silicon, gallium nitride (GaN) has a wider band gap, higher thermal conductivity and higher electron mobility in AlGaN/GaN heterostructures. At present, GaN-based power and radio frequency high-electron-mobility transistors (HEMTs) show excellent performance and are widely applied in the field of base station communication, aerospace and radar systems [1–4]. As a core component of the power amplifier (PA) in the radio remote unit of a base station, the failure of an AlGaN/GaN HEMT reduces the product yield, work efficiency and service life, further aggravating the failure rate of PAs.

AlGaN/GaN HEMTs usually work under high-frequency and high-voltage conditions. On the one hand, low-frequency dispersion is induced by traps and has an impact on high-frequency performance [5–7]. On the other hand, these devices need to withstand extremely high drain voltages. The electric potential lines gather at either the gate or drain of the device and result in an electric field peak. When the electric field peak is higher than the critical breakdown electric field, avalanche ionization occurs and the device breaks down [8,9]. Therefore, it is necessary to optimize the electric field distribution by designing the device structure so as to improve the breakdown voltage (V_{br}). The electrical characteristics of AlGaN/GaN HEMTs can be enhanced by introducing a field plate structure, which brings in a uniform space charge distribution in a lateral direction and, therefore, increases the V_{br} [10–15]. In a failure analysis and reliability study of AlGaN/GaN HEMTs, cracks are always found at the junctions of the source field plate (SFP). It is reported that these cracks may come from the process of field plate metal deposition on uneven surfaces because of the stresses at these junctions [16–18]. However,

SFP cracks are potentially risky, and their effects on a device's electrical performance are still unclear; therefore, further research is urgently needed.

In this work, firstly, an AlGaN/GaN HEMT device with an SFP structure is characterized experimentally after its breakdown, and cracks are found at the junctions of the SFP. Secondly, the effects of these SFP cracks on the device's electrical performance are studied systematically based on TCAD simulations. Various device structures with and without an SFP and with differing crack widths are simulated and compared. Finally, a beveled SFP architecture is proposed to provide a feasible solution for avoiding cracks and increasing the reliability of AlGaN/GaN HEMTs.

2. Failure Analysis Experiments

During the failure analysis of a commercial AlGaN/GaN HEMT device, a large number of focused ion beam (FIB, Thermo Scientific Helios G4 PFIB UXe DualBeam system) characterizations were carried out after breakdown. The images were acquired at 5 kV and 0.17 nA under secondary electron imaging mode. From scanning electron microscope (SEM) surface morphology, the gate metal at the breakdown points was broken or partially damaged, as shown in Figure 1a. FIB cross-section characterization was carried out at position A, near the breakdown point as shown in Figure 1b. The morphology of the gate metal was partially damaged, and there was a crack in the SFP near the drain side. The SFP broke at the crack and was lifted near the source side. Position B, with a better surface morphology below the breakdown point, was characterized as shown in Figure 1c. The SFP was disconnected, and there were cracks at the junctions above the gate cap. Another position, C (not shown in Figure 1a), on the same gate where no breakdown occurred was also characterized. As shown in Figure 1d, the SFP was still connected, but there was also a trend of cracking at the junctions. It is inferred that the cracks were not caused by breakdown but caused by the SFP deposition process, since there was no crack in the secondary passivation layer between the SFP and the gate cap. In the SFP deposition process, the metal is relatively prone to fracture due to stress [16–18]. According to the FIB characterization results, there is a certain correlation between gate breakdown and SFP cracking; however, a further study of the internal mechanism is still needed.

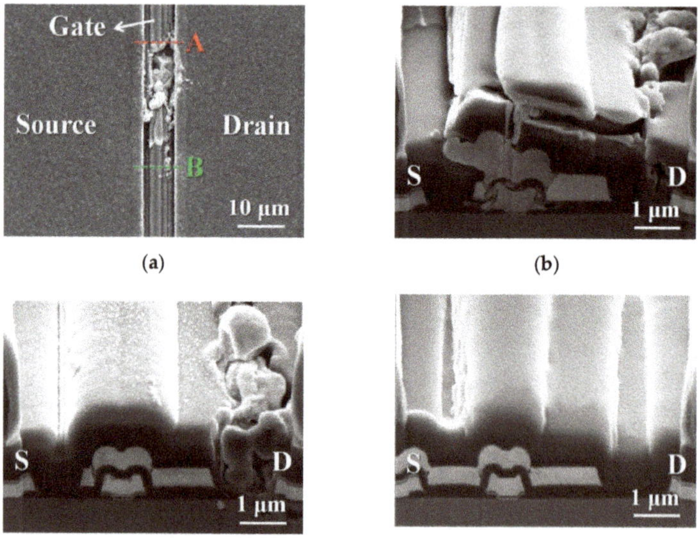

(a) (b)

(c) (d)

Figure 1. (**a**) SEM surface image of the AlGaN/GaN HEMT after breakdown; (**b–d**) SEM cross-section images of different breakdown and non-breakdown points along the same gate as marked in (**a**).

3. Materials and Methods

APSYS (Crosslight, Vancouver, Canada) device simulations were carried out to analyze the detailed electrical characteristics [19]. The simulated AlGaN/GaN HEMT structure with a T-gate is illustrated in Figure 2. The epitaxial layer structure consisted of an n-doped (2×10^{16} cm^{-3}) semi-insulating SiC substrate layer, AlN nucleation layer, GaN buffer layer and GaN channel layer. Above these were an unintentionally doped $Al_{0.3}Ga_{0.7}N$ barrier layer and a SiN passivation layer. Traps with a maximum concentration of 1×10^{17} cm^{-3} and a relative energy level of 0.45 eV were assigned to model Fe doping in the buffer layer [20,21]. Two-dimensional electron gas was generated using interface and polarization statements, and the electron mobility of 1400 $cm^2V^{-1}s^{-1}$ was used in the simulation. In order to simulate the breakdown mechanism, the Chenoweth law for impact ionization with user-definable parameters was used [22]. The V_{br} was defined as the drain voltage when the drain-to-source current reached 1 mA/mm [10,23].

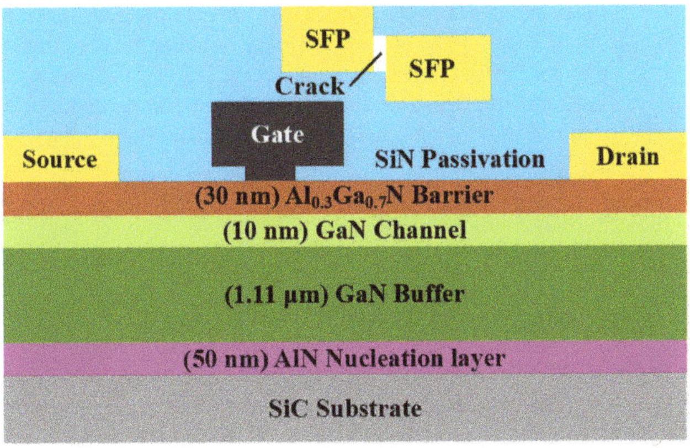

Figure 2. Cross-sectional illustration of the AlGaN/GaN HEMT with a cracked SFP.

A Schottky contact with a work function of 5.4 eV was used as a gate electrode, and an ohmic contact with a work function of 4.0 eV was used for the source and drain electrodes. For the AlGaN/GaN HEMT with a cracked SFP, the crack height was set as 200 nm, and the width was varied from 0–50 nm in 10 nm step. In addition, the material within the crack was set as air in the simulation. The SiC substrate and source electrode were grounded. For the breakdown simulation, a gate voltage of $V_g = -6$ V was applied to imitate the off-state situation. The SFP and source electrode maintained the same potential in the simulation. The other main dimension parameters used are provided in Table 1.

Table 1. Dimension parameters for simulation.

Name	Value	Name	Value
Source-to-gate distance	1.7 μm	Gate foot thickness	0.05 μm
Gate-to-drain distance	4.5 μm	Gate head thickness	0.55 μm
Source/drain length	1 μm	Source/drain thickness	0.2 μm
Gate foot length	0.45 μm	SFP length	1.7 μm
Gate head length	1.25 μm	SFP thickness	0.55 μm

4. Results and Discussion

Cracks of different shapes often appear at the junctions of a SFPs, and it is necessary to investigate the effect of different crack widths on a device's electrical performance. Therefore, simulations of different AlGaN/GaN HEMT structures were carried out. When w = 0 nm, it means an HEMT with a conventional SFP but without a crack. For comparison, an HEMT without an SFP was also investigated. The transfer and transconductance characteristics of these devices are shown in Figure 3a. The drain voltage (V_d) was set to 10 V, and the gate voltage (V_g) was swept from −6 V–2 V. Then, the threshold voltage (V_{th}) was extracted from a linear extrapolation of the transfer characteristics at the point of maximum transconductance. The presence of an SFP and SFP cracks had little effect on the transfer and transconductance properties of the device. The V_{th} and the maximum transconductance were about −2.53 V and 227 mS/mm, respectively. As shown in Figure 3b, the V_{br} was 160 V at V_g = −6 V for the HEMT without an SFP. For the HEMT with an SFP, when w increased from 0 nm to 50 nm, the V_{br} decreased from 261 V to 181 V. This decreasing trend slowed down when w increased to 20 nm. It was seen that the SFP had a positive effect, improving the V_{br}. However, once the SFP had a crack, the V_{br} considerably decreased. Furthermore, the relationship between V_{br} and V_{th} depending on crack width is depicted in Figure 3c. Figure 3d shows simulated pulsed I-V characteristics with a 1 μs pulse width and a 1 ms pulse period. The quiescent bias point was set as ($V_{g,q}$ = −6 V, $V_{d,q}$ = 80 V). In contrast to the static results, the HEMT with a conventional SFP showed a maximum saturated drain current, while the HEMT without an SFP showed a minimum one. The reason for the discrepancy between the dynamic and static I-V characteristics is that the SFP alleviates the high electric field and restrains the trapping in the buffer [24,25]. The appearance of a crack affected the modulation effect of the SFP on the electric field. As a result, the trapping process increased, and the saturated drain current decreased.

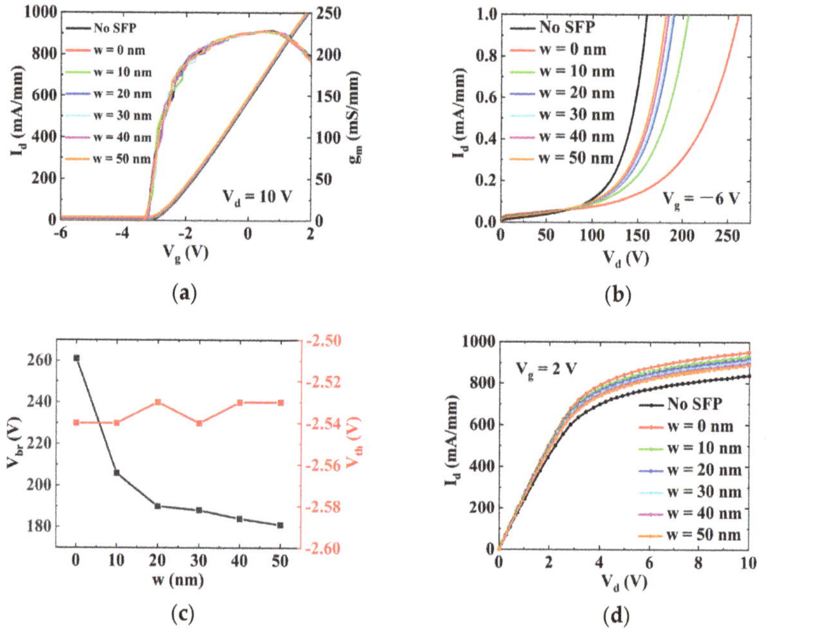

Figure 3. (**a**) Drain current and transconductance versus gate voltage; (**b**) off-state drain current versus drain voltage; (**c**) breakdown voltage and threshold voltage versus crack width; (**d**) pulsed I-V characteristics for AlGaN/GaN HEMTs without SFP and with SFP of different crack widths.

In order to find the weak point where the device is prone to breakdown, the impact ionization rates were simulated at their respective V_{br} points. Cross-sections of the impact ionization rate distribution for three different AlGaN/GaN HEMT structures are shown in Figure 4. For the HEMT without an SFP and the HEMT with a 20 nm wide cracked SFP, the max impact ionization rate was located at the gate edge of the drain side. However, if there was no crack in the SFP (w = 0 nm), the position of the max impact ionization rate moved to the SFP edge of the drain side, as shown in Figure 4b. High impact ionization rates lead to enhanced carrier generation and early breakdowns of a device [8,9], and the appearance of an SFP decreases the impact ionization rate and changes the distributions at the same time [13,14].

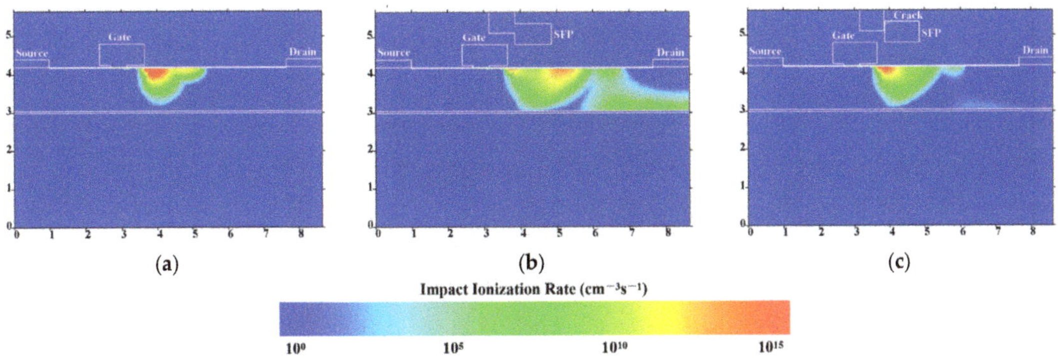

Figure 4. Impact ionization rate contours for AlGaN/GaN HEMT (**a**) without an SFP; (**b**) with a conventional SFP; (**c**) with a 20 nm wide cracked SFP at their respective V_{br} points.

The potential distributions for the three different aforementioned AlGaN/GaN HEMT structures were simulated at $V_g = -6$ V and $V_d = 160$ V. The contours for a closeup region, including the gate and SFP, are shown in Figure 5. It was observed that for the HEMT without an SFP, most of the potential lines crowded at the drain side of the gate, as well as in the barrier and buffer layers. Crowded potential lines usually lead to a high electric field and consequently a high impact ionization rate. For the HEMT with a conventional SFP, high potential line density appeared at both the drain side of the gate and the drain side of the SFP, which is attributed to the modulation effect of the SFP [12,26,27]. The potential line densities in the barrier and buffer layers were also lower than those for the HEMT without an SFP. Once the SFP had cracked, some potential lines appeared in the cracked region. The right part of the SFP did not share the same potential as the left part anymore, and its modulation effect finally weakened.

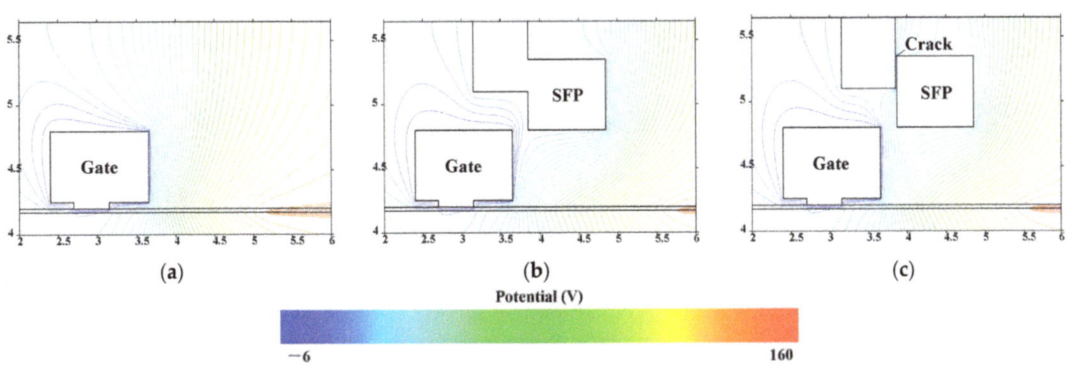

Figure 5. Potential contours for AlGaN/GaN HEMT (**a**) without an SFP; (**b**) with a conventional SFP; (**c**) with a 20 nm wide cracked SFP at $V_g = -6$ V, $V_d = 160$ V.

The electric field is also an important parameter which determines the V_{br} of an HEMT device [28–30]. Therefore, the electric field along the channel before breakdown ($V_g = -6$ V, $V_d = 160$ V) is shown in Figure 6. It was found that the electric field peaks all appeared at the gate edge of the drain side. The HEMT without an SFP showed the maximum electric field peak of 1.82 MV/cm. As the w increased, the electric field peak increased from 1.38 MV/cm to 1.70 MV/cm. It is obvious that the V_{br} is inversely proportional to the electric field peak. The field plate plays a major role in the electric field, impact ionization rate distribution and the breakdown of the AlGaN/GaN HEMT; however, the occurrence of a crack will reduce the electric field peak inhibition by the field plate. Given this, the wider the SFP crack becomes, the poorer the electrical performance.

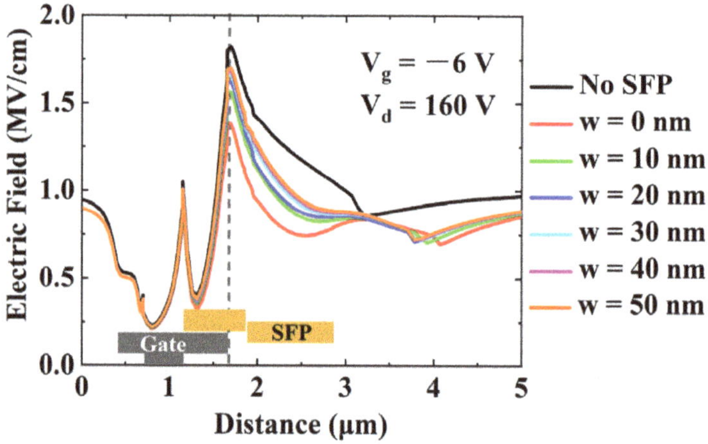

Figure 6. Electric field along the channel for various AlGaN/GaN HEMTs.

As discussed above, the appearance of SFP cracks degrades the electrical performance of AlGaN/GaN HEMTs. In the process of SFP metal deposition, cracks form easily at these junctions due to the steep angle and high stresses [16–18]. High stress usually appears at the right-angled corner. When the angle increases, the stress reduces, and, consequently, the cracking tendency decreases [31,32]. Accordingly, we propose a beveled SFP structure (see Figure 7a) since increasing the bevel angle at this junction may relieve the stress and reduce the probability of cracking. A further simulation study investigating whether the rise of junction angle affects the electrical performance of a device was carried out. For convenience, the angle is named θ as illustrated in Figure 7a. When θ = 90°, it means the HEMT with a conventional SFP of which the junction is a right angle. In the following simulations, the total length of SFP was kept unchanged and θ changed from 90° to 162° (90°, 108°, 135°, 149°, 157° and 162°). The same electrical parameter simulations were carried out as in the above study. As an example, a proposed HEMT with an SFP bevel angle of 108° achieved a V_{br} of 256 V, which is very close to the HEMT with a conventional SFP. The channel electric field and the cross-sectional impact ionization rate distributions also showed similar results. As shown in Figure 7b, when θ increased from 90° to 162°, the V_{br} decreased from 261 V to 201 V. Furthermore, when θ was increased to 135°, the rate of decline in V_{br} increased. Even so, the V_{br} of 201 V was still higher than that of the HEMT with a 20 nm wide cracked SFP. The channel electric field peak was extracted before breakdown ($V_g = -6$ V, $V_d = 200$ V), and it showed an inverse trend compared with the V_{br}. With an increase in θ, the distance from the lower edge of the SFP to the channel becomes longer, so the modulation effect of the SFP on the electric field is weakened [12,33,34]. As a result, the beveled SFP reduces the stress and the risk of cracking, but it also reduces the electrical performance of the device.

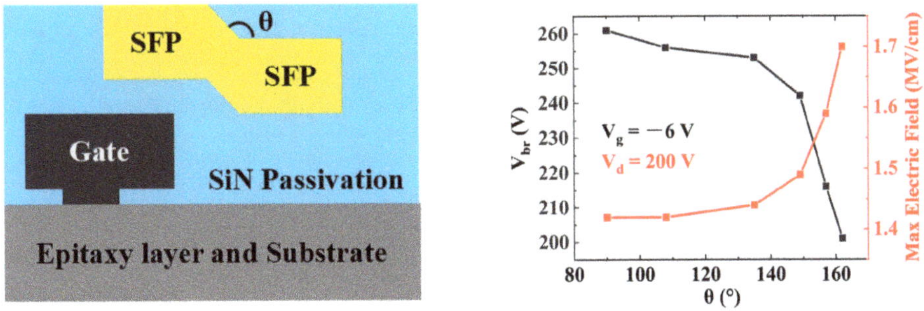

(a) (b)

Figure 7. (**a**) Cross-sectional illustration of AlGaN/GaN HEMT with a bevel SFP; (**b**) breakdown voltage and channel electric field peak versus the bevel angle.

5. Conclusions

To investigate the effects of cracks in SFPs on a device's electrical performance, AlGaN/GaN HEMTs with different SFP structures were studied by using simulations. The results show that compared with the HEMT without an SFP, the HEMT with a conventional SFP showed better breakdown and dynamic I-V performances. However, we found that once a crack appeared in the SFP, the device's electrical performance degraded, and the decreasing trend slowed down when the crack width increased to 20 nm. The physical mechanism was investigated from the aspects of impact ionization rate, potential and channel electric field. The SFP structure effectively alleviated the peak electric field at the gate edge of the drain side, but the crack reduced the effect of SFP undesirably. The beveled SFP structure is a good strategy to reduce the risk of cracking; however, this is at the expense of a device's electrical indicators. When the bevel angle increases to 135°, the breakdown voltage begins to drop rapidly. Thus, it is recommended to balance these factors according to actual needs, and a bevel angle of no more than 135° is suggested. Field plate technology is simple and compatible with other electric field optimization technologies. These results can be used for fabricating future HEMT devices with SFPs to achieve a high breakdown voltage and high reliability.

Author Contributions: Conceptualization, Y.-N.B., C.-L.D. and X.-L.C.; characterization, C.-L.D.; simulation, Y.-N.B. and C.-L.D.; data curation and analysis, Y.-N.B., C.-L.D., X.-L.C. and R.Y.; writing—original draft preparation, Y.-N.B. and C.-L.D.; writing—review and editing, Y.-N.B., C.-L.D., X.-L.C., R.Y., H.-J.L., Y.Z., X.-Y.D. and J.-J.Z. All authors have read and agreed to the published version of the manuscript.

Funding: This research was funded by the Fundamental Research Funds for the Central Universities under Grant QTZX22022.

Institutional Review Board Statement: Not applicable.

Informed Consent Statement: Not applicable.

Data Availability Statement: Not applicable.

Conflicts of Interest: The authors declare no conflict of interest.

References

1. Fletcher, A.S.; Nirmal, D. A survey of gallium nitride HEMT for RF and high power application. *J. Superlattice Microst.* **2017**, *109*, 519–537. [CrossRef]
2. Amano, H.; Baines, Y.; Beam, E.; Matteo, B. The 2018 GaN power electronics roadmap. *J. Appl. Phys.* **2018**, *51*, 163001. [CrossRef]
3. Nakajima, S. GaN HEMTs for 5G Base Station Applications. In Proceedings of the 2018 IEEE International Electron Devices Meeting (IEDM), San Francisco, CA, USA, 1–5 December 2018; pp. 14.2.1–14.2.4.
4. Huang, L.H.; Chiou, H.K. An Ultra-compact 14.9-W X-Band GaN MMIC Power Amplifier. In Proceedings of the 2020 IEEE Asia-Pacific Microwave Conference (APMC), Hong Kong, China, 8–11 December 2020; pp. 257–259.

5. Jardel, O.; Groote, F.D.; Reveyrand, T.; Jacquet, J.-C.; Charbonniaud, C.; Teyssier, J.-P.; Floriot, D.; Quere, R. An Electrothermal Model for AlGaN/GaN Power HEMTs Including Trapping Effects to Improve Large-Signal Simulation Results on High VSWR. *IEEE Trans. Microw. Theory Tech.* **2007**, *55*, 2660–2669. [CrossRef]
6. Raffo, A.; Vadala, V.; Schreurs, D.M.M.-P.; Crupi, G.; Avolio, G.; Caddemi, A.; Vannini, G. Nonlinear Dispersive Modeling of Electron Devices Oriented to GaN Power Amplifier Design. *IEEE Trans. Microw. Theory Tech.* **2010**, *58*, 710–718. [CrossRef]
7. Potier, C.; Martin, A.; Campovecchio, M.; Laurent, S.; Quere, R.; Jacquet, J.C.; Jardel, O.; Piotrowicz, S.; Delage, S. Trap characterization of microwave GaN HEMTs based on frequency dispersion of the output-admittance. In Proceedings of the 2014 9th European Microwave Integrated Circuit Conference, Rome, Italy, 6–7 October 2014; pp. 464–467.
8. Gaudenzio, M.; Matteo, M.; Enrico, Z. Breakdown mechanisms in AlGaN/GaN HEMTs: An overview. *Jpn. J. Appl. Phys.* **2014**, *53*, 100211.
9. Saito, W.; Suwa, T.; Uchihara, T.; Naka, T.; Kobayashi, T. Breakdown behaviour of high-voltage GaN-HEMTs. *Microelectron. Reliab.* **2015**, *55*, 1682–1686. [CrossRef]
10. Kwak, H.T.; Chang, S.B.; Jung, H.G.; Kim, H.S. Thermal Analysis of AlGaN/GaN High-Electron-Mobility Transistor and Its RF Power Efficiency Optimization with Source-Bridged Field-Plate Structure. *J. Nanosci. Nanotechnol.* **2018**, *18*, 5860–5867. [CrossRef]
11. Godfrey, D.; Nirmal, D.; Arivazhagan, L.; Roy, B.; Chen, Y.L.; Yu, T.H.; Roy, B.; Yeh, W.K.; Godwinraj, D. Investigation of AlGaN/GaN HEMT Breakdown analysis with Source field plate length for High power applications. In Proceedings of the 2020 5th International Conference on Devices, Circuits and Systems (ICDCS), Coimbatore, India, 5–6 March 2020; pp. 244–246.
12. Liao, B.Y.; Zhou, Q.B.; Qin, J.; Wang, H. Simulation of AlGaN/GaN HEMTs' Breakdown Voltage Enhancement Using Gate Field-Plate, Source Field-Plate and Drain Field Plate. *Electronics* **2019**, *8*, 406. [CrossRef]
13. Soni, A.; Singh, A.; Shrivastava, M. Novel Drain-Connected Field Plate GaN HEMT Designs for Improved V_{BD}–R_{ON} Tradeoff and RF PA Performance. *IEEE Trans. Electron Devices* **2020**, *67*, 1718–1725. [CrossRef]
14. Augustine Fletcher, A.S.; Nirmal, D.; Ajayan, J.; Arivazhagan, L. Analysis of AlGaN/GaN HEMT using discrete field plate technique for high power and high frequency applications. *AEU Int. J. Electron. Commun.* **2019**, *99*, 325–330. [CrossRef]
15. Kwak, H.K.; Jang, K.W.; Kim, H.J.; Lee, S.H.; Lim, J.W.; Kim, H.S. DC Characteristics of AlGaN/GaN High-Electron Mobility Transistor with a Bottom Plate Connected to Source-Bridged Field Plate Structure. *J. Nanosci. Nanotechnol.* **2019**, *19*, 2319–2322. [CrossRef] [PubMed]
16. Zhou, L.; San, Z.W.; Hua, Y.J.; Lin, L.; Zhang, S.; Zhao, Z.G.; Zhou, H.J.; Yin, W.Y. Investigation on Failure Mechanisms of GaN HEMT Caused by High-Power Microwave (HPM) Pulses. *IEEE Trans. Electromagn. Compat.* **2017**, *59*, 902–909. [CrossRef]
17. Zhang, S.; Wei, K.; Ma, X.H.; Zhang, Y.C.; Asif, M.; Liu, G.G.; Huang, S.; Zheng, Y.K.; Wang, X.H.; Niu, J.B.; et al. Millimeter-wave AlGaN/GaN HEMTs breakdown voltage enhancement by an air-bridge recessed source field plate (RSFP). *Solid State Electron.* **2019**, *160*, 107629. [CrossRef]
18. Bothe, K.M.; Ganguly, S.; Guo, J.; Liu, Y.Y.; Niyonzima, A.; Tornblad, O.; Fisher, J.; Gajewski, D.A.; Sheppard, S.T.; Noori, B. Improved X-Band Performance and Reliability of a GaN HEMT with Sunken Source Connected Field Plate Design. *IEEE Electron Device Lett.* **2022**, *43*, 354–357. [CrossRef]
19. User's Manual. 2019. Available online: www.crosslight.com (accessed on 25 June 2022).
20. Sharma, K.; Dupouy, E.; Bouslama, M.; Sommet, R.; Nallatamby, J.-C. Impact of the Location of Iron Buffer Doping on Trap Signatures in GaN HEMTs. In Proceedings of the 2020 International Workshop on Integrated Nonlinear Microwave and Millimetre-Wave Circuits (INMMiC), Cardiff, UK, 16–17 July 2020; pp. 1–3.
21. Raja, P.V.; Subramani, N.K.; Gaillard, F.; Bouslama, M.; Sommet, R.; Nallatamby, J.-C. Identification of Buffer and Surface Traps in Fe-Doped AlGaN/GaN HEMTs Using Y21 Frequency Dispersion Properties. *Electronics* **2021**, *10*, 3096. [CrossRef]
22. Chynoweth, A.G. Ionization rates for electrons and holes in silicon. *Phys. Rev.* **1958**, *109*, 1537–1540. [CrossRef]
23. Zhang, J.L.; He, L.; Li, L.A.; Yang, F.; Shen, Z.; Zhou, D.Q.; Chen, Z.J.; Zhang, X.R.; He, L.; Wu, Z.S.; et al. The breakdown behavior of GaN epitaxial material on silicon. In Proceedings of the 2016 13th China International Forum on Solid State Lighting: International Forum on Wide Bandgap Semiconductors China (SSLChina: IFWS), Beijing, China, 15–17 November 2016; pp. 81–84.
24. Saito, Y.; Tsurumaki, R.; Noda, N.; Horio, K. Analysis of reduction in lag phenomena and current collapse in field-plate AlGaN/GaN HEMTs with high acceptor density in a buffer layer. *IEEE Trans. Device Mater. Rel.* **2018**, *18*, 46–53. [CrossRef]
25. Meneghini, M.; Rossetto, I.; Bisi, D.; Stocco, A.; Chini, A.; Pantellini, A.; Lanzieri, C.; Nanni, A.; Meneghesso, G.; Zanoni, E. Buffer traps in Fe-doped AlGaN/GaN HEMTs: Investigation of the physical properties based on pulsed and transient measurements. *IEEE Trans. Electron Devices* **2014**, *61*, 4070–4077. [CrossRef]
26. Mao, W.; Fan, J.S.; Du, M.; Zhang, J.F.; Zheng, X.F.; Wang, C.; Ma, X.H.; Zhang, J.C.; Hao, Y. Analysis of the modulation mechanisms of the electric field and breakdown performance in AlGaN/GaN HEMT with a T-shaped field-plate. *Chin. Phys. B* **2016**, *25*, 127305. [CrossRef]
27. Wu, H.; Fu, X.J.; Hu, S.D. A 650V Enhancement Mode GaN HEMT Device with Field Plate for Power Electronic Applications. In Proceedings of the 2021 International Conference on Electrical, Communication, and Computer Engineering (ICECCE), Kuala Lumpur, Malaysia, 12–13 June 2021; pp. 1–5.
28. Lashway, C.R.; Berzoy, A.; Mohammed, O. Field-plate length variation on GaN devices for BV and on-resistance characterization. In Proceedings of the 2018 International Applied Computational Electromagnetics Society Symposium (ACES), Denver, CO, USA, 25–29 March 2018; pp. 1–2.

29. Berzoy, A.; Lashway, C.R.; Moradisizkoohi, H.; Mohammed, O.A. Breakdown voltage improvement and analysis of GaN HEMTs through field plate inclusion and substrate removal. In Proceedings of the 2017 IEEE 5th Workshop on Wide Bandgap Power Devices and Applications (WiPDA), Albuquerque, NM, USA, 30 October–1 November 2017; pp. 138–142.
30. Soni, A.; Shrivastava, M. Implications of Various Charge Sources in AlGaN/GaN Epi-Stack on the Drain & Gate Connected Field Plate Design in HEMTs. *IEEE Access* **2022**, *10*, 74533–74541.
31. Ackaert, J.; Vanderstraeten, D.; Vandevelde, B. On the impact of the edge profile of interconnects on the occurrence of passivation cracks of plastic-encapsulated electronic power devices. In Proceedings of the 2011 IEEE International Conference on IC Design & Technology, Kaohsiung, Taiwan, 2–4 May 2011; pp. 1–4.
32. Huang, X.K.; Zhu, L.P.; Nguyen, B.; Tran, V.; Isom, H.S. Passivation Stress versus Top Metal Profiles by 3 D Finite Element Modeling. In Proceedings of the 2013 International Conference on Compound Semiconductor Manufacturing Technology, New Orleans, LA, USA, 13–16 May 2013; pp. 323–326.
33. Chen, T.; Zhou, Q.; Wei, D.; Dong, C.; Chen, W.; Zhang, B. Physics-Based 2-D Analytical Model for Field-Plate Engineering of AlGaN/GaN Power HFET. *IEEE Trans. Electron Devices* **2019**, *66*, 116–125. [CrossRef]
34. Chugh, N.; Kumar, M.; Haldar, S.; Bhattacharya, M.; Gupta, R.S. Applicability of Field Plate in Double Channel GaN HEMT for Radio-Frequency and Power-Electronic Applications. *Silicon* **2022**, *14*, 1029–1038. [CrossRef]

crystals

MDPI

Article

Numerical Study of Growth Rate and Purge Time in the AlN Pulsed MOCVD Process

Wei-Jie Lin and Jyh-Chen Chen *

Department of Mechanical Engineering, National Central University, Zhongli 320, Taiwan
* Correspondence: jcchen@ncu.edu.tw

Abstract: The relationship between the purge time and the overall growth rate in pulsed injection metal–organic chemical vapor deposition with different V/III ratios is studied by numerical analysis. The transport behavior of TMAl and $TMAlNH_3$ during the process is studied to understand the effect of the adductive reaction. The results show that, as the V/III ratio increases, there is a significant reduction in the average growth rate per cycle, without the addition of a purging H_2 pulse between the III and V pulses, due to stronger adductive reaction. The adductive reaction can be reduced by inserting a purging pulse of pure H_2 between the III and V pulses, but there is a decrease in the overall growth rate due to the longer cycle time. At smaller V/III ratios, the growth rate decreases with increasing purge times, since the gain in reducing the adductive reaction is offset by the detrimental effect of extending the cycle time. The degree of reduction in the adductive reaction is higher for larger V/III ratios. When the benefit of reducing the adductive reaction overcomes the deficiency of the extending cycle time, a remarkable enhancement of the growth rate can be obtained at high V/III ratios by inserting a pure H_2 purge pulse between the III and V pulses. A higher overall growth rate can be achieved at higher V/III ratios by choosing an appropriate purge time.

Keywords: MOCVD; nitride; pulsed injection; growth rate

Citation: Lin, W.-J.; Chen, J.-C. Numerical Study of Growth Rate and Purge Time in the AlN Pulsed MOCVD Process. *Crystals* **2022**, *12*, 1101. https://doi.org/10.3390/cryst12081101

Academic Editors: Peng Chen and Zhizhong Chen

Received: 18 July 2022
Accepted: 4 August 2022
Published: 5 August 2022

1. Introduction

III-nitride materials are widely used in power devices, high-frequency communications, and UV applications because of their excellent properties. Among the many III-nitride materials, aluminum nitride (AlN) is one of the most important, often used as a buffer layer between the sapphire substrate and other III semiconductor materials, because the lattice mismatch between AlN and sapphire is minor. However, the small migration of Al atoms on the substrate makes it difficult to grow high-quality AlN film [1]. One way to increase the migration of the Al atoms is to increase the temperature of the substrate. However, the high-temperature growth conditions not only make the parasitic reaction more intense, they also induce the problem of thermal expansion. Another way to increase the migration of Al atoms is to apply the pulsed injection method (PI method) during the metal–organic chemical vapor deposition (MOCVD) process. During the pulsed process, the supply of TMAl and NH3 is carried out separately. The TMAl follows the pyrolysis reaction pathway instead of adducting with NH_3 in the gas phase before being deposited on the surface. It changes the dominant species on the surface from Al-N to Al. The migration of Al atoms increases because the number of N atoms is much lower during the pulsed process [2]. Smoother AlN film can be achieved due to the lateral growth of film caused by the higher Al migration. Demir et al. [3] grew high-quality AlN by pulsed MOCVD (PIMOCVD) at a growth temperature of 1443 K with III and V pulse durations of 4 s and 2 s, respectively.

Many studies have pointed out the need for a hydrogen purge between the TMAl pulse (III pulse) and NH_3 pulse (V pulse) to further suppress the adductive reactions and achieve better surface roughness. Khan et al. [4] fixed the duration of the different pulses to 1 s with a 1 s hydrogen purge time in between. The crystalline quality of the material grown by the pulsed process at 723 K was as good as the material grown by the steady-flow process at 1273 K. Takeuchi et al. [5] compared the growth rates of AlN film grown by the steady-flow process and pulsed process with a 1 s purge time. In the steady-flow process, the TMAl and NH3 were supplied at the same time. At a V/III ratio of 20.8, it was found that the growth rate of the pulsed process was 5.56 times that of the steady-flow process.

The effects caused by different TMAl and NH3 flow rates have also been investigated. Kröncke et al. [6] varied the effective V/III ratio between 200 and 700 and found that increasing the purge time after supplying TMAl could reduce the density of pits. The growth rate of the AlN film for these processes was about 1 μm/h. Rahman et al. [7,8] grew AlN films using NH_3 flow rates ranging from 0.2 to 2.2 slm, with a fixed TMAl flow rate of 15 slm. The durations of the III and V pulses were 4 s and 2 s, respectively, and the growth temperature was 1453 K. The results showed that the best AlN film quality was obtained with 0.6 slm NH_3. The growth rate of the AlN films was between 0.225 and 0.249 μm/h.

The potential of the pulsed method to improve the quality of AlN has been confirmed in numerous studies. However, there has been great divergence in the growth conditions used in different works. The pulsed process includes more external control parameters than a steady-flow process, making the transport phenomenon during the growth process more complicated. In order to obtain optimal growth conditions, it is important to have an in-depth understanding of its transport phenomena. The transport phenomena during the pulsed process have been investigated numerically in several works. Nakamura et al. [9] performed numerical experiments and observed the generation of AlN particles with different purge times (0, 1, and 2 s) and a constant V/III ratio of about 4000. The results showed the mole concentration of AlN particles to be eliminated with a purge time of 2 s. However, there was a decrease in the overall growth rate per cycle as the purge time increased. Endres et al. [10] also carried out numerical experiments. They showed that for a V/III ratio of 20.8, the optimal purge duration would be 5–6 times the duration of the III and V pulses, resulting in the highest growth rate. Lin and Chen [11] conducted a numerical investigation to explore the physical mechanisms in the evolution of flow motion and heat and mass transfer during the pulse process. They found that the duration of the III and V pulses does not affect the particle generation if the duration time is longer than a critical value (a pulse steady state is reached). The diffusive mass transport of NH_3 in H_2 is greater than that of TMAl in H_2 due to the higher mass diffusion coefficient of NH_3 in H_2. Therefore, the V pulses easily exceed the III pulses before reaching the substrate when the III/V ratio is 1. More AlN particles occur in the high-temperature region during the H_2 pulse after the III pulse. To reduce the occurrence of AIN particles, the duration of the H_2 pulse after the III pulse should be longer than that after the V pulse. However, the relationship between the purge time and the V/III ratio during the pulsed process and the transport phenomena is still unclear. When the V/III ratio is increased, the purge time should be extended, since more NH_3 will remain in the chamber, leading to stronger parasitic reactions. Furthermore, at a higher V/III, the transient mass transport process becomes more complex, since the main species of the V pulse changes from H_2 to NH_3.

In this study, the numerical scheme developed in our previous study [11] is extended to first study the transport phenomenon and overall growth rate of the cases without purging and with different V/III ratios. Then, the relationship between the purge time, V/III ratio and overall growth rate is investigated and the maximum overall growth rates for different V/III ratios are investigated.

2. Modeling

Figure 1 shows a schematic diagram of the simplified horizontal MOCVD reactor built in our previous study [11] based on the study of Chen et al. [12]. The temperature of the substrate surface was maintained at 1200 K and the operating pressure was 85 Torr. For pulsed processes, the total flow rate for all pulse sequences was fixed at 7 slm to maintain the stability of the flow field. The flow rates of the III and V pulses are denoted by $Q_{TMAl,p}$ and $Q_{NH3,p}$, respectively. The molar flow rate of TMAl for the III pulse was kept at 30 μmol/min. The V/III ratio was varied from 1 to 10,000. In the previous study, the concentration of precursors was found to remain stable at 0.15 s after the start of the III or V pulse, which is close to the effective residence time (about 0.13 s). Therefore, in this study, the supply time of the TMAl and NH3 was fixed at 0.15 s and the pure hydrogen purge time between III and V pulses was changed from 0 s to 0.1 s. The working fluid in the chamber was considered as an ideal gas. The Reynolds number for the cases investigated in this study was below 400. Compressible laminar flow was applied to describe the flow motion during the process. The equations for the conservation of mass, momentum, heat energy, species, and the detail of material properties are listed in the previous work [10]. The reaction model follows the work performed by Mihopoulos [13]. The commercial software COMSOL was applied to solve the MOCVD processes coupling the fluid dynamics, the heat and mass transfer, and the chemical reactions by the finite element method. The discretization for the momentum and heat equation was linear and that for mass transport and reaction model was quadratic. Due to the rapid chemical reaction near the inlet, the distribution of the width of the element was arranged as arithmetic distribution and the size of the smallest element was 1 μm. The distribution of the height of the elements near the chamber wall also followed arithmetic distribution to calculate the boundary layer effects and the mass flux caused by surface reactions. The backward differentiation formula (BDF) was used to solve the transient behavior and the maximum order for the BDF was 2. To investigate the transport behavior of different species during the process, the maximum time step was limited by 0.001 s. The simulation results obtained in the steady-flow process with this scheme in our previous study [11] are in good agreement with previous results [10,12,13].

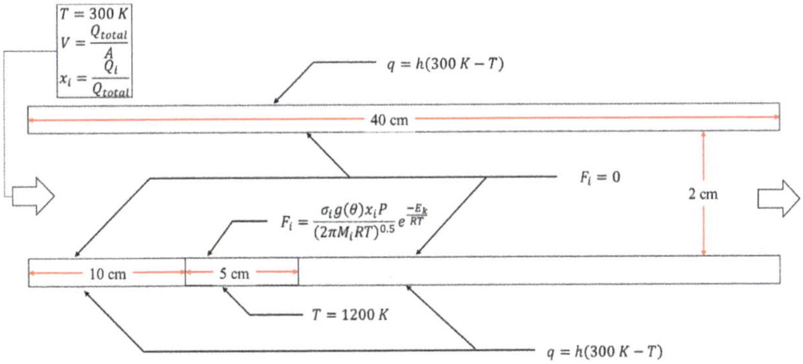

Figure 1. Schematic diagram and boundary conditions of a simplified horizontal MOCVD reactor.

3. Results and Discussion

Before studying the pulsed process, it is necessary to understand the effect of the V/III ratios during the steady-flow process. Figure 2 shows the relationship between the average growth rate along the entire wafer surface and the V/III ratio when the operating pressure is 40 Torr. As the V/III ratio increases, the growth rate decreases due to the higher generation of Al particles caused by higher amounts of NH_3. This trend is similar to the numerical and experimental results presented in [14] for a close-coupled showerhead (CCS) reactor during the steady-flow process. However, the growth rate obtained in this study is higher than in the CCS results. This may be due to the difference in growth area between the CCS reactor used in ref. [14] (60.8 cm^2) and the channel reactor (25 cm^2) used for the present study.

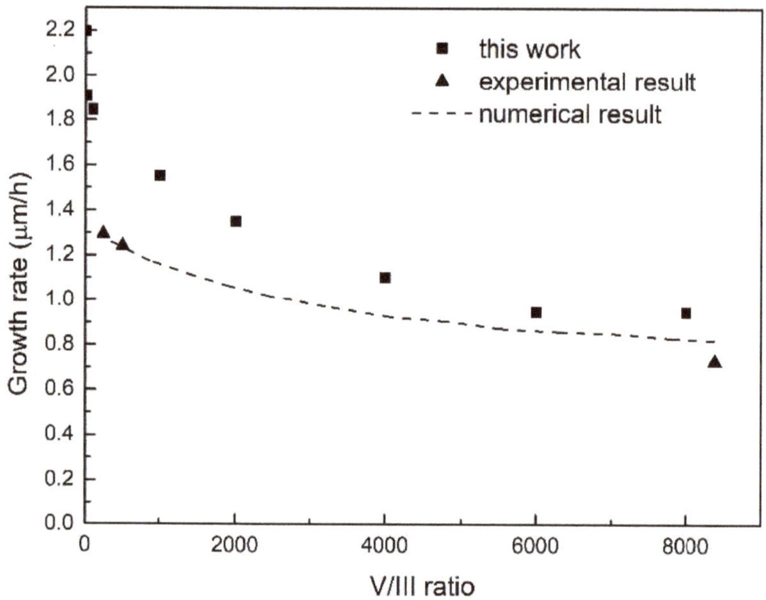

Figure 2. Relationship between the average growth rate and the V/III along the entire wafer surface. The square points represent the results obtained in this work and the dashed line and triangles indicate the results from a previous numerical and experimental study [14].

The pulse process considered here first starts with a III pulse, followed by a V pulse, with no purge between the III and V pulses, after which the entire process is cycled repeatedly. Figure 3 shows the average growth rate along the wafer surface during the pulsed process and Figures 4–6 show the mass fraction distributions of TMAl and $TMAlNH_3$ at t = 0.18 s, 0.35 s, and 0.46 s for different V/III ratios. The III pulse reaches the hot substrate surface at approximately t = 0.05 s and the AlN film grows through the pyrolysis reaction [13]. Therefore, there is significant increase in the average growth rate around t = 0.05 s. The V pulse is injected into the chamber at 0.15 s and then the adductive reaction occurs at the rear interface of the III pulse. The adductive reaction between the TMAl and NH_3 leads to the formation of Al particles. Figure 4 shows that at t = 0.18 s, the front interface of TMAlNH3 is still noted to reach the region close to the hot substrate. The movement of the front interface of $TMAlNH_3$ is faster with a higher V/III ratio. Due to the stronger adductive reaction, more $TMAlNH_3$ is produced at a higher V/III ratio (Figure 4b). It affects the substrate at approximately t = 0.2 s. In Figure 3, the beginning of an abrupt drop at around t = 0.2 s can be observed, which occurs earlier and faster with a higher V/III ratio.

The second III pulse starts at 0.3 s and the growth rate increases again at approximately t = 0.36 s (Figure 3). However, as the V/III ratio increases, the magnitude of the maximum growth rate becomes smaller. As can be seen from Figure 5a, for higher V/III ratios, the movement of the front interface of the III pulse is slower because more NH_3 remains in the chamber from the previous cycle which reacts with the TMAl to form $TMAlNH_3$. The production of TMAlNH3 increases when the V/III ratio is higher (Figure 5b). This results in a greater V/III ratio, and a later and slower increase in the growth rate after t = 0.35 s (Figure 3). Figure 6 shows the mass fraction distributions of TMAl and TMAlNH3 at t = 0.46 s for different V/III ratios, which is 0.01 s after the start of the second V pulse. When the V/III ratio is higher, the rear interface of the III pulse moves faster (Figure 6a), similar to the movement of the rear interface shown in Figure 4a. However, the front interface of the III pulse is affected by the remaining NH_3 in the chamber. The larger the V/III ratio, the greater the formation of $TMAlNH_3$ near the front and rear interfaces (Figure 6b). This makes the region occupied by the TMAl smaller. Therefore, the maximum value of the average growth rate for the second cycle decreases as the V/III ratio increases (Figure 3). Figure 7 shows the mass fraction distribution of TMAl at t = 0.49 s for different V/III ratios. The movement of the rear interface over the wafer with a V/III ratio equal to 10000 can be seen. This results in a rapid drop in the average growth rate because the rear interface causes a rapid drop in TMAl above the wafer. The phenomena in the third and subsequent cycles are similar to those in the second cycle. It is clear that the decrease in the average growth rate for higher V/III ratios is due to greater $TMAlNH_3$ formation. The insertion of a pure H_2 pulse between the III and V pulses prevents the NH_3 from reacting with the TMAl to form $TMAlNH_3$.

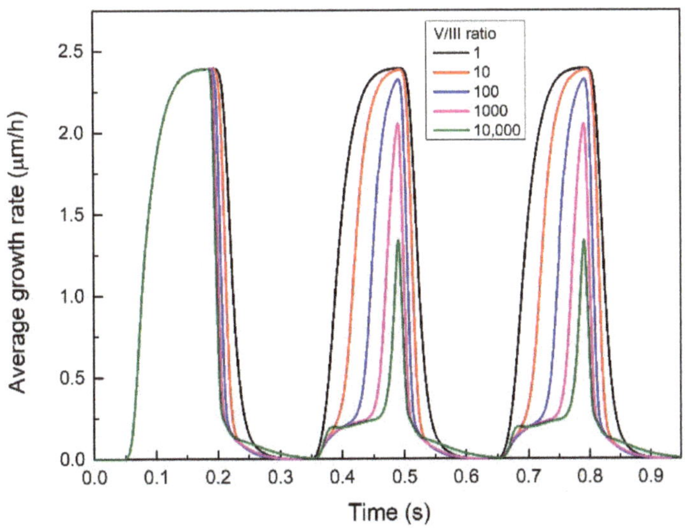

Figure 3. Average growth rate along the wafer surface during the pulsed process.

According to our previous results [11], the formation of $TMAlNH_3$ can be minimized by inserting a pure H_2 purge pulse between the III and V pulses. On the other hand, there is a decline in the overall growth rate due to a wider cycle time. As can been seen in Figure 3, the growth rate after the first cycle is significantly affected by changes in the V/III ratios. In practice, the pulsed process time is usually greater than several minutes. Therefore, the differential impact of the first cycle on the overall growth rate will be very small. To investigate the effect of the purge time on the overall growth rate, the average growth rate for the second cycle is calculated. Figure 8 shows the average growth rate throughout the second cycle for different purge times and different V/III ratios. The decrease in the average growth rate as the V/III ratio increases occurs because of the formation of TMAlNH3. The adductive reaction is weak when the V/III ratio is equal to 1. The benefit of reducing the formation of $TMAlNH_3$ is outweighed by the detrimental effect of the increasing cycle time which causes the average growth rate to decrease with an increase in purge time. When the V/III ratio is equal to 100, there is an increase in the average growth rate as the purge time increases from 0 s to 0.05 s, because the benefit of the reduction in $TMAlNH_3$ outweighs the disadvantage of the increasing cycle time. However, there is a decrease in the average growth rate when the purge time increases from 0.05 s to 0.1 s. In this case, the generation of particles (formation of $TMAlNH_3$) is very low. The benefit of reducing particle generation is outweighed by the disadvantage of an increase in the cycle time. The maximum growth rate is about 0.75 μm/h. When the V/III ratio is equal to 10,000, the average growth rate increases as the purge time increases. However, the average growth rate may not increase with longer purge times, because the growth rate at higher V/III ratios does not exceed that obtained at lower V/III ratios. The maximum average growth rate can be expected to be around 0.7 μm/h.

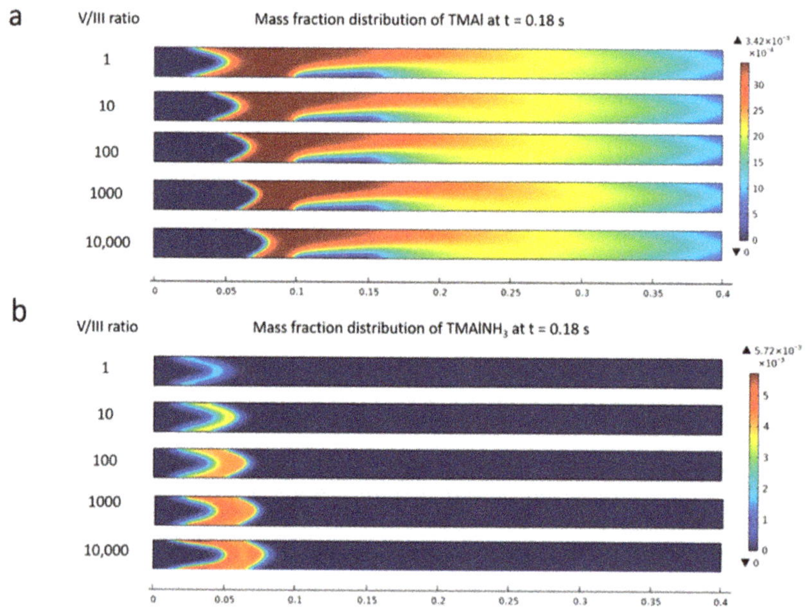

Figure 4. Mass fraction distributions of (**a**) TMAl and (**b**) $TMAlNH_3$ at 0.18 s for different V/III ratios.

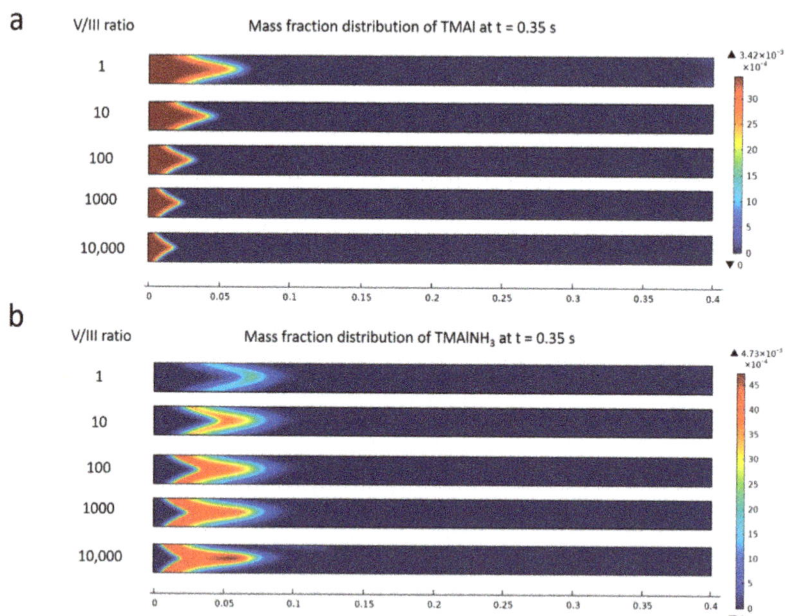

Figure 5. Mass fraction distributions of (**a**) TMAl and (**b**) $TMAlNH_3$ at 0.35 s for different V/III ratios.

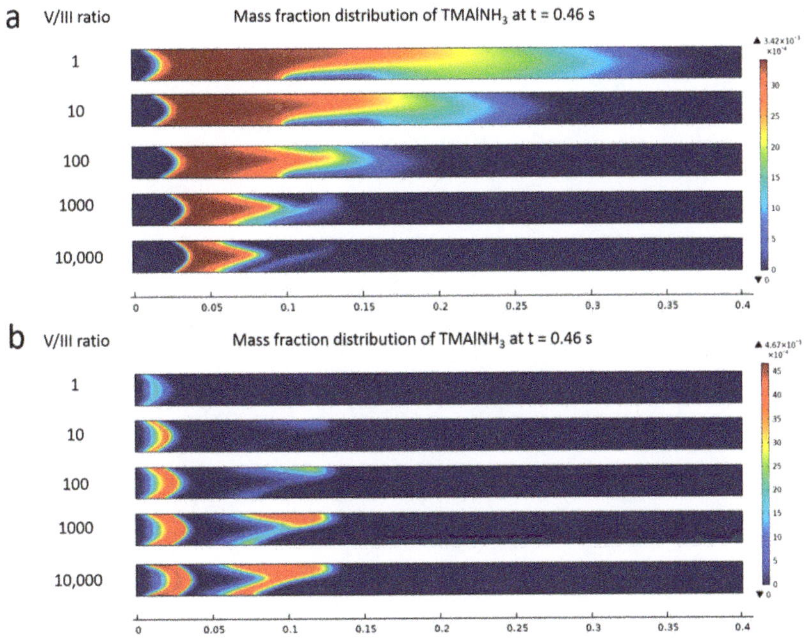

Figure 6. Mass fraction distributions of (**a**) TMAl and (**b**) $TMAlNH_3$ at 0.46 s for different V/III ratios.

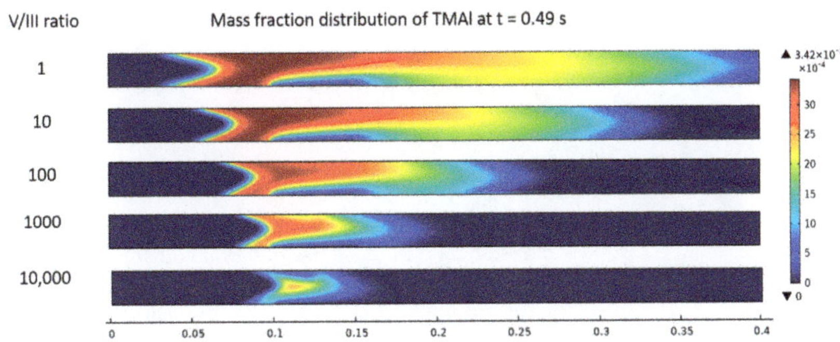

Figure 7. Mass fraction distributions of TMAl at 0.49 s for different V/III ratios.

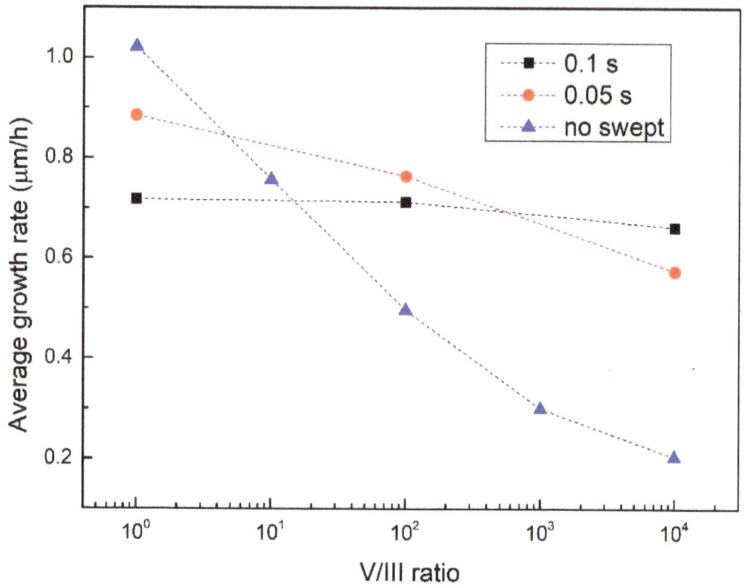

Figure 8. Average growth rate in the second cycle for different purge times and V/III ratios.

4. Conclusions

A numerical analysis is performed to investigate the effects of the V/III ratio on the growth rate during the pulse injection process. The evolution of TMAl and $TMAlNH_3$ during the process is studied in detail. The adductive reaction can be reduced by inserting a pure H_2 purge pulse between III and V pulses, but there may be a decrease in the overall growth due to the longer cycle time. The trade-off between the benefits of reducing particle generation and the disadvantages of increasing the process time can be understood by investigating the relationship between the purge time, the III/V ratio, and the overall growth rate. When the V/III ratio is increased, there is a significant decrease in the average growth rate per cycle, when there is no purge pulse between the III and V pulses due to the stronger adductive reaction (the higher Al particle generation). For smaller V/III ratios, there is a decrease in the growth rate as the purge time increases because the benefit of reducing the adductive reaction is offset by the detrimental effect of the increasing cycle time. This causes the average growth rate to decrease as the purge time increases. The benefit of reducing the adductive reaction is more significant for higher V/III ratios. With increasing purge times, the decline in the growth rate becomes smaller as the V/III ratio increases. Clearly, inserting a pure H_2 purge pulse between the III and V pulses can significantly improve the growth rate when the V/III ratios are high. Higher overall growth rates can be obtained at higher V/III ratios by selecting an appropriate purge time.

Author Contributions: W.-J.L., conceptualization, methodology, formal analysis, writing—original draft; J.-C.C., conceptualization, writing—review and editing, supervision. All authors have read and agreed to the published version of the manuscript.

Funding: The authors would like to thank the Ministry of Science and Technology, R.O.C, for their support of this study through grant number MOST-106-2221-E-008-052-MY3.

Institutional Review Board Statement: Not applicable.

Informed Consent Statement: Not applicable.

Data Availability Statement: Not applicable.

Conflicts of Interest: The authors declare no conflict of interest.

References

1. Wang, T.Y.; Liang, J.H.; Fu, G.W.; Wuu, D.S. Defect annihilation mechanism of AlN buffer structures with alternating high and low V/III ratios grown by MOCVD. *CrystEngComm* **2016**, *18*, 9152–9159. [CrossRef]
2. Nakamura, K.; Hirako, A.; Ohkawa, K. Analysis of pulsed injection of precursors in AlN-MOVPE growth by computa-tional fluid simulation. *Phys. Status Solidi C* **2010**, *7*, 2268–2271. [CrossRef]
3. Demir, I.; Li, H.; Robin, Y.; McClintock, R.; Elagoz, S.; Razeghi, M. Sandwich method to grow high quality AlN by MOCVD. *J. Phys. D Appl. Phys.* **2018**, *51*, 085104. [CrossRef]
4. Khan, M.A.; Kuznia, J.N.; Skogman, R.A.; Olson, D.T.; Mac Millan, M.; Choyke, W.J. Low pressure metalorganic chemical vapor deposition of AIN over sapphire substrates. *Appl. Phys. Lett.* **1992**, *61*, 2539–2541. [CrossRef]
5. Takeuchi, M.; Shimizu, H.; Kajitani, R.; Kawasaki, K.; Kumagai, Y.; Koukitu, A.; Aoyagi, Y. Improvement of crystalline quality of N-polar AlN layers on c-plane sapphire by low-pressure flow-modulated MOCVD. *J. Cryst. Growth* **2006**, *298*, 336–340. [CrossRef]
6. Kröncke, H.; Figge, S.; Aschenbrenner, T.; Hommel, D. Growth of AlN by pulsed and conventional MOVPE. *J. Cryst. Growth* **2013**, *381*, 100–106. [CrossRef]
7. Rahman, M.N.A.; Talik, N.A.; Khudus, M.I.M.A.; Sulaiman, A.F.; Allif, K.; Zahir, N.M.; Shuhaimi, A. Ammonia flux tailoring on the quality of AlN epilayers grown by pulsed atomic-layer epitaxy techniques on (0 0 0 1)-oriented sapphire substrates via MOCVD. *CrystEngComm* **2019**, *21*, 2009–2017. [CrossRef]
8. Rahman, M.N.B.A.; Yusuf, Y.; Anuar, A.; Mahat, M.R.; Chanlek, N.; Talik, N.A.; Khudus, M.I.M.B.A.; Zainal, N.; Majid, W.H.A.; Shuhaimi, A. Agglomeration enhancement of AlN surface diffusion fluxes on a (0 0 0 1)-sapphire substrate grown by pulsed atomic-layer epitaxy techniques via MOCVD. *CrystEngComm* **2020**, *22*, 3309–3321. [CrossRef]
9. Nakamura, K.; Makino, O.; Tachibana, A.; Matsumoto, K. Quantum chemical study of parasitic reaction in III–V nitride semi-conductor crystal growth. *J. Organomet. Chem.* **2000**, *611*, 514–524. [CrossRef]
10. Endres, D.; Mazumder, S. Numerical investigation of pulsed chemical vapor deposition of aluminum nitride to reduce particle formation. *J. Cryst. Growth* **2011**, *335*, 42–50. [CrossRef]
11. Lin, W.J.; Chen, J.C. A numerical study of the effect of pulse duration on preventing particle generation during the AlN pulsed MOCVD process. *Mater. Sci. Semicond. Process.* **2022**, *148*, 106816. [CrossRef]
12. Chen, C.H.; Liu, H.; Steigerwald, D.; Imler, W.; Kuo, C.P.; Craford, M.G.; Ludowise, M.; Lester, S.; Amano, J. A study of parasitic reactions between NH3 and TMGa or TMAl. *J. Electron. Mater.* **1996**, *25*, 1004–1008. [CrossRef]
13. Mihopoulos, T.G.; Gupta, V.; Jensen, K.F. A reaction-transport model for AlGaN MOVPE growth. *J. Cryst. Growth* **1998**, *195*, 733–739. [CrossRef]
14. Lobanova, A.; Mazaev, K.; Talalaev, R.; Leys, M.; Boeykens, S.; Cheng, K.; Degroote, S. Effect of V/III ratio in AlN and AlGaN MOVPE. *J. Cryst. Growth* **2005**, *287*, 601–604. [CrossRef]

crystals

MDPI

Article

Efficiency Droop and Degradation in AlGaN-Based UVB Light-Emitting Diodes

Yi-Tsung Chang [1], Mu-Jen Lai [2,*], Rui-Sen Liu [3], Shu-Chang Wang [4], Xiong Zhang [5], Lin-Jun Zhang [1], Yu-Hsien Lin [6], Shiang-Fu Huang [7,8], Lung-Chien Chen [6] and Ray-Ming Lin [7,9,*]

1 Department of Physics, School of Science, Jimei University, Xiamen 361021, China; oliver_ytchang@jmu.edu.cn (Y.-T.C.); linjun_zhang_liana@163.com (L.-J.Z.)
2 Jiangxi Litkconn Academy of Optical Research Co., Ltd., Ganzhou 341700, China
3 Guangdong Institute of Semiconductor Micro-Nano Manufacturing Technology, Foshan 528200, China; rsliu2021@sinanogd.ac.cn
4 Changshu Institute of Technology, College of Electronics and Information Engineering, Changshu 215500, China; scw@cslg.edu.cn
5 Advanced Photonics Center, School of Electronic Science and Engineering, Southeast University, Nanjing 210096, China; esezx@seu.edu.cn
6 Department of Electro-Optical Engineering, National Taipei University of Technology, Taipei 10608, Taiwan; lyulin.tw@yahoo.com.tw (Y.-H.L.); ocean@ntut.edu.tw (L.-C.C.)
7 Department of Otolaryngology-Head and Neck Surgery, Chang Gung Memorial Hospital, Taoyuan 33302, Taiwan; shiangfu.huang@gmail.com
8 Department of Public Health, Chang Gung University, Taoyuan 33302, Taiwan
9 Department of Electronic Engineering and Institute of Electronics Engineering, Chang Gung University, Taoyuan 33302, Taiwan
* Correspondence: 15007070418@139.com (M.-J.L.); rmlin@mail.cgu.edu.tw (R.-M.L.)

Abstract: In this study, we found that the current droop (J-droop) in AlGaN-based UVB light-emitting diodes was more obvious at higher temperatures, despite both the main and parasitic peaks undergoing monotonic decreases in their intensity upon an increase in the temperature. The slower temperature droop (T-droop) did not occur when the forward current was increased to temperatures greater than 298 K. After an aging time of 6000 h, the emission wavelengths did not undergo any obvious changes, while the intensity of the parasitic peak barely changed. Thus, the degradation in the light output power during long-term operation was not obviously correlated to the existence of the parasitic peak.

Keywords: current droop; temperature droop; AlGaN; ultraviolet-B; light-emitting diodes; external quantum efficiency

Citation: Chang, Y.-T.; Lai, M.-J.; Liu, R.-S.; Wang, S.-C.; Zhang, X.; Zhang, L.-J.; Lin, Y.-H.; Huang, S.-F.; Chen, L.-C.; Lin, R.-M. Efficiency Droop and Degradation in AlGaN-Based UVB Light-Emitting Diodes. *Crystals* **2022**, *12*, 1082. https://doi.org/10.3390/cryst12081082

Academic Editor: Peng Chen

Received: 22 June 2022
Accepted: 11 July 2022
Published: 1 August 2022

1. Introduction

AlGaN-based ultraviolet-B light-emitting diodes (UVB-LEDs) that display emission wavelengths of 280–315 nm have attracted a great deal of attention since the implementation of the Minamata Convention on Mercury and due to the global health crisis arising from the severe acute respiratory syndrome coronavirus 2 (SARS-CoV-2) [1,2]. This is because there is much potential for using such UVB-LEDs in medical and agricultural applications [3–7]. Although several reports have disclosed that high external quantum efficiencies (EQEs) are possible [8–12], current droop (J-droop) and temperature droop (T-droop) processes, which commonly occur with visible-light LEDs, are also severe problems in AlGaN-based UVB-LEDs, even for emission wavelengths of less than 280 nm [13–20]. In addition to problems related to efficiency droop, accompanying parasitic peaks are often observed in electroluminescence (EL) spectra plotted on a logarithmic intensity scale [13,14,16,17]. The shapes of the parasitic peaks of AlGaN-based UVB-LEDs possess a few characteristic features, with a closed-form model having been developed to obtain more insight into their origin [16]. One emission peak, near 340 nm, which likely originated from an intra-band-gap radiative transition in the last quantum barrier (LQB) next to the electron-blocking layer (EBL), was influenced at high currents by enhanced hole injection upon an increase in the temperature. The other emission peak, near 410 nm, is possibly attributable to a radiative transition on the p-side. In other words, at lower driving currents, significant decreases have occurred in the intensity of both the signals from multiple quantum wells (MQWs) and the parasitic emissions when the temperature is increased to above 300 K [16]. In addition, two main parasitic peaks (310 and 400 nm) have been observed extending the low-energy side of the main peak at 275 nm [17]. The suggested origins of these parasitic peaks are deep-level radiative centers in the active region, or they are associated with other centers located in the p-region. Despite a few recent reports discussing the phenomena of T-droop, J-droop, and current-induced degradation in AlGaN-based UVB-LEDs [16,21,22], the mechanisms through which the parasitic peaks affect the degradation behavior during aging operation remain unclear—although they have attracted great interest.

In this study, we fabricated AlGaN-based UVB-LEDs that displayed a main emission peak wavelength of approximately 306 nm and an EL full width at half maximum (FWHM) of approximately 9 nm. We examined the EL characteristics of these UVB-LEDs over a temperature range of 298 to 348 K and a forward current (I_f) range from 10 to 130 mA to obtain insight into the behavior of the EL emission and the efficiency droop. Furthermore, we evaluated the impact of the parasitic peaks on the aging lifetimes of these UVB-LEDs, and we suggest a mechanism for the degradation that occurred during the aging operation.

2. Materials and Methods

All epitaxial layers were obtained using low-pressure metal–organic chemical vapor deposition (LP-MOCVD). First, a 2.2 μm thick AlN buffer layer was grown on a 2 inch (0001)-oriented sapphire substrate. Next, a strain-relieving interlayer consisting of 30 periods of an AlN/AlGaN superlattice was grown on the AlN buffer layer. A 1.5 μm thick layer of undoped $Al_{0.6}Ga_{0.4}N$ was then grown on the interlayer, followed by a 2 μm thick Si-doped n-$Al_{0.5}Ga_{0.5}N$ layer as the n-contact layer. The active region of the MQWs included five pairs of 2 nm thick $Al_{0.35}Ga_{0.65}N$ quantum wells (QWs) and 8 nm thick $Al_{0.45}Ga_{0.55}N$ quantum barriers (QBs), followed by a two-fold Mg-doped p-$Al_{0.55}Ga_{0.45}N$ (10 nm)/Mg-doped p-$Al_{0.4}Ga_{0.6}N$ (2 nm) structure as the EBL. Subsequently, a 50 nm thick Mg-doped p-AlGaN layer, with the Al content decreasing from 30% to 0%, was grown on the EBL. Finally, a 20 nm thick Mg-doped p-GaN layer was deposited, serving as the p-contact layer. After the epitaxial layers had been grown, the sample was annealed in situ (700 °C, 15 min) under a N_2 ambient to activate the Mg dopants. UVB-LED chips having dimensions of 550×550 μm^2 were fabricated using standard flip-chip processing technologies [23].

The prepared UVB-LED chips were flip-bonded onto AlN-based direct plating copper ceramic (AlN-DPC) lead frames (LFs) using the eutectic method, then attached to a quartz glass as the hermetic cover. The packaged samples were soldered onto an Al core printed circuit board (MCPCB). The light output power (LOP), current–voltage characteristics, and EL spectra of the samples were measured using an ATA-5000 LED photoelectric measurement system (Everfine) equipped with a 30 cm diameter integrating sphere. During measurement, the temperature of the heat-sink enclosing the samples was controlled (at 298, 323, or 348 K) while the values of I_f varied from 10 to 130 mA. Aging tests were performed over 6000 h at a value of I_f of 40 mA at room temperature (RT) to investigate the UVB-LED characteristics.

3. Results and Discussion

Figure 1a reveals the dependence of the EQEs and LOPs of the fabricated UVB-LED on the three tested temperatures and the various forward currents. The LOP increased monotonically when the forward current was increased to 130 mA, reaching maximum values of 8.6, 7.8, and 6.9 mW at 298, 323, and 348 K, respectively. The maximum EQEs at 298, 323, and 348 K (1.77, 1.69, and 1.52%, respectively) occurred when the value of

I_f was 10 mA; at 130 mA, they decreased to 1.62, 1.48, and 1.32%, respectively. After applying Equation (1), we obtained J-droops of 8.47, 12.42, and 13.16% at 298, 323, and 348 K, respectively.

$$\text{J} - \text{droop}\ (\%) = \frac{\text{Peak EQE @10mA} - \text{Peak EQE @forward current}}{\text{Peak EQE @10mA}} \times 100\% \quad (1)$$

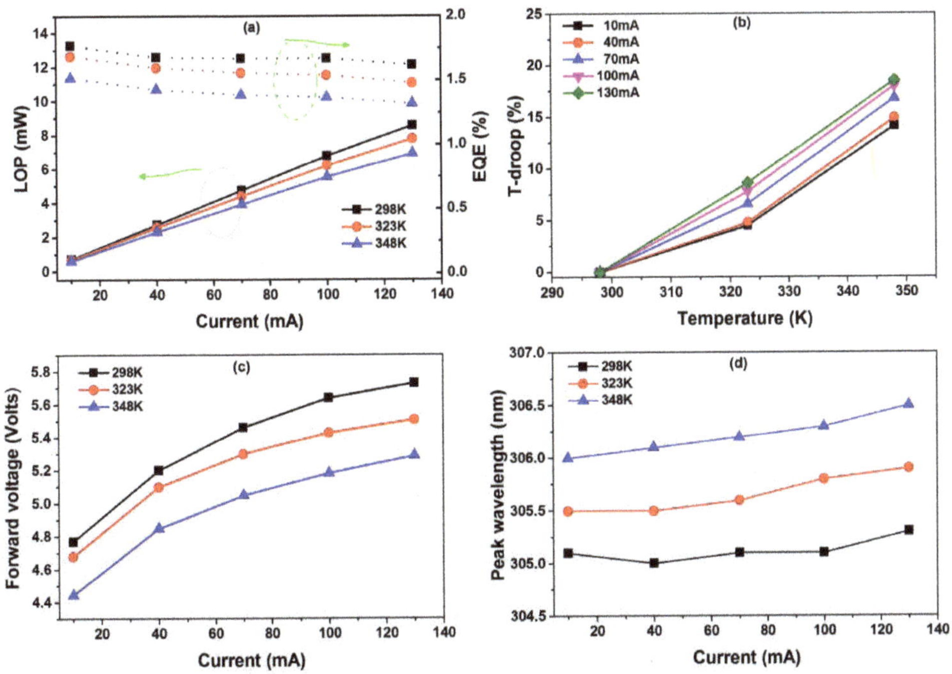

Figure 1. (**a**) EQEs (dotted lines) and LOPs (solid lines) of the AlGaN-based UVB-LED plotted with respect to the forward current, measured at three different temperatures. (**b**) T-droop plotted with respect to temperature, measured at various forward currents. (**c**) Forward voltage and (**d**) peak wavelength plotted with respect to forward current, measured at the three temperatures.

For a given value of I_f, the EQEs decreased monotonically upon an increase in the temperature, and then decreased further upon an increase in the value of I_f. These findings suggested that the efficiency droop was probably related to a mechanism of enhanced non-radiative recombination, with carrier leakage or Auger recombination being exacerbated [16–21]. Figure 1b presents the T-droops with respect to temperature, which was quantified as follows:

$$\text{T} - \text{droop}\ (\%) = \frac{\text{Peak EQE @298K} - \text{Peak EQE @T}}{\text{Peak EQE @298K}} \times 100\% \quad (2)$$

Interestingly, after applying Equation (2), the T-droops measured at 10, 40, 70, 100, and 130 mA were 4.52, 4.76, 6.59, 7.78, and 8.64%, respectively, when the value of T was 323 K; at 348 K, they increased to 14.12, 14.88, 16.77, 17.96, and 18.52%, respectively. We observed monotonically increasing T-droops upon an increase in the temperature from 298 to 348 K, with higher T-droops occurring at higher values of I_f at a constant temperature. This behavior is not consistent with a previous finding that the T-droop decreased upon an increase in the current from 0.5 to 100 mA at a temperature of 340 K [20]. We suggest that the mechanism of T-droop in this study originated from Auger recombination, with carrier leakage becoming dominant when the temperature reached 348 K.

Figure 1c,d displays the effects of the forward voltage (V_f) and peak wavelength on the forward currents of the fabricated UVB-LED at the three tested temperatures. The forward voltages decreased and the peak wavelengths increased upon an increase in the temperature at the same forward current due to shrinkage of the band gap; the peak wavelength underwent a slight blue-shift (<0.5 nm) upon an increase in the current from 10 to 40 mA, when the temperature was 298 or 323 K. This result implies the absence of a band-filling effect or coulomb screening of the quantum confinement Stark effect (QCSE) in the MQWs of the fabricated UVB-LED. When the temperature was 348 K, the peak wavelength underwent a slight red-shift when the current was increased from 10 to 130 mA due to the effect of self-heating. Thus, suppression of the temperature-dependent recombination mechanism in the MQWs of the fabricated UVB-LED presumably occurred at temperatures of less than 348 K.

Figure 2a–c presents the normalized EL emission spectra plotted on a logarithmic scale, measured at 298, 323, and 348 K, respectively, for forward currents (I_f) supplied in the range from 10 to 130 mA. Each EL emission spectrum featured a main peak wavelength near 306 nm, with a parasitic peak appearing in the range from 330 to 370 nm. The relative intensities of the parasitic peaks decreased when the value of I_f and the temperature increased. Figure 3a–d displays the spectral distributions and the peak intensities of the EL recorded at various forward currents and temperatures of 298, 323, and 348 K. At a constant temperature, the intensity ratios between the parasitic and main peaks decreased when the value of I_f increased. However, at a constant value of I_f, the intensity ratios of the parasitic and main peaks decreased upon an increase in the temperature (Figure 3d). This behavior is different from that expected when considering the previously suggested hypothesis that the parasitic peak emission is related to a defect-assisted emission in the MQW active region [13]. In addition, Wu et al. reported that the band-edge emission in the EL spectra surged, while the increases in the intensities of two parasitic peaks stalled upon an increase in the value of I_f at various temperatures [17]. They obtained maximum EQEs of 1.19 and 0.61% at temperatures of 20 and 300 K, respectively; when the current increased from 1 to 7 mA, these values decreased to approximately 0.82 and 0.53%, respectively [17]. In other words, the decrease in the EQE at a temperature of 20 K (ca. 31%) was more pronounced than that at 300 K (ca. 13.1%). In contrast, in this study, we found that the J-droop was more obvious at higher temperatures (Figure 1a). This finding is similar to that for the InGaN-based blue LED [18], despite both the main and parasitic peaks undergoing monotonic decreases in intensity upon an increase in the temperature (Figure 3b,c).

Figure 4 provides the normalized intensities of the peak wavelengths recorded at temperatures of 298, 323, and 348 K with forward currents applied in a range from 10 to 130 mA. In a previous study, Santi et al. observed a parasitic peak near 340 nm for their UVB-LED and attributed its origin to radiative transitions through deep levels in the QB next to the EBL and not inside the QWs [16].

However, their paper reveals that a slower T-droop occurred when the temperature was greater than 300 K upon an increase in the value of I_f [16]. Nevertheless, in the present study, we did not observe a similar T-droop phenomenon when the temperature was 348 K. Furthermore, we found that the normalized intensities of the peak wavelengths underwent a steeper decline upon an increase in the temperature when increasing the value of I_f. According to the dependence of the emission intensity on temperature, Figure 4 indicates that at a higher forward current, the AlGaN-based UVB-LED exhibited weaker temperature stability. This trend is dramatically different from that reported previously [19]. Figure 5 presents the relative LOP measured over time for the UVB-LED operated at 40 mA (current density: ca. 28.5 A/cm^2) and RT, as well as a photograph of the UVB-LED during operation. A rapid degradation in the LOP, to 79.7% of its initial value, occurred in the first 500 h of the aging operation; it then decreased more slowly to 70.6% of its initial value when the aging time reaching 6000 h (the inset in Figure 5 displays the LOP plotted with respect to a logarithmic time scale). Such behavior has been observed many times previously [21–28]. This performance is close to the specifications of commercial solid-state lighting products and the estimated 70% lifetime (L70) is comparable with values reported by Ruschel et al. for a 310 nm UVB-LED operated at a current density of 33.5 A/cm^2 [21]. Figure 6a–c presents the time-dependence of the forward voltage, the EL peak wavelength, and the EL FWHM, respectively, which were all measured during aging tests performed with an I_f value of 40 mA. The forward voltage decreased dramatically at the onset of operation but remained stable during operation between 500 and 6000 h.

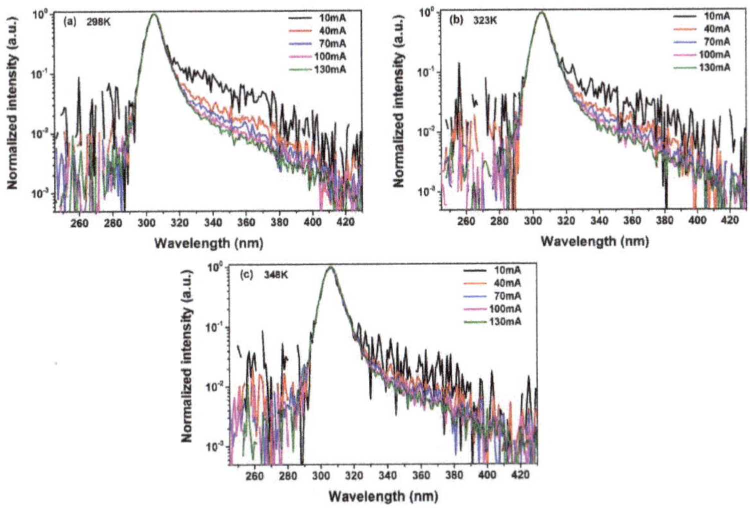

Figure 2. Normalized EL emission spectra of the AlGaN-based UVB-LED, recorded at various forward currents (**a**) 298, (**b**) 323, (**c**) 348 K.

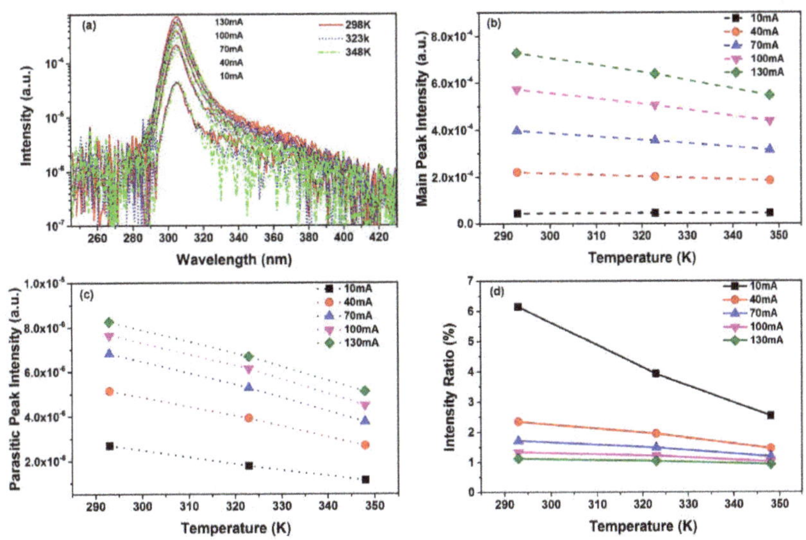

Figure 3. EL spectral data recorded at various forward currents and temperatures of 298, 323, and 348 K. (**a**) Spectral distributions, (**b**) main peak intensities, (**c**) parasitic peak intensities, and (**d**) peak intensity ratios.

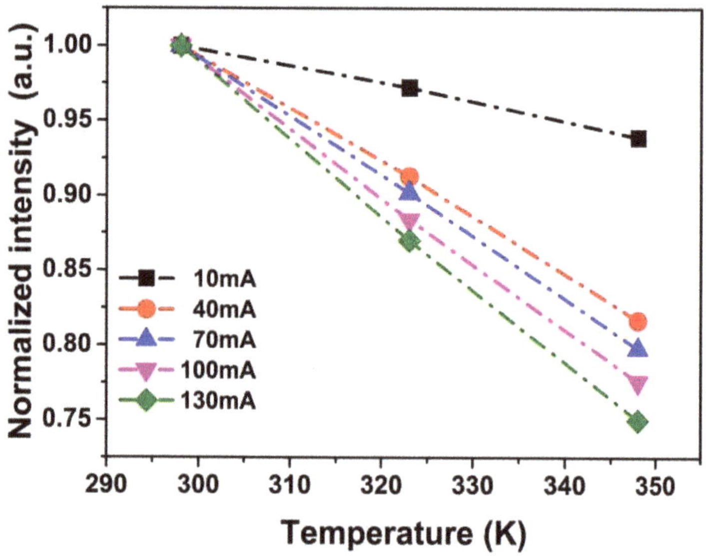

Figure 4. Normalized QW intensities of the peak wavelengths of the AlGaN-based UVB-LED measured at temperatures of 298, 323, and 348 K while applying forward currents from 10 to 130 mA at a step of 30 mA.

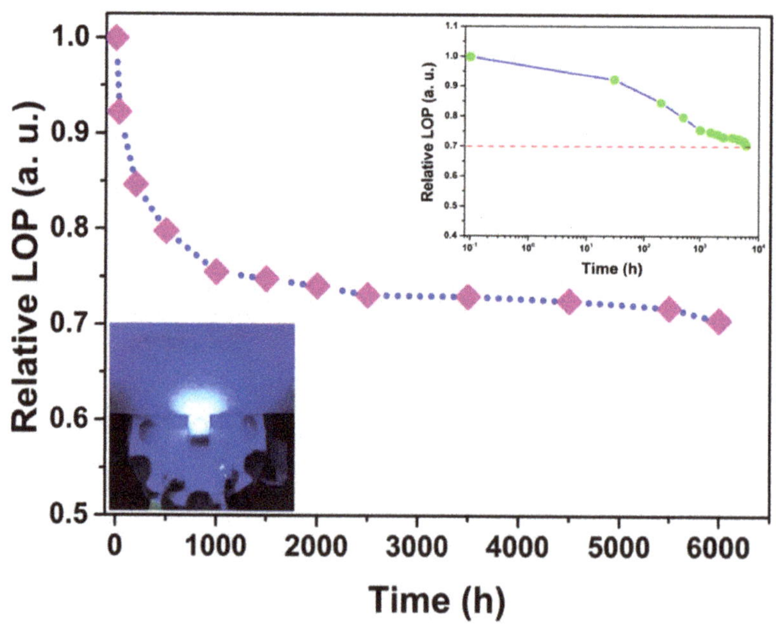

Figure 5. Relative LOP of the AlGaN-based UVB-LED, measured during aging tests performed at 40 mA. Insets: (left) photograph of the UVB-LED during operation; (right) LOP plotted on a logarithmic time scale.

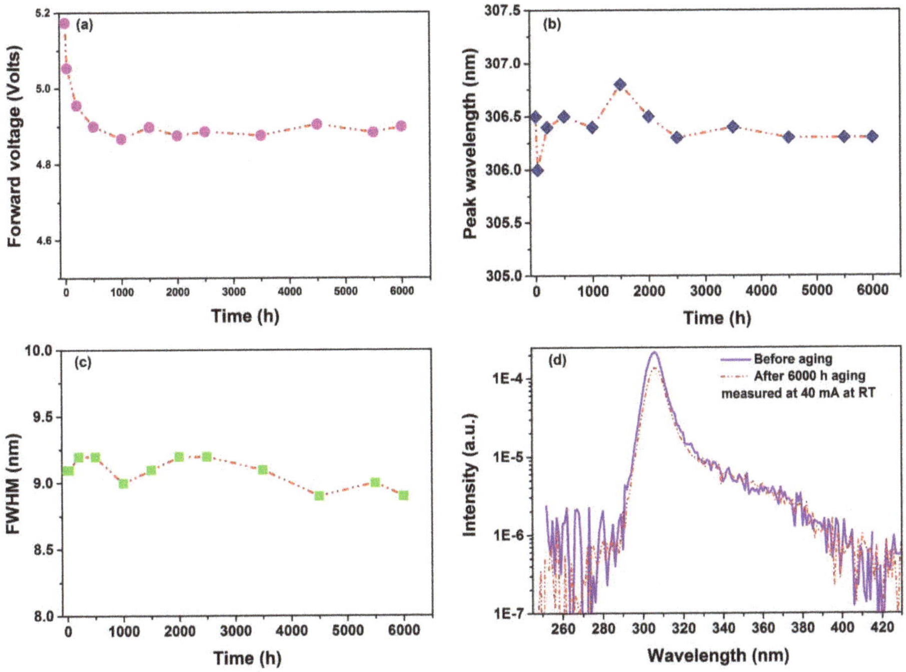

Figure 6. (**a**) Forward voltages, (**b**) EL peak wavelengths, and (**c**) EL FWHMs of the AlGaN-based UVB-LED measured at a value of I_f of 40 mA during the aging tests. (**d**) EL emission spectra of the AlGaN-based UVB-LED measured at a value of I_f of 40 mA before and after performing the aging tests.

Interestingly, while the LOPs and forward voltages decreased simultaneously when the aging time was increased to 6000 h, no significant changes occurred in the peak wavelengths or the FWHMs. This behavior suggested that the crystal quality of the MQWs did not undergo serious deterioration over time. Moreover, Figure 6d displays the EL emission spectra measured at a value of I_f of 40 mA before and after aging for 6000 h. Although the emission wavelengths of both the main and parasitic peaks did not undergo any obvious changes before and after aging, the intensity of the main peak decreased to 63.6% of its initial value, whereas the intensity of the parasitic peak barely changed. The behavior of the parasitic peak in this case was different from that reported by Trivellin et al. during the degradation of a 280-nm UVC-LED [13]. Thus, further investigations will be necessary to determine whether the mechanism responsible for the parasitic peak observed in our EL spectra involved radiative recombination in the EBL or p-layer through electron leakage. In addition, the degradation of the LOP during long-term operation was not obviously correlated with the existence of the parasitic peak. Therefore, we infer that the rapid decrease in the LOP and the decrease in the value of V_f that occurred simultaneously at the onset of the current stress were due to intrinsic defects in the MQWs and extrinsic defects dynamically induced by the current stress; those defects were occupied by injection carriers, resulting in a decrease in the width of the space-charge region. Because the number of defects in the MQWs gradually became saturated, the rate of degradation decreased during further operation. Previous reports have suggested that decreases in LOPs arise as a result of the diffusion of dopants [24,26–28]. We will investigate this phenomenon in future studies.

4. Conclusions

We found that the J-droop in an AlGaN-based UVB-LED was more obvious at higher temperatures, despite both the main and parasitic peaks undergoing monotonic decreases in their intensity upon increases in the temperature. A higher T-droop occurred upon increases in both the value of I_f and the temperature. Although the emission wavelengths did not undergo obvious changes after 6000 h of aging, the intensity of the main peak decreased to 63.6% of its initial value, whereas the intensity of the parasitic peak barely changed. This result suggested that the mechanism responsible for the origin of the parasitic emission peak did not also lead to serious deterioration in the crystal quality; furthermore, the degradation of the LOP during long-term operation was not obviously correlated with the existence of the parasitic peak. At the onset of current stress, the intrinsic defects in the MQWs and the extrinsic defects dynamically induced by the current stress (occupied by injection carriers) resulted in a decrease in the width of the space-charge region. Moreover, the decelerating rate of degradation during further operation was due to the gradual saturation of the number of defects in the MQWs.

Author Contributions: Conceptualization, M.-J.L. and Y.-T.C.; methodology, M.-J.L., Y.-T.C. and R.-S.L.; formal analysis, M.-J.L. and Y.-T.C.; investigation, M.-J.L., R.-S.L. and R.-M.L.; data curation, L.-J.Z., Y.-H.L. and M.-J.L.; writing—original draft preparation, M.-J.L. and Y.-T.C.; writing—review and editing, S.-F.H., S.-CW., X.Z., L.-C.C. and R.-M.L.; visualization, X.Z., L.-C.C. and R.-M.L.; supervision, L.-C.C. and R.-M.L.; project administration, M.-J.L., L.-C.C., S.-F.H. and R.-M.L.; funding acquisition, Y.-T.C., S.-C.W., S.-F.H. and R.-M.L. All authors have read and agreed to the published version of the manuscript.

Funding: Jimei University Research Project (contract no. Z91956/4412), National Natural Science Foundation of China (grant no. 62005026), Natural Science Foundation of Jiangsu Province (grant no. BK20191027), and Suzhou Science and Technology Project (grant no. SZS2020313), Ministry of Science and Technology (MOST) of Taiwan (contract MOST 109-2221-E-182-060), Chang Gung Memorial Hospital (grant BMRP 591).

Institutional Review Board Statement: Not applicable.

Informed Consent Statement: Not applicable.

Data Availability Statement: Not applicable.

Acknowledgments: We thank Keh-Yung Cheng, Liann-Be Chang, and Chieh-Hsiung Kuan for the helpful discussions.

Conflicts of Interest: The authors declare no conflict of interest.

References

1. Amano, H.; Collazo, R.; De Santi, C.; Einfeldt, S.; Funato, M.; Glaab, J.; Hagedorn, S.; Hirano, A.; Hirayama, H.; Ishii, R.; et al. The 2020 UV emitter roadmap. *J. Phys. D Appl. Phys.* **2020**, *53*, 503001. [CrossRef]
2. Raeiszadeh, M.; Adeli, B. A Critical Review on Ultraviolet Disinfection Systems against COVID-19 Outbreak: Applicability, Validation, and Safety Considerations. *ACS Photon.* **2020**, *7*, 2941–2951. [CrossRef]
3. Kneissl, M.; Rass, J. *III-Nitride Ultraviolet Emitters: Technology and Applications*; Springer: New York, NY, USA, 2016.
4. Matsuura, S.; Ishikura, S. Suppression of Tomato mosaic virus disease in tomato plants by deep ultraviolet irradiation using light-emitting diodes. *Lett. Appl. Microbiol.* **2014**, *59*, 457–463. [CrossRef] [PubMed]
5. Middelkamp-Hup, M.A.; Bos, J.D.; Rius-Diaz, F.; Gonzalez, S.; Westerhof, W. Treatment of vitiligo vulgaris with narrow-band UVB and oral Polypodium leucotomos extract: A randomized double-blind placebo-controlled study. *J. Eur. Acad. Dermatol. Venereol.* **2010**, *21*, 942–950. [CrossRef] [PubMed]
6. Beani, J.C.; Jeanmougin, M. Narrow-band UVB therapy in psoriasis vulgaris: Good practice guideline and recommendations of the French Society of Photodermatology. *Ann. Dermatol. Vénéréologie* **2010**, *137*, 21–31. [CrossRef]
7. Verhaeghe, E.; Lodewick, E.; van Geel, N.; Lambert, J. Intrapatient Comparison of 308-nm Monochromatic Excimer Light and Localized Narrow-Band UVB Phototherapy in the Treatment of Vitiligo: A Randomized Controlled Trial. *Dermatology* **2011**, *223*, 343–348. [CrossRef]
8. Khan, M.A.; Matsumoto, T.; Maeda, N.; Kamata, N.; Hirayama, H. Improved external quantum efficiency of 293 nm AlGaN UVB LED grown on an AlN template. *Jpn. J. Appl. Phys.* **2018**, *58*, SAAF01. [CrossRef]

9. Khan, M.A.; Maeda, N.; Jo, M.; Akamatsu, Y.; Tanabe, R.; Yamada, Y.; Hirayama, H. 13 mW operation of a 295–310 nm AlGaN UV-B LED with a p-AlGaN transparent contact layer for real world applications. *J. Mater. Chem. C* **2018**, *7*, 143–152. [CrossRef]
10. Susilo, N.; Enslin, J.; Sulmoni, L.; Guttmann, M.; Zeimer, U.; Wernicke, T.; Weyers, M.; Kneissl, M. Effect of the GaN:Mg contact layer on the light-output and current–voltage characteristic of UVB LEDs. *Phys. Status Solidi A* **2017**, *215*, 1700643. [CrossRef]
11. Guttmann, M.; Hermann, M.; Enslin, J.; Graupeter, S.; Sulmoni, L.; Kuhn, C.; Wernicke, T.; Kneissl, M. Improved light extraction and quantum efficiencies for UVB LEDs with UV-transparent p-AlGaN superlattices. *Proc. SPIE* **2017**, *10104*, 101041S. [CrossRef]
12. Khan, M.A.; Itokazu, Y.; Maeda, N.; Jo, M.; Yamada, Y.; Hirayama, H. External quantum efficiency of 6.5% at 300 nm emission and 4.7% at 310 nm emission on bare wafer of AlGaN-based UVB LEDs. *ACS Appl. Electron. Mater.* **2020**, *2*, 1892–1907. [CrossRef]
13. Trivellin, N.; Monti, D.; Piva, F.; Buffolo, M.; De Santi, C.; Zanoni, E.; Meneghesso, G.; Meneghini, M. Degradation processes of 280 nm high power DUV LEDs: Impact on parasitic luminescence. *Jpn. J. Appl. Phys.* **2019**, *58*, SCCC19. [CrossRef]
14. Monti, D.; Santi, C.D.; Ruos, S.D.; Piva, F.; Glaab, J.; Rass, J.; Einfeldt, S.; Mehnke, F.; Enslin, J.; Wernicke, T.; et al. High-current stress of UV-B (In)AlGaN-based LEDs: Defect-generation and diffusion processes. *IEEE Trans. Electron Dev.* **2019**, *66*, 3387–3392. [CrossRef]
15. Meneghini, M.; De Santi, C.; Tibaldi, A.; Vallone, M.; Bertazzi, F.; Meneghesso, G.; Zanoni, E.; Goano, M. Thermal droop in III-nitride based light-emitting diodes: Physical origin and perspectives. *J. Appl. Phys.* **2020**, *127*, 211102. [CrossRef]
16. Santi, C.D.; Meneghini, M.; Monti, D.; Glaab, J.; Guttmann, M.; Rass, J.; Einfeldt, S.; Mehnke, F.; Enslin, J.; Wernicke, T.; et al. Recombination mechanisms and thermal droop in AlGaN-based UV-B LEDs. *Photonics Res.* **2017**, *5*, A44–A51. [CrossRef]
17. Wu, T.; Lin, Y.; Peng, Z.; Chen, H.; Shangguan, Z.; Liu, M.; Chen, S.-W.H.; Lin, C.-H.; Kuo, H.-C.; Chen, Z. Interplay of carriers and deep-level recombination centers of 275-nm light-emitting diodes—Analysis on the parasitic peaks over wide ranges of temperature and injection density. *Opt. Express* **2019**, *27*, A1060–A1073. [CrossRef]
18. Oh, C.H.; Shim, J.I.; Shin, D.S. Current- and temperature-dependent efficiency droops in InGaN-based blue and AlGaInP-based red light-emitting diodes. *Jpn. J. Appl. Phys.* **2019**, *58*, SCCC08. [CrossRef]
19. Ploch, N.L.; Einfeldt, S.; Frentrup, M.; Rass, J.; Wernicke, T.; Knauer, A.; Kueller, V.; Weyers, M.; Kneissl, M. Investigation of the temperature dependent efficiency droop in UV LEDs. *Semicond. Sci. Technol.* **2013**, *28*, 125021. [CrossRef]
20. Peng, Z.; Guo, W.; Wu, T.; Guo, Z.; Lu, Y.; Zheng, Y.-C.; Lin, Y.; Chen, Z. Temperature-Dependent Carrier Recombination and Efficiency Droop of AlGaN Deep Ultraviolet Light-Emitting Diodes. *IEEE Photonics J.* **2020**, *12*, 8200108. [CrossRef]
21. Ruschel, J.; Glaab, J.; Beidoun, B.; Ploch, N.L.; Rass, J.; Kolbe, T.; Knauer, A.; Weyers, M.; Einfeldt, S.; Kneissl, M. Current-induced degradation and lifetime prediction of 310 nm ultraviolet light-emitting diodes. *Photonics Res.* **2019**, *7*, B36–B40. [CrossRef]
22. Ruschel, J.; Glaab, J.; Brendel, M.; Kneissl, M. Localization of current-induced degradation effects in (InAlGa)N-based UV-B LEDs. *J. Appl. Phys.* **2018**, *124*, 084504. [CrossRef]
23. Liu, T.-Y.; Huang, S.-M.; Lai, M.-J.; Liu, R.-S.; Zhang, X.; Chang, Y.-T.; Zhang, L.-J.; Lin, R.-M. Narrow-band AlGaN-based UVB light-emitting diodes. *ACS Appl. Electron. Mater.* **2021**, *3*, 4121. [CrossRef]
24. Glaab, J.; Ploch, C.; Kelz, R.; Stölmacker, C.; Lapeyrade, M.; Ploch, N.L.; Rass, J.; Kolbe, T.; Einfeldt, S.; Mehnke, F.; et al. Degradation of (InAlGa)N-based UV-B light emitting diodes stressed by current and temperature. *J. Appl. Phys.* **2015**, *118*, 094504. [CrossRef]
25. Nippert, F.; Mazraehno, M.T.; Davies, M.J.; Hoffmann, M.P.; Lugauer, H.-J.; Kure, T.; Kneissl, M.; Hoffmann, A.; Wagner, M.R. Auger recombination in AlGaN quantum wells for UV light-emitting diodes. *Appl. Phys. Lett.* **2018**, *113*, 071107. [CrossRef]
26. Glaab, J.; Haefke, J.; Ruschel, J.; Brendel, M.; Rass, J.; Kolbe, T.; Knauer, A.; Weyers, M.; Einfeldt, S.; Guttmann, M.; et al. Degradation effects of the active region in UV-C light-emitting diodes. *J. Appl. Phys.* **2018**, *123*, 104502. [CrossRef]
27. Zhu, X.; Su, M.; Chen, Z.; Deng, S.; Yao, H.; Wang, Y.; Chen, Z.; Deng, J.; Sun, W. Degradation analysis with characteristics and simulations of 265 nm UV-C LED. *J. Mater. Sci. Mater. Electron.* **2021**, *32*, 17115–17122. [CrossRef]
28. Su, M.; Zhu, X.; Guo, Q.; Chen, Z.; Deng, S.; Chen, Z.; Wang, Y.; Deng, J.; Sun, W. Characterization and simulation of 280 nm UV-LED degradation. *AIP Adv.* **2021**, *11*, 035315. [CrossRef]

crystals MDPI

Article

The Sensing Mechanism of InAlN/GaN HEMT

Yanli Liu [1,*], Xiao He [1], Yan Dong [2], Su Fu [1], Yuhui Liu [1] and Dunjun Chen [3,*]

[1] School of Information and Electronic Engineering, Shandong Technology and Business University, Yantai 264005, China; 2019420095@sdtbu.edu.cn (X.H.); 2020420067@sdtbu.edu.cn (S.F.); 2020420014@sdtbu.edu.cn (Y.L.)
[2] School of Electronic Engineering, Heilongjiang University, Harbin 150080, China; 2021001@hlju.edu.cn
[3] School of Electronic Science and Engineering, Nanjing University, Nanjing 210093, China
* Correspondence: yanliliu@sdtbu.edu.cn (Y.L.); djchen@nju.edu.cn (D.C.)

Abstract: The sensing mechanism of InAlN/GaN high electron mobility transistors (HEMTs) is investigated systematically by numerical simulation and theoretical analysis. In detail, the influence of additional surface charge on device performance and the dependence of surface sensing properties on InAlN barrier thickness are studied. The results indicate that the saturation output drain current I_{dsat} and two-dimensional electron gas (2DEG) concentration increase with the increase of positive surface charge density, which decrease with the increase of negative surface charge. The influence of negative surface charge on device performance is more remarkable than that of positive surface charge. Additionally, the modulation ability of surface charge on device performance increases with the decrease ofInAlN barrier thickness. The modulation of surface charge on device performance and the influence of barrier thickness on surface sensing sensitivity are mainly attributed to the variation of the energy band structure, surface potential and 2DEG concentration in the HEMT heterostructure. This work provides important support for structural optimization design of GaN-based HEMT sensors.

Keywords: surface charge; InAlN/GaN; HEMT; surface sensitivity

Citation: Liu, Y.; He, X.; Dong, Y.; Fu, S.; Liu, Y.; Chen, D. The Sensing Mechanism of InAlN/GaN HEMT. *Crystals* **2022**, *12*, 401. https://doi.org/10.3390/cryst12030401

Academic Editors: Peng Chen, Zhizhong Chen and Cheljong Choi

Received: 16 January 2022
Accepted: 13 March 2022
Published: 16 March 2022

1. Introduction

Due to the superior properties such as wide bandgap, high breakdown electric field, high saturation drift velocity and high thermal conductivity, GaN-based high electron mobility transistors (HEMTs) exhibit important applications in the fields of high temperature, high frequency and high power devices [1–4]. Meanwhile, two dimensional electron gas (2DEG) with a sheet concentration of as much as 10^{13} cm^{-2} is formed because of the polarization discontinuity at AlGaN (or InAlN) /GaN heterointerface. The 2DEG density strongly depends on surface states of GaN-based heterostructure. Any changes of surface state will result in changes of energy band structure and surface potential of the heterostructure. Moreover, 2DEG responds quite sensitively to any changes of the free surface charge situation. To a first approximation, for each positive or negative ion adsorbed at the free surface, accordingly, one electron is gained or lost in the 2DEG. Since the mobilities of adsorbed ions at the free surface and electrons in the 2DEG differ by many orders of magnitude, changes in the surface charge will be visible as changes in the 2DEG conductivity, however, they will be amplified by the mobility ratio. This is the basic mechanism which makes GaN-based HEMT attractive for sensor applications [5–7]. Recently, there are tremendous reports about AlGaN/GaN HEMT sensors for detecting gas, liquid, chemical and biomaterial, which can be used in the fields of environmental monitoring, food safety analysis and biomedicine [8–12]. In the past few years, researchers mainly focused on surface modification of AlGaN/GaN HEMTs for the detection of various target substances. For different targets to be tested, the best obtained result is different. For example, the highest sensing sensitivities for a pH sensor [13] and a Hg^{2+} ion sensor [14] are 167.71 mV/pH and 0.3 μA/ppb, respectively. Besides, there are a few studies about the methods to improve the sensing performance by optimizing HEMT device structure such as gateless structure [15,16], gate-recessed structure [17,18] and extended-gate structure [19,20]. For gateless HEMT, the sensing distance between the surface sensing area and the 2DEG channel is scaled compared to the conventional GaN-based HEMT, which makes sensing sensitivity enhanced. Due to the progress of controlling ability from the gate-recessed region, the sensing performance of the gate-recessed HEMT sensors obviously enhanced compared with the planar gate HEMT. For extended-gate HEMT, by extending the gate area and separating the sensing area from the channel of the HEMT, the device has a larger sensing area. Therefore, all of the sensing sensitivity, stability and detection limit of the HEMT sensors are improved.

In fact, to further improve the sensing performance of GaN-based HEMTs effectively, it is necessary to investigate the sensing mechanism comprehensively. Until now, there are few reports about the sensing mechanism of GaN-based HEMT. Besides, owing to the relatively mature epitaxial and manufacturing technology, the heterostructure mainly used for HEMT sensors in the past few years is AlGaN/GaN. Compared with conventional AlGaN/GaN heterostructure, the lattice-matched $In_{0.17}Al_{0.83}N$/GaN heterostructure, has higher carrier concentration and better 2DEG confinement [21,22]. In fact, lattice-matched $In_{0.17}Al_{0.83}N$/GaN heterostructure is more suitable for HEMT sensors. However, there are only a few reports about InAlN/GaN HEMT sensors [23,24]. In this work, the modulation law of surface charge on the device performance and the effect of InAlN barrier layer thickness on the sensing performance of InAlN/GaN HEMT are studied systematically, which is considered to provide a broad-version route map for designing InAlN based HEMT sensors.

2. Device Structure and Theoretical Models

The InAlN/GaN HEMT device structure used in this study is shown in Figure 1. The HEMT heterostructure on sapphire substrate consists of a 2-μm-thick GaN buffer layer and a thin $In_{0.17}Al_{0.83}N$ barrier layer. A SiN passivation layer was applied between the gate sensing area and ohmic contacts (source and drain electrodes). To be close to the real experiment, the n-type doping densities in both GaN buffer layer and InAlN barrier layer used in this work were set to be 1×10^{16} cm^{-3}, due to the fact that the unintentionally doping InAlN/GaN heterostructures grown in the experiment exhibit n-type doping with a concentration of 1×10^{16} cm^{-3}. The distance between the source and drain electrodes is 3 μm. The length of the sensing area for gate region is 1.5 μm.

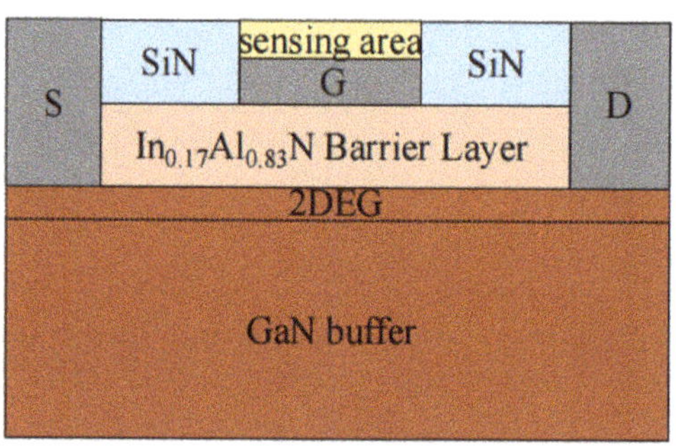

Figure 1. Schematic diagram of the InAlN/GaN HEMT.

To obtain the I_{dsat}, 2DEG concentration and the energy band profile of the HEMT with varying surface charge, the Schrödinger equation, Poisson's equation and the carrier continuity equation are solved self-consistently [25,26]. The equations can be expressed as follows, respectively.

Schrödinger equation:

$$\left[\frac{\hbar^2}{2m^*}\frac{d^2}{dz^2} + V(z)\right]\psi(z) = E\psi(z) \tag{1}$$

where $\hbar$ is the Planck's constant, m^* is the effective mass of the electron, $V(z)$ is the potential energy of the electron, E is the energy of the electron and ψ is the wave function. Poisson's equation:

$$\frac{d}{dz}\left(\varepsilon\frac{d}{dz}\right)\phi(z) = -q\left[N_d^+(z) - n(z)\right] \tag{2}$$

where ε is the dielectric constant, $\phi(z)$ is the electrostatic potential, q is the charge, N_d^+ is the dissociative sender concentration and n is the electron concentration. Since the contribution of the acceptor charge as well as the holes on the total charge are very small, the acceptor charge and holes are neglected in the calculation.

Carrier continuity equation:

$$\frac{\partial n}{\partial t} = \frac{1}{q}\mathrm{div}\boldsymbol{J}_n + G_n - R_n \tag{3}$$

where J_n is the electron current density, G_n is the electron production rate and R_n is the electron complex rate.

The self-consistent procedure is that the potential V is obtained using Poisson's equation from an initial guess of the mobile charge concentration, and then inserted into the Schrödinger equation, which is solved to obtain the energy levels and wavefunctions of the systems. The new electron charge density is then calculated by applying Fermi statistics

$$n_{2D}(z) = \frac{m^* K_B T}{\pi\hbar^2}\sum_i |\psi_i(z)|^2 \ln\left[1 + e^{(E_F - E_i)/m^* K_B T}\right] \tag{4}$$

where E_F is Fermi level, E_i is energy of the ith quantized level, T is temperature and K_B is Boltzmann's constant.

The calculated density is then plugged into Poisson's equation and the iteration repeated until convergence is achieved.

The polarization model, Shockley–Read-Hall recombination model and Albrecht model incorporated for concentration dependent low field mobility were incorporated in the simulations study through Silvaco TCAD [27]. We applied a fixed temperature of 300 K due to the neglecting self-heating effect when the device is operated at low biases. The electron transport was also included based on a drift-diffusion model [28,29]. In the simulation, we set the value of saturation velocity and the parameters of electron and hole recombination lifetimes (taun0 and taup0) in SRH recombination model. Besides, we adopted the parameters for GaN, InN, and AlN that Braga et al. [30] have reported such as energy band structure, spontaneous polarization, mobilities and saturation velocities. Linear interpolations are adopted to compute parameter values as a function of mole fraction in InAlN.

To verify the validity of the simulation results in this work, we compare the simulation results with previous experimental report. As shown in Figure 2, simulated DC current of InAlN/GaN HEMT without additional surface charge is basically matched to the experimental results of the same structure [31,32].

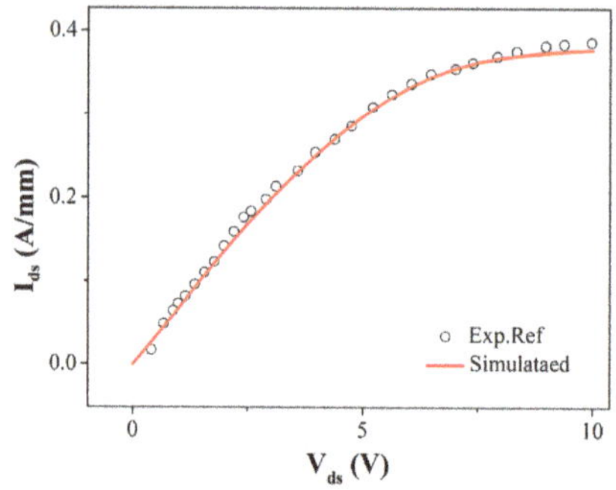

Figure 2. Output characteristics of InAlN/GaN HEMT from simulation and experimental results.

3. Results and Discussion

3.1. Influence of Surface Charge on Device Performance of InAlN/GaN HEMTs

Figure 3 shows the output characteristics and carrier concentration distributions of the InAlN/GaN HEMT with different surface charge density. As shown in Figure 3a, the saturation drain current I_{dsat} increases with the increase of positive surface charge density and decreases with the increase of negative surface charge density. The influence of negative surface charge on device performance is more significant than that of positive charge. In fact, the effect of surface charge on the I_{dsat} mainly resulted from the changes of 2DEG concentration. The carrier concentration distributions of the HEMT under gate region with different surface charge density are shown in Figure 3b. Corresponding to the change of I_{dsat}, the 2DEG concentration increases with the increase of positive surface charge, which decreases with the increase of negative surface charge.

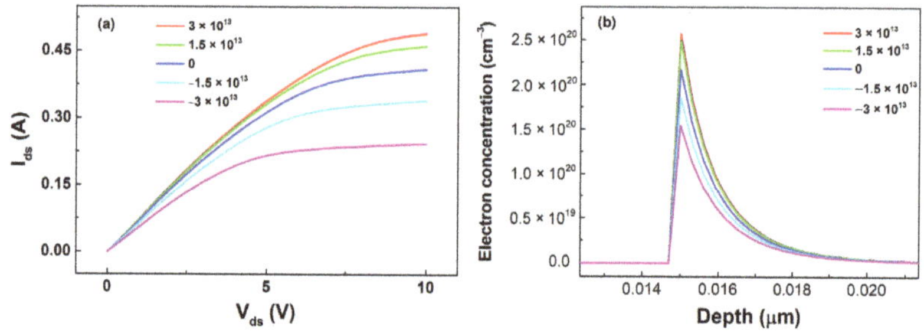

Figure 3. (**a**) The output characteristics and (**b**) carrier concentration distributions of the HEMT with different surface charge density at $V_g = 0$ V.

The intrinsic physical reason for the changes of I_{dsat} and 2DEG concentration with surface charge can be mainly attributed to the effect of the energy band structure and surface potential of the InAlN/GaN HEMT heterostructure. The conduction band diagram of the HEMT under gate region with different surface charge density is shown in Figure 3a. The corresponding surface potential representing energy difference between the conduction band edge of surface and Fermi level E_F is shown in Figure 3b. Considering the energy band diagram of the HEMT as sketched in Figure 3, the sheet charge density N_S can be written as [33]

$$N_s = \sigma_{pol,interface} - [\frac{\varepsilon_{barrier}}{d_{barrier} \cdot q}(\Phi_s + (E_F - \Delta E_c)] \tag{5}$$

where $d_{barrier}$ is the barrier thickness, q is the elementary charge, Φ_S is the surface potential, E_F is the Fermi level at the interface relative to the bottom of the GaN conduction band and ΔE_C is the conduction band offset between InAlN barrier layer and GaN buffer.

As shown in Figure 4, the surface potential of InAlN/GaN HEMT heterostructure under gate region increases slightly with positive surface charge. Correspondingly, the depth of triangular quantum potential well at InAlN/GaN interface becomes deeper, which enhanced the 2DEG confinement. So, the 2DEG concentration and I_{dsat} increase with the increase of positive surface charge density. As the negative surface charge density increases, the surface potential decreases and the depth of the triangular potential well becomes shallower, which weakens the confinement of the 2DEG at InAlN/GaN interface. So, the 2DEG concentration and I_{dsat} will decrease. In fact, the simulated sensing rule in this work is consistent with a previous experiment report that the output current of HEMT device decreases gradually as the pH value increases [11]. For HEMT pH sensor, the drain current decreased with increased pH value, because the adsorbed negative charges on HEMT sensing area increases with pH value.

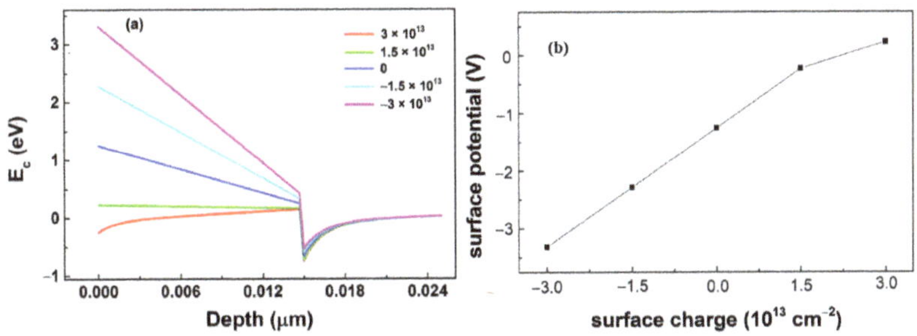

Figure 4. (**a**) The conduction band diagram and (**b**) surface potential of the HEMT under the gate region with different surface charge density.

3.2. The Effect of InAlN Barrier Thickness on the Sensing Performance of InAlN/GaN HEMT Device

To further study the surface sensing properties of HEMT, the effect of the InAlN barrier thickness on the surface sensing performance of InAlN/GaN HEMT devices is investigated by comparing the device performance before and after adding the same density of surface charges ($\sigma_S = -3 \times 10^{13}$ cm^{-2}). The InAlN barrier thicknesses ranged from 5 to 25 nm. Figure 5 shows the output characteristics and 2DEG concentration distributions of HEMT for varying InAlN thickness with and without additional surface charge. As shown in Figure 5, with the increase of barrier layer thickness, the I_{dsat} and 2DEG concentration increase when the additional surface charge is zero, which can be deduced from the energy band diagram [34,35]. Correspondingly, the energy drop in the barrier layer increases, leading to a deeper triangular quantum well at the hetero-interface. In further, the confinement of the quantum well on 2DEG is enhanced. Consequently, the 2DEG concentration and drain current also increase with the increase of barrier layer thickness.

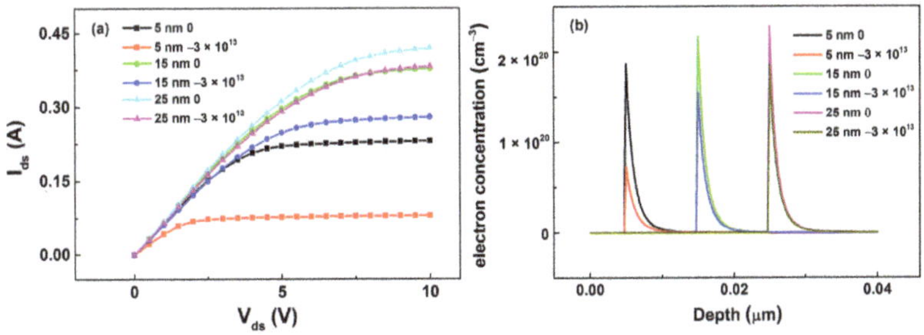

Figure 5. (**a**) The output characteristics and (**b**) carrier concentration distributions of HEMT with surface charge on the structure at different InAlN barrier layer thicknesses.

Additionally, from Figure 5 we can get the variation quantity of the saturation current ΔI_{dsat} and 2DEG concentration Δn before and after adding the same density surface charges for various InAlN thicknesses. The change rates of the $\Delta I_{dsat}/I_{dsat}$ and the $\Delta n/n$ are shown in Figure 6. In contrast to the variation of the I_{dsat} and 2DEG concentration, the ΔI_{dsat}, $\Delta I_{dsat}/I_{dsat}$, Δn and $\Delta n/n$ decrease gradually with the increase of InAlN thickness for HEMT. The change rates of the $\Delta I_{dsat}/I_{dsat}$ are approximately in the range of 0.1–0.65.

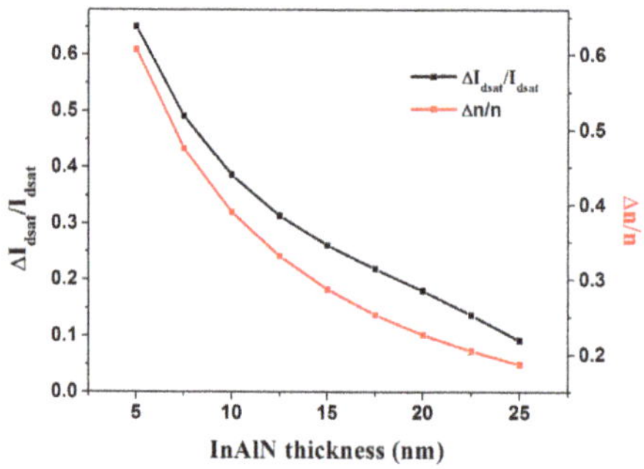

Figure 6. The change rates of the saturation drain current and the 2DEG concentration of the HEMT with different InAlN barrier thicknesses.

From the above results, it can be confirmed that the influence of surface charge on device performance increases with the decrease of the InAlN barrier thickness, which means that the modulation ability of surface charge on device performance will be increased by decreasing the InAlN thickness. So, we can conclude that the surface sensing sensitivity can be enhanced by thinning the barrier layer thickness in a certain range, when GaN-based HEMT is used as sensor.

The intrinsic physical mechanism for the above phenomenon and results is still resulted from the changes of the energy band structure and surface potential of the InAlN/GaN HEMT heterostructure. Figure 7 shows the energy band diagram of the HEMT under the gate region with and without additional surface charge for various InAlN barrier thicknesses. It can be seen from the diagram that with the increase of InAlN barrier thickness, the depth difference of the triangular potential well before and after additional surface charge decreases due to the enlarged distance between the sensing area and the 2DEG channel. Therefore, the influence of the same density surface charge on the interface potential well increases with the decrease of InAlN barrier thickness, which in turn affects the output characteristics and 2DEG concentration of the HEMT. So, the modulation ability of surface charge on device I_{dsat} and the 2DEG concentration increase with the decrease of InAlN thickness, which means that the sensing sensitivity of GaN-based HEMT sensor can be improved by thinning the barrier layer.

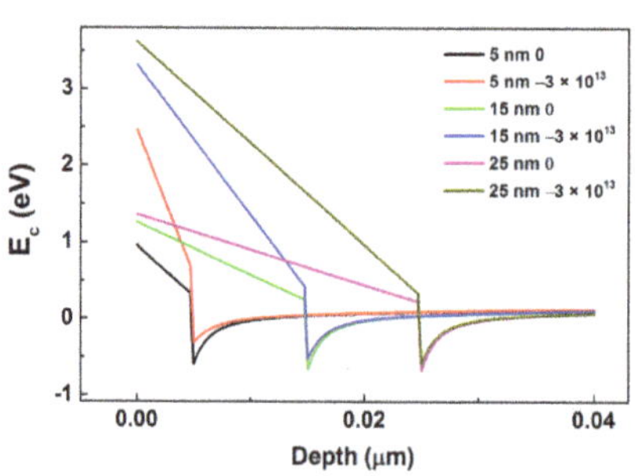

Figure 7. The conduction band diagram of HEMT for different InAlN barrier layer thicknesses with and without additional surface charge density.

4. Conclusions

In this work, the sensing mechanism of InAlN/GaN HEMT and the dependence of the sensing sensitivity on the InAlN barrier layer thickness were studied by solving the Schrodinger equation, Poisson's equation and the carrier continuity equation self-consistently. The results exhibit that both the I_{dsat} and 2DEG concentration increase with the increase of positive surface charge, which decrease with the increasing negative charge. The influence of negative surface charge on device performance is more significant than that of positive surface charge. This can be attributed to the change of the energy band structure and surface potential of HEMT with varying surface charge density. Moreover, the I_{dsat} and 2DEG concentration increase with the increase of InAlN barrier thickness, while the modulation ability of surface charge on device performance will decrease due to the enlarged distance between the sensing area and 2DEG channel. This means that the sensing sensitivity will decrease. So, when GaN-based HEMT is used as surface ion sensors, the barrier layer thickness should be as thin as possible.

Author Contributions: Conceptualization, Y.L. (Yanli Liu) and D.C.; formal analysis, X.H. and Y.D.; investigation, S.F. and Y.L. (Yuhui Liu); writing—original draft preparation, Y.L. (Yanli Liu); writing—review and editing, D.C. All authors have read and agreed to the published version of the manuscript.

Funding: This research was funded by the National Nature Science Foundation of China, grant number 61804089; Shandong Provincial Science and Technology Support Program of Youth Innovation Team in College, grant numbers 2019KJN041 and 2020KJN005; Basic Research Project of the Basic Research Business of the Provincial University in Heilongjiang Province, grant number 2021-KYYWF-0036; and the PhD Start-up Fund of Shandong Technology and Business University, grant number (BS201608).

Institutional Review Board Statement: Not applicable.

Informed Consent Statement: Not applicable.

Data Availability Statement: Not applicable.

Conflicts of Interest: The authors declare no conflict of interest.

References

1. Palacios, T.; Chakraborty, A.; Rajan, S.; Poblenz, C.; Keller, S.; DenBaars, S.P.; Speck, J.S.; Mishra, U.K. High-power AlGaN/GaN HEMTs for ka-band applications. *IEEE Electron. Device Lett.* **2005**, *26*, 781–783. [CrossRef]
2. Driad, S.B.; Maher, H.; Defrance, N.; Hoel, V.; De Jaeger, J.C.; Renvoise, M.; Frijlink, P. AlGaN/GaN HEMTs on Silicon Substrate With 206-GHz FMAX. *IEEE Electron. Device Lett.* **2013**, *34*, 36–38. [CrossRef]
3. Xu, R.; Chen, P.; Liu, M.H.; Zhou, J.; Li, Y.M.; Cheng, K.; Liu, B.; Chen, D.J.; Xie, Z.L.; Zhang, R.; et al. 3.4-kV AlGaN/GaN Schottky Barrier Diode on Silicon Substrate With Engineered Anode Structure. *IEEE Electron. Device Lett.* **2021**, *42*, 208–211. [CrossRef]
4. Fitch, R.C.; Walker, D.E.; Green, A.J.; Tetlak, S.E.; Gillespie, J.K.; Gilbert, R.D.; Sutherlin, K.A.; Gouty, W.D.; James, P.; Theimer, J.P.; et al. Implementation of High-Power-Density X-Band AlGaN/GaN High Electron Mobility Transistors in a Millimeter-Wave Monolithic Microwave Integrated Circuit Process. *IEEE Electron. Device Lett.* **2015**, *36*, 1004–1007. [CrossRef]
5. Sharma, N.; Mishra, S.; Singh, K.; Chaturvedi, N.; Chauhan, A.; Periasamy, C.; Kharbanda, D.K.; Parjapat, P.; Khanna, P.K.; Chaturvedi, N. High-resolution AlGaN/GaN HEMT-based electrochemical sensor for biomedical applications. *IEEE Trans. Electron. Devices* **2019**, *67*, 289–295. [CrossRef]
6. Ibbetson, J.P.; Fini, P.T.; Ness, K.D.; DenBaars, S.P.; Speck, J.S.; Mishra, U.K. Polarization effects, surface states, and the source of electrons in AlGaN/GaN heterostructure field effect transistors. *Appl. Phys. Lett.* **2000**, *77*, 250–252. [CrossRef]
7. Ould-Abbas, A.; Zeggaia, O.; Bouchaour, M.; Zeggai, H.; Sahouane, N.; Madani, M.; Trari, D.; Boukais, M.; Chabane-Sari, N.E. Study on functionalizing the surface of AlGaN/GaN high electron mobility transistor based sensors. *J. Optoelectron. Adv. Mater.* **2013**, *15*, 1323–1327.
8. Ahn, J.; Kim, D.; Park, K.H.; Yoo, G.; Heo, J. Pt-decorated graphene gate AlGaN/GaN MIS-HEMT for ultrahigh sensitive hydrogen gas detection. *IEEE Trans. Electron. Devices* **2021**, *68*, 1255–1261. [CrossRef]
9. Chaturvedi, N.; Singh, K.; Kachhawa, P.; Lossy, R.; Wuerfl, J. AlGaN/GaN HEMT based sensor and system for polar liquid detection. *Sens. Actuators A Phys.* **2020**, *302*, 111799. [CrossRef]

10. Woo, K.; Kang, W.; Lee, K.; Lee, P.; Kim, Y.; Yoon, T.S.; Cho, C.Y.; Park, K.H.; Ha, M.W.; Lee, H.H. Enhancement of cortisol measurement sensitivity by laser illumination for AlGaN/GaN transistor biosensor. *Biosens. Bioelectron.* **2020**, *159*, 112186. [CrossRef]
11. Dong, Y.; Son, D.H.; Dai, Q.; Lee, J.H.; Won, C.H.; Kim, J.G.; Chen, D.J.; Lee, J.H.; Lu, H.; Zhang, R.; et al. High sensitive pH sensor based on AlInN/GaN heterostructure transistor. *Sensors* **2018**, *18*, 1314. [CrossRef] [PubMed]
12. Varghese, A.; Periasamy, C.; Bhargava, L. Fabrication and charge deduction based sensitivity analysis of GaN MOS-HEMT device for glucose, MIG, C-erbB-2, KIM-1, and PSA detection. *IEEE Trans. Nanotechnol.* **2019**, *18*, 747–755. [CrossRef]
13. Cho, S.K.; Cho, W.J. High-Sensitivity pH sensor based on coplanar gate AlGaN/GaN metal-oxide-semiconductor high electron mobility transistor. *Chemosensors* **2021**, *9*, 42. [CrossRef]
14. Nigam, A.; Sharma, N.; Tripathy, S.; Kumar, M. Development of semiconductor based heavy metal ion sensors for water analysis: A review. *Sens. Actuators A Phys.* **2021**, *330*, 112879. [CrossRef]
15. Sokolovskij, R.; Zheng, H.Z.; Li, W.M.; Zhou, G.N.; Wang, Q.; Zhang, G.Q.; Yu, H.Y. Application of a gateless AlGaN/GaN HEMT sensor for diesel soot particulate detection. *Sens. Actuators B Chem.* **2021**, *349*, 130811. [CrossRef]
16. Jia, X.L.; Chen, D.J.; Liu, L.; Lu, H.; Zhang, R.; Zheng, Y.D. Highly selective and sensitive phosphate anion sensors based on AlGaN/GaN high electron mobility transistors functionalized by ion imprinted polymer. *Sci. Rep.* **2016**, *6*, 27728. [CrossRef] [PubMed]
17. Sokolovskij, R.; Zhang, J.; Zheng, H.; Li, W.M.; Jiang, Y.; Yang, G.Y.; Yu, H.Y.; Sarro, P.M.; Zhang, G.Q. The Impact of Gate Recess on the H_2; Detection Properties of Pt-AlGaN/GaN HEMT Sensors. *IEEE Sens. J.* **2020**, *20*, 8947–8955. [CrossRef]
18. Lee, C.T.; Chiu, Y.S. Gate-recessed AlGaN/GaN ISFET urea biosensor fabricated by photoelectrochemical method. *IEEE Sens. J.* **2015**, *16*, 1518–1523. [CrossRef]
19. Ding, X.Z.; Miao, B.; Gu, Z.Q.; Wu, B.J.; Hu, Y.M.; Wang, H.; Zhang, J.; Wu, D.M.; Lu, W.H.; Li, J.D. Highly sensitive extended gate-AlGaN/GaN high electron mobility transistor for bioassay applications. *RSC Adv.* **2017**, *7*, 55835–55838. [CrossRef]
20. Pyo, J.Y.; Jeon, J.H.; Koh, Y.; Cho, C.Y.; Park, H.H.; Park, K.H.; Lee, S.W.; Cho, W.J. AlGaN/GaN high-electron-mobility transistor pH sensor with extended gate platform. *AIP Adv.* **2018**, *8*, 85106. [CrossRef]
21. Hao, Y.; Xue, J.S.; Zhang, J.C. Research progress in the group-III nitride InAlN semiconductorheterostructures. *Sci. China Inform. Sci.* **2012**, *42*, 1577–1587.
22. Tülek, R.; Ilgaz, A.; Gökden, S.; Teke, A.; Öztürk, M.K.; Kasap, M.; Özçelik, S.; Arslan, E.; Özbay, E. Comparison of the trans-port properties of high quality AlGaN/AlN/GaN and AlInN/AlN/GaN two-dimensional electron gas heterostructures. *J. Appl. Phys.* **2009**, *105*, 13707. [CrossRef]
23. Jia, X.L.; Huang, X.Y.; Yin, T.; Yang, L.H.; Chen, D.J. Ultrasensitive Detection of Phosphate Using Ion-Imprinted Polymer Functionalized AlInN/GaN High Electron Mobility Transistors. *IEEE Electron. Device Lett.* **2016**, *37*, 913–915. [CrossRef]
24. Chapin, C.A.; Miller, R.A.; Dowling, K.M.; Chen, R.Q.; Senesky, D.G. InAlN/GaN high electron mobility micro-pressure sensors for high-temperature environments. *Sens. Actuator A Phys.* **2017**, *263*, 216–223. [CrossRef]
25. Kong, Y.C.; Zheng, Y.D.; Chu, R.M.; Gu, S.L. Influence of Al-content on the property of the two-dimensional electron gases in AlxGa1-xN/GaN heterostructures. *Acta Phys. Sin.* **2003**, *52*, 1756–1760. [CrossRef]
26. Amarnath, G.; Panda, D.K.; Lenka, T.R. Modeling and simulation of DC and microwave characteristics of AlInN (AlGaN)/AlN/GaN MOSHEMTs with different gate lengths. *Int. J. Numer. Modell. Electron. Netw. Devices Fields* **2019**, *32*, e2456. [CrossRef]
27. Mohammadzamani, M.; Fathipour, M. An investigation of highly scaled III-nitride Gallium-face and Nitrogen-face HEMTs. *Superlattices Microstruct.* **2015**, *78*, 125–133. [CrossRef]
28. Albrecht, J.D.; Wang, R.P.; Ruden, P.P.; Farahmand, M.; Brennan, K.F. Electron transport characteristics of GaN for high temperature device modeling. *J. Appl. Phys.* **1998**, *83*, 4777–4781. [CrossRef]
29. Chang, Y.; Tong, K.Y.; Surya, C. Numerical simulation of current–voltage characteristics of AlGaN/GaN HEMTs at high temperatures. *Semicond. Sci. Technol.* **2005**, *20*, 188. [CrossRef]
30. Braga, N.; Mickevicius, R.; Gaska, R.; Hu, X.; Shur, M.S.; Khan, M.A.; Simin, G.; Yang, J. Simulation of hot electron and quantum effects in AlGaN/GaN heterostructure field effect transistors. *J. Appl. Phys.* **2004**, *95*, 6409–6413. [CrossRef]
31. Higashiwaki, M.; Matsui, T. MBE growth and device characteristics of InAlN/GaN HFETs. *Phys. Status Solidi C* **2005**, *2*, 2598–2601. [CrossRef]
32. Kanaga, S.; Dutta, G.; Kushwah, B.; DasGupta, N.; DasGupta, A. Low Temperature and High Pressure Oxidized Al_2O_3 as Gate Dielectric for AlInN/GaN MIS-HEMTs. *IEEE Trans. Device Mater. Reliab.* **2020**, *20*, 613–621. [CrossRef]
33. Ambacher, O.; Smart, J.; Shealy, J.R.; Weimann, N.G.; Chu, K.; Murphy, M.; Schaff, W.J.; Eastman, L.F.; Dimitrov, R.; Wittmer, L.; et al. Two-dimensional electron gases induced by spontaneous and piezoelectric polarization charges in N-and Ga-face AlGaN/GaN heterostructures. *J. Appl. Phys.* **1999**, *85*, 3222–3233. [CrossRef]
34. Jogai, B. Influence of surface states on the two-dimensional electron gas in AlGaN/GaN heterojunction field-effect transistors. *J. Appl. Phys.* **2003**, *93*, 1631–1635. [CrossRef]
35. Behmenburg, H.; Khoshroo, L.R.; Eickelkamp, M.; Mauder, C.; Fieger, M.; Ketteniss, N.; Woitok, J.; Wamwangi, D.; Wuttig, M.; Hernandez, S.E.; et al. Influence of barrier thickness on AlInN/AlN/GaN heterostructures and device properties. *Phys. Status Solidi C* **2009**, *6*, S1041–S1044. [CrossRef]

crystals

Review

Recent Progress on AlGaN Based Deep Ultraviolet Light-Emitting Diodes below 250 nm

Chunyue Zhang [1,2], Ke Jiang [1,2,*], Xiaojuan Sun [1,2] and Dabing Li [1,2,*]

1 State Key Laboratory of Luminescence and Applications, Changchun Institute of Optics, Fine Mechanics and Physics, Chinese Academy of Sciences, Changchun 130033, China
2 Center of Materials Science and Optoelectronics Engineering, University of Chinese Academy of Sciences, Beijing 100049, China
* Correspondence: jiangke@ciomp.ac.cn (K.J.); lidb@ciomp.ac.cn (D.L.); Tel.: +86-0431-8617-6345 (K.J.); +86-0431-8617-6345 (D.L.)

Abstract: AlGaN based deep ultraviolet (DUV) light-emitting diodes (LEDs), especially with a wavelength below 250 nm, have great application potential in the fields of sterilization and disinfection, gas sensing, and other aspects. However, with the decrease of emission wavelength, performance collapse occurs and the external quantum efficiencies (EQE) of sub-250 nm LEDs are usually below 1% for a long time. Low efficiencies are resulted from problem accumulation of all aspects, including n/p-type doping and contacts, carrier confinements and transports, light extraction, etc. To achieve high EQE of sub-250 nm LEDs, problems and solutions need to be discussed. In this paper, the research progress, development bottlenecks, and corresponding solutions of sub-250 nm LEDs are summarized and discussed in detail.

Keywords: UVC-LED; AlGaN; doping; carrier confinement; light extraction

Citation: Zhang, C.; Jiang, K.; Sun, X.; Li, D. Recent Progress on AlGaN Based Deep Ultraviolet Light-Emitting Diodes below 250 nm. *Crystals* **2022**, *12*, 1812. https://doi.org/10.3390/cryst12121812

Academic Editors: Peng Chen, Zhizhong Chen and Shin-Tson Wu

Received: 21 November 2022
Accepted: 9 December 2022
Published: 13 December 2022

Publisher's Note: MDPI stays neutral with regard to jurisdictional claims in published maps and institutional affiliations.

1. Introduction

The ultraviolet (UV) spectrum can be divided into UVA (315–400 nm), UVB (280–315 nm), UVC (200–280 nm), and VUV (10–200 nm) of which the far-UVC band is from 200 to 250 nm. Mercury lamp is currently the most important UV light source on the market. However, the toxicity of mercury makes it urgent to find alternative UV light sources. AlGaN based LEDs have many advantages, including the light weight, small size, long lifetime, low power consumption, easy integration, fast switching, and tunable emission wavelength from 365 to 210 nm through adjusting the alloy composition. These characteristics make AlGaN based LED the unique and most potential light source in the UV range. Some applications of AlGaN based UV LEDs are summarized in Figure 1a. In addition to the ordinary applications, such as medical treatment, gas sensing, non-line-of-sight communication, and charge management system [1–6], as shown in Figure 1b, AlGaN based sub-250 nm LED is more suitable for safe and highly efficient sterilization due to its high radiation energy and strong absorption by the stratum corneum that does not contain cell nuclei [7], as shown in Figure 1c. Recent research proved that at the biocidal doses for multi-resistant pathogens at 40 to 60 mJ/cm^2 irradiation of 222 and 233 nm irradiation, no relevant DNA damage and radical formation occurred in the skin [8,9].

However, AlGaN based sub-250 nm LEDs are faced with many bottlenecks at present. It has been reported that the EQE drops from 0.6% to 0.00013% when the emission wavelength decreases from 264 to 220 nm [10]. Additionally, when the emission wavelength decreases from 239 to 217 nm, the EQE drops by more than three orders of magnitude [11]. There are many reasons for the sharp drop in quantum efficiency in this wave band, most of which are the effects brought about by the increase of Al composition. Generally, to achieve an emission below 250 nm, the Al composition of quantum well must be higher than 62%. Inevitably, the other function layers, such as the buffer layer, n-type electron injection layer, quantum barrier layers, electron block layer, and p-type hole injection layer, must provide even higher Al composition, usually higher than 70% [12]. The super-high Al composition introduces many more serious problems than UVA-, UVB-, and above-250 nm UVC-LEDs. This paper first summarizes the development status of sub-250 nm LEDs and then introduces the problems that limit the efficiency improvement from the aspects of n/p-type doping, carrier confinement and transport, and light extraction and solutions that have been provided nowadays.

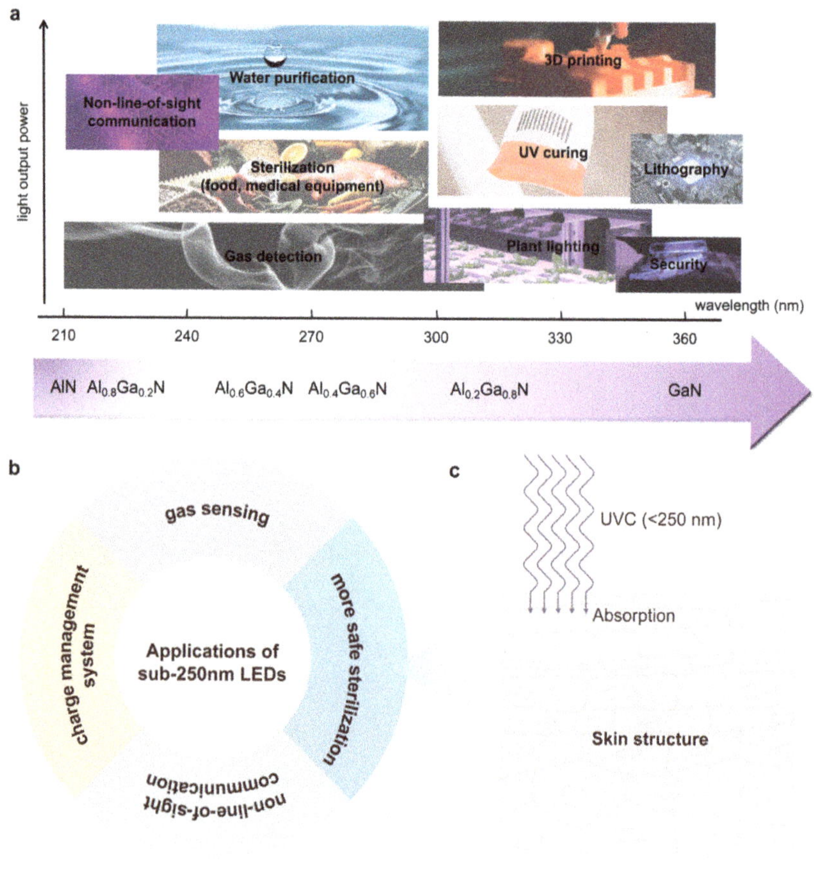

Figure 1. Partial applications of AlGaN-based LEDs of (**a**) UV band and (**b**) below 250 nm. Images are partially from internet. (**c**) Schematic of sterilization on the skin by high UVC below 250 nm.

2. Advances in DUV-LEDs below 250 nm

On the whole, the performance level of sub-250 nm LEDs is much lower than that of the above-250 nm LED. The internal quantum efficiency (IQE), EQE, wall-plug efficiency (WPE), and light out-put power (LOP) have all dropped significantly. Up to now, the maximum EQE of 275 nm LED is 20.3% [13], the maximum EQE of 265 nm LED is 10% [14], while the majority of reported EQEs of sub-250 nm LEDs are below 1%. In 2006, Taniyasu et al. [15] obtained both n- and p-type AlN, and developed p-AlN/undoped AlN/n-AlN pin LED structure with n-AlN/AlGaN and p-AlN/AlGaN superlattices as the electron and hole lateral conduction layers, realizing emission at 210 nm. This is the first report of successful realization of AlGaN based 210 nm emission, confirming the possibility of full band coverage of AlGaN based LEDs. However, the device performance is extremely low, with the operation voltage of 45 V at 20 mA, LOP and EQE of 0.02 μW and 10^{-6}% at 40 mA, respectively. In 2009, Hirayama et al. [16] fabricated an LED on a multilayer AlN buffer layer. They used a thin quantum well of 1.3–2 nm to suppress the polarization field in the well. At 100 mA, the peak wavelength located at 227 nm and the LOP and EQE were 0.15 mW and 0.2%, respectively. Next year, they [17] further improved the LED efficiency by introducing multi-quantum-barrier (MQB) electron blocking layers (EBL), providing an important scheme for preventing electron leakage of sub-250 nm LEDs. They achieved an LOP of 4.8 mW at 135 mA under CW operation for the 250 nm LED, corresponding to an EQE of 1.18%, which is still the highest EQE reported for 250 nm and sub-250 nm LEDs.

Recent years, the performance of sub-250 nm LEDs has been further improved. In 2020, Lobo-Ploch et al. [18] optimized the growth and fabrication processes including using epitaxially laterally overgrown (ELOG) sapphire substrates and vanadium-based low resistance n-contacts. They fabricated a 233 nm LED with an EQE of 0.36% at 100 mA. Recently, Jo et al. [19] demonstrated a 228 nm LED on sapphire substrates with an LOP of 1.4 mW at 150 mA by modifying the multiple quantum well (MQW) and spacer layer structures.

The method of using non-polar plane have also been explored [20]. Although the light extraction efficiency (LEE) was estimated to be 25 times higher than that on the polar plane, the EL intensity was approximately 70 times lower than that on the polar plane, which probably resulted from the higher defect density. New design concepts may break through the transparency and conductivity problem of sub-250 nm LEDs. Liu D. et al. [21] demonstrated a 229 nm LED adopting a p-type Si nanomembrane as both a p-contact and a hole injection layer. LOP was measured to be 0.16 mW under 100 mA current and EQE was calculated to be 0.03%. Next year, they demonstrated a 226 nm LED with a p-Si nanomembrane and achieved an EQE of 0.2% [22], which was the highest value among LEDs with emission wavelength below 230 nm in the CW mode.

Strong quantum-confinement GaN/AlN heterostructures can produce optical emission far above the bulk bandgap of GaN (3.4 eV), and offer another path to achieve DUV emission below 250 nm. Furthermore, it has been reported that delta-QW design could resolve the band-mixing issue and ensure large transverse electric (TE) spontaneous radiation rate [23]. S.M. Islam et al. [24] demonstrated that ultrathin (1–2 monolayers) GaN dots/disks embedded in an AlN matrix can achieve controllable emission in the 219–235 nm range, showing the shortest wavelength emission of 219 nm with an IQE of 40% from binary GaN active regions till date.

AlGaN based far-UVC LEDs grown by homogeneous epitaxy were also reported due to development of AlN single-crystal substrate. Toshio Nishida et al. [25] fabricated UV LED structures on AlN substrates as early as 2004, and compared the performance with LEDs on sapphire substrates. They found that the saturation injection current of LEDs on AlN substrate was two times higher than that of LEDs on sapphire substrate. This work confirmed the huge potential of bulk AlN substrates for UV-LEDs. In 2020, Yoshikawa et al. [26] reported the fabrication of 230–237 nm LEDs on AlN substrates. Thanks to the high-quality single crystal AlN substrate, the 233 nm LED exhibited EQE of 0.25% at 233 nm. Compared to sapphire and AlN substrates, Si substrate has several advantages [27], including low cost, good electrical and thermal conduction, and easy integration to Si-based electronics, etc. However, to obtain DUV LEDs on Si substrate has some problems that AlGaN epilayers grown on Si possess poor material quality owing to the large lattice and thermal mismatches between AlGaN and Si. The nanowire technology may solve the problem [28]. In 2015, S. Zhao et al. [29] achieved the first 210 nm emitting AlN nanowire LEDs with a turn on voltage of approximately 6 V by utilizing N-polar AlGaN nanowires directly grown on Si substrate. The IQE of the N-polar AlN nanowires LED can reach 80%.

Another important aspect of an LED are the reliability and lifetime. Lifetime in the order of 10,000 h for the long-wavelength UV LED (UVA and UVB) have been commercialized [30], while the lifetime for the long-wavelength UVC LED (>250 nm) is still limited to thousands of hours [31,32] although Ruschel et al. [31] demonstrated a 265 nm UVC LED with L70 lifetime more than 10,000 h. Meneghini et al. [33] pointed out that the stress induced nonradiative recombination increase and point defects in or around the active region might cause output power degradation. However, at present, there are few studies on the reliability and lifetime of sub-250 nm LEDs. The reported record lifetime of sub-250 nm LEDs was created by Yoshikawa et al. [26], which was expected to exceed 3600 h. The authors speculated that the possible reason for the long lifetime was the lower oxygen content in the novel structure. Glaab et al. [34], in detail, analyzed two possible degradation mechanisms of sub-250 nm LEDs. The first mechanism may be the activation of initially H-passivated defect complexes in or around the quantum well active region by hot carriers from Auger recombination, which may be responsible for the rapid decrease of LOP during the initial 50 to 100 h. The second mechanism may be the carrier injection efficiency reduction triggered by hot carriers from Auger recombination, which may be responsible for the LOP degradation after working more than 50 h.

The EQE, LOP and WPE of 250 nm and sub-250 nm LEDs are summarized in Figure 2a–c, respectively. Up to now, the research on AlGaN based sub-250 nm LED is still relatively few, compared to other UV wave bands. There are even no reports that demonstrated a WPE over 1%, implying a long way to realize large-scale commercial application. However, as it can been seen, the performance of recently reported sub-250 nm LEDs have increased by more than one magnitude over those reported several years ago, indicating the future of the commercialization of AlGaN based sub-250 nm LED is expectable.

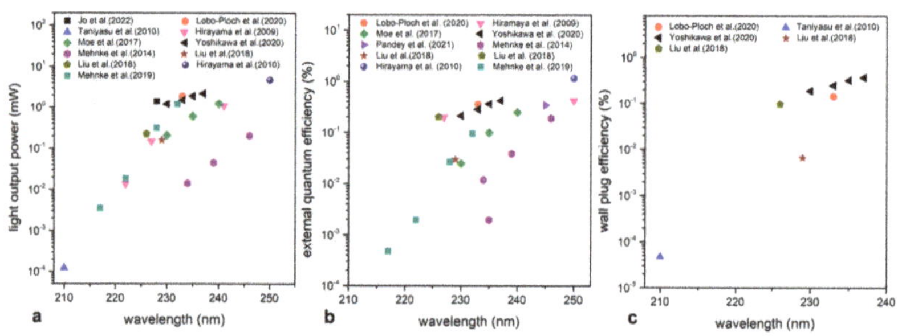

Figure 2. Summary of the reported (**a**) EQE, (**b**) LOP, and (**c**) WPE of 250 nm and sub-250 nm LEDs in dependence of emission wavelength in recent literature [11,16–22,26,35–37].

3. Key Issues and Challenges

To analyze the key issues and challenges limiting the performance of AlGaN-based sub-250 nm LEDs, it is usually necessary to discuss from the two aspects of material growth and device physics. Generally, the technology roadmap of AlGaN-based sub-250 nm LEDs follows the ordinary UVC LEDs, which means the typical device structure and fabrication process of a sub-250 nm LED are basically identical to an UVC one. Figure 3a is a schematic diagram of the basic structure of an AlGaN-based UV LED. Typically, the epitaxial structure from bottom to top includes (1) substrate, generally sapphire or single crystal AlN, (2) AlN or AlGaN buffer layer, to reduce defects density and relax stress during growth, (3) n-AlGaN layer, to form n-type ohmic contact and provide electrons to active region for recombination, (4) active region, the main light emission region and usually made up of 3–5 periods of undoped $Al_xGa_{1-x}N/Al_yGa_{1-y}N$ quantum wells ($x < y$), (5) EBL, with a wider bandgap than all other function layers to suppress the electrons to overflow to p-side, (6) p-AlGaN, to inject holes into active region for recombination, and (7) p-GaN cap layer, to form a p-type ohmic contact. From the perspective of materials, the bottom buffer layer quality that basically determines the upper epi-layer quality and the n/p doping efficiency that determine the carrier injection efficiency will significantly influence the device performance. Figure 3b is the schematic energy band structure of an AlGaN-based UV LED. The electrons and holes transport to the active region and recombine to each other to emit UV photons. If the carriers cannot effectively transport to the active region and be confined, the radiative recombination must be very weak. To say the least, even though the carrier recombination generates a large number of photons, the photons cannot be effectively extracted, and the device performance will be very low, too.

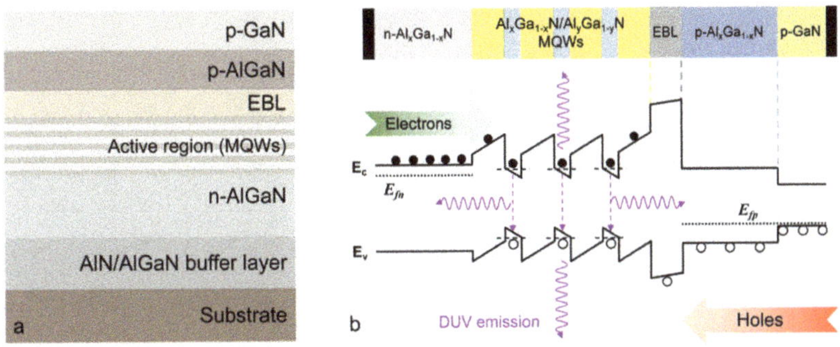

Figure 3. Schematic diagram for (**a**) epitaxial structure and (**b**) energy band structure of a typical AlGaN-based UV LED, which are suitable for sub-250 nm LEDs, too.

Quantitatively, if we choose the WPE as the key parameter to assess the performance of an LED, the relationship between the device performance and the main parameters can be expressed as the following Formula (1):

$$\mathrm{WPE} = \frac{\mathrm{LOP}}{\mathrm{I \cdot V}} = \mathrm{EQE \cdot EE} = \mathrm{IQE \cdot LEE \cdot EE} = \mathrm{RRE \cdot CIE \cdot LEE \cdot EE} \quad (1)$$

where I and V are the operation current and voltage, EE is the electrical efficiency, RRE is the radiative recombination efficiency, CIE is the carrier injection efficiency, respectively. The ABC model [38] is often used to estimate the RRE by the following Formula (2):

$$\mathrm{RRE} = \frac{\mathrm{B}n^2}{\mathrm{A}n + \mathrm{B}n^2 + \mathrm{C}n^3} \quad (2)$$

where the coefficients A, B, and C represent the SRH non-radiative recombination coefficient (Shockley-Read-Hall, SRH), radiative recombination coefficient, and Auger recombination coefficient, and n is the carrier concentration, respectively. Apparently, the material crystal quality and carrier distribution will influence the performance through RRE, the doping efficiency and contact quality can mainly determine the performance through EE and CIE, and furthermore the LEE will directly determine the device performance. Consequently, the key issues and challenges including material crystal quality, n/p-type doping, carrier transport and confinement, and light extraction will be discussed in this part.

3.1. Material Crystal Quality

Guttmann et al. [10] have studied the reasons for the decrease of EQE at 264–220 nm. As shown in Figure 4a, with the wavelength decreasing, the trend of EQE is consistent with the trend of IQE, implying to improve the IQE becomes even more urgent for short wavelength range. As can be seen from the ABC model, IQE and non-radiative recombination process are inseparable. Due to the mismatch of lattice constants and thermal expansion coefficients between the AlGaN epilayer and substrates, dislocation density is generally high. Ban et al. [39] explored the relationship between IQE and dislocation densities (DDs) in the $Al_xGa_{1-x}N/Al_yGa_{1-y}N$ active region with different Al compositions, and found that under the condition of the same dislocation density and injected carrier density, the active region has the same IQE regardless of the emission wavelength. The functional relationship between IQE and DDs is depicted in Figure 4b. It can be seen from the figure that the dislocation density needs to be less than 10^7 cm^{-2} to achieve an IQE of over 95%.

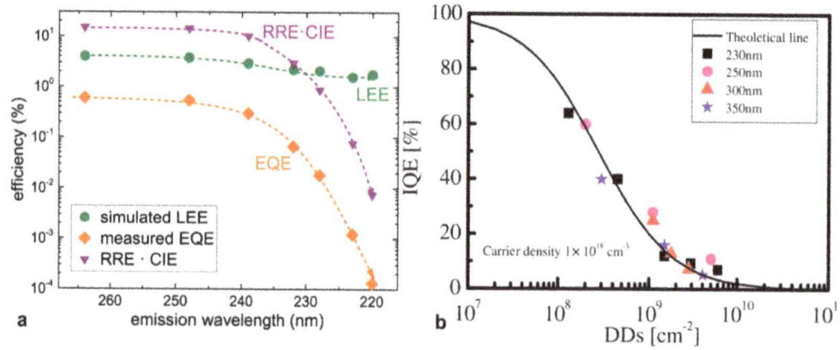

Figure 4. (**a**) Measured on-wafer EQE, simulated on-wafer LEE and corresponding product of RRE and CIE as a function of the emission wavelength [10]. (**b**) IQE as the function of DD at a weak excitation with the carrier density of 1×10^{18} cm^{-3} [39].

Before growing the LED structure, an AlN buffer layer is first grown on the sapphire substrate. The AlN buffer layer quality largely determines the quality of the upper LED structure. At the present stage, several technologies are commonly used to reduce dislocation density in the AlN/Sapphire template.

1. Two-step growth method [40–43]. It is used to alleviate the mismatch between AlN and sapphire substrate and improve the AlN quality. The first step is to form AlN crystal islands under high V/III ratio and low temperature. The second step is to grow AlN buffer layer with low threading dislocation density (TDD) and smooth surface through enhanced Al atom surface migration and lateral growth. The mechanism for reducing TDD is the transition from 3D to 2D growth mode that facilitates dislocation bending and annihilation along the growth direction. Hasan et al. [40] recently reported 4 um-thick AlN layers with total dislocation density of 1.1×10^9 cm^{-2}. Two-step growth exists the issue that the crystal islands may change their crystal orientation under the actions of airflow and rotation, etc., and thus generate threading screw dislocations. Three-step growth based on two-step growth was proposed [42]. The additional growth step is inserting a morphology control layer between the buffer layer and the step-flow-growth layer by adjusting the ammonia gas flow rate to a certain value. The twist angle of island could be controlled and the threading edge dislocation lines are more likely to form a loop and annihilate. The edge dislocation can drop to the order of 10^7 cm^{-2};
2. Epitaxial lateral overgrowth (ELOG) method [44–50]. To be specific, the low DD is caused by filtration related to patterns on sapphire substrate and annihilation related to the formation of a loop structure during growth [47]. Nakano et al. [49] first demonstrated the ELOG-AlN layers on trenched-patterned sapphire substrates and achieved a dislocation density of 6.7×10^8 cm^{-2}. To further enhanced the coalescence of ELOG-AlN, Imura et al. [48] grew ELOG-AlN layers on patterned AlN templates grown on sapphire substrates and achieved a TDD value of less than 10^7 cm^{-2}. Recently, Liu et al. [50] proposed a preset strain modulation method on nano-patterned AlN templates by introducing HTA pretreatment. The XRC FWHM values of the (0002) and $(10\bar{1}2)$ reflections were significantly dropped from 190 and 311 arcsec to 140 and 228 arcsec, respectively. At present, ELOG combined with HTA is believed to be the most effective approach for the growth of AlN layers in terms of crystal quality, cost and stability;
3. High-temperature annealing (HTA) of sputtered AlN on Sapphire [51–56]. Miyake et al. [55] first investigated the annealing of an AlN layer grown on a sapphire substrate and achieved a TDD of 4.7×10^8 cm^{-2}. Annealing at temperatures above 1500 °C is effective in improving the quality of AlN layers. Due to the evolution mechanism

of strains induced by HTA, it possesses high repeatability and stability and exhibits great potential in realizing mass production. The mechanism for reducing TDD after HTA was discussed in detail by Ben et al. [52]. Under TH, adjacent dislocations with different Burgers Vectors are more likely to form voids, which provide the inner surface for dislocations to terminate. The TDD below 10^7 cm^{-2} can be achieved through this method;

4. Migration-enhanced epitaxy (MEE) method [57–61]. MEE is achieved by alternative supply of the TMAl and NH_3 sources. The Al atom migration can be enhanced and the gas phase reaction can be prevented in the meantime. When NH_3 is not supplied, there are few N atoms on the epitaxial surface and the migration of Al atoms can be enhanced. Based on this route, NH_3 pulsed-flow method and modified MEE were proposed [58–60]. Hirayama et al. [58] fabricated a high-quality AlN buffer by NH_3 pulse-flow growth technique for the first time. The NH_3 pulse-flow growth is realized by firstly injecting continuous TMAl flow during the NH_3 pulse-flow sequence, secondly introducing continuous-flow mode to reduce the surface roughness, and then repeating the pulse- and continuous-flow modes. Li et al. [56] recently realized growing AlN layers under a relatively fast growth rate of 0.98 μm/h and achieved flat surface morphology (RMS = 0.5 nm) by NH_3-pulse flow mode with a small duty ratio of NH_3. A modified MEE is characterized by a combination of simultaneous source supply and conventional MEE, consisting of three different growth sequences including simultaneous, alternating supply and a combination of the two. Banal et al. [61] utilized the modified MEE to control the initial nucleation and achieved the XRC FWHM values of ~45 and ~250 arcsec for the (0002) and $(10\bar{1}2)$ planes;
5. Interlayer method [62–68]. The inserting layers work by stress management and dislocation filter. Common insertion layers include medium-temperature (MT) interlayer [62–65], superlattices [66,67], graphene [68], etc. The introduction of a MT interlayer could lead to the formation of high-density self-organized nano-voids, which can alleviate tensile stress and reduce TDD due to promoting dislocation termination and bending at macrosteps at the same time. He et al. [64] recently successfully fabricated a 5.6 μm-thick AlN film by employing a MT interlayer (930 °C). The tensile stress in the AlN film had a reduction of 64% through this method and the TDD showed an extremely low value of 4.7×10^7 cm^{-2}. Wang et al. [66] grew high quality thick AlGaN on sapphire by inserting ten periods AlN (8 nm)/$Al_{0.2}Ga_{0.8}N$ superlattices (SLs), and the TDD was reduced by two orders of magnitude.

For substrate selection, AlN single-crystal substrate is an ideal substrate for AlN and Al-rich AlGaN growth with low extended defect density and low non-radiative recombination centers [69–71]. The DD of AlGaN grown on AlN single crystal can be lower than 10^3 cm^{-2}, which provides a significant progress for the application of ultrawide-bandgap AlGaN. However, with the development of ELOG and HTA technologies, the AlN template DD grown on sapphire substrate has reached very low magnitude of 10^8 cm^{-2} [48], allowing the better performance of sub-250 nm LEDs grown on cost-effective sapphire substrate. Although, at present, the DD of LEDs grown on sapphire substrates is four orders of magnitude larger than that of LEDs grown on single-crystal AlN, the EQE of sub-250 nm LEDs grown on sapphire are similar to those grown on single-crystal AlN [18].

3.2. n-AlGaN with High Al Composition

The contact resistivity and sheet resistivity of the n-AlGaN layer, which influence the electron injection efficiency, are depended on the Si doping efficiency. The ionization energy (E_a) of Si is approximately 15 meV when Al composition is smaller than 80% [72], therefore the free electron concentration can achieve the order of 10^{19} cm^{-3} and the contact resistance and sheet resistance are within acceptable ranges. However, when the Al content is higher than 80% which is essential for sub-250 nm LED, Si doping faces severe difficulties. First, the Si E_a sharply increases with Al composition and can reached to 250 meV in

AlN [72,73], as shown in Figure 5a. In principle, E_a can be reduced by increasing the doping concentration to form impurity bands, yet self-compensation effect will reduce the free electron concentration when the Si doping concentration exceeds the critical doping dose, and the conductivity of the film will drop drastically, which is called a "knee behavior" [74], as shown in Figure 5b. Furthermore, it has been reported that the rapid decrease in conductivity also correlates with the transformation of Si from shallow donor to *DX* center [74]. The name of *DX* center derives from the fact that it is originally thought to be a deep donor (*D*) level associated with an unknown impurity (*X*) [75]. It is a highly localized center of negative charge related to two donor atoms, whose generation can be described by a chemical-like Equation (3)

$$2d^0 \rightarrow DX^- + d^+ \tag{3}$$

where d^0 and d^+ represent neutral and ionized substituted shallow donor atoms, respectively. The *DX* centers are thought to resulting from the breaking and displacement of covalent bonds of Si atoms and nearest neighbor metal atoms in wurtzite crystals.

Low self-compensation can be achieved via V_{Al}-nSi_{Al} complexes by defect quasi-Fermi level control (dQFL), chemical potential control (CPC), and the optimization of growth conditions including reactor pressure and temperature, etc. [76–78]. Non-equilibrium process such as Si implantation can inhibit the formation of *DX* centers, providing higher free carrier concentrations. In 2020, Breckenridge et al. [79] implanted Si into homoepitaxial AlN and performed a low thermal budget damage recovery and activation process, achieving an n-AlN with conductivity of 0.05 $\Omega^{-1}\cdot cm^{-1}$ at room temperature, as shown in Figure 5c. Later, in 2021, Breckenridge et al. [76] further realized Si-implanted n-AlN with higher conductivity (>1 $\Omega^{-1}\cdot cm^{-1}$) and carrier concentration (5×10^{18} cm^{-3}) on low TDD AlN substrate, maintaining the Si dopant in a shallow state via a non-equilibrium annealing process and dQFL control method.

For the purpose of enhancing the WPE, the contact resistivity of the n-electrode must be reduced. However, the n-type ohmic contact with high Al composition faces with physical limitations that the electron affinity of AlGaN decreases from 3.18 eV for GaN to 1.01 eV for AlN [80], giving rise to the Schottky barrier increase at the metal semiconductor (MS) interface with Al content. Figure 5d summarized the electrical characteristics of n-contacts by carefully optimizing the annealing conditions on n-AlGaN with Al content from 0.75 to 0.95 [81]. At present, suitable low work function metals have not been found. However, vanadium can form thermal stable nitrides with low work functions and is considered as an alternative metal to Ti. When Al mole fraction is greater than 0.6, the common Ti/Al-based n-electrode should be replaced by V/Al-based contacts [82]. Moe et al. [35] adopted Ti/Al/Ni/Au on n-$Al_{0.87}Ga_{0.13}N$ as the n-electrode and fabricated an LED with emission wavelength of 233 nm. The device had an operation voltage of 10.4 V at 100 mA. Sulmoni et al. [80] used V/Al-based n-type contacts on n-$Al_{0.87}Ga_{0.13}N$ and fabricated an LED. The operation voltage was 9.9 V at 100 mA. In addition, Nagata et al. [83] reported that a thin SiN_x interlayer can reduce the n-type contact resistivity by more than one order of magnitude in the Al mole fraction range from 0.62 to 0.87.

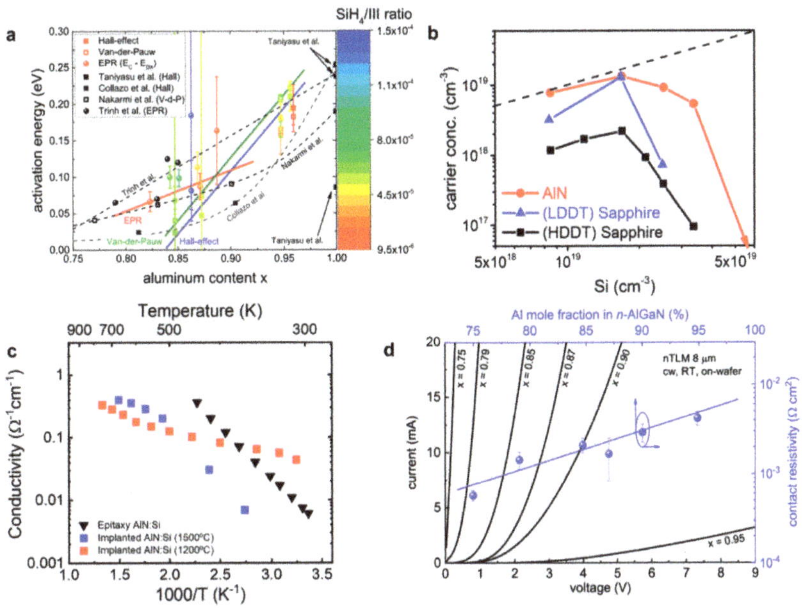

Figure 5. (**a**) The dependence of Si E_a on Al composition [73]. (**b**) The conductivity of n-AlGaN on different substrate with increasing Si doping concentration [74]. (**c**) Temperature dependent conductivity for the epitaxially doped AlN film, the Si-implanted AlN film on sapphire, and the Si-implanted AlN film on AlN substrate [79]. (**d**) IV curves and contact resistivity for n-contacts on *n*-AlGaN as a function of Al mole fraction [81].

3.3. Carrier Confinement and Transport

Driven by the external electric field, electrons and holes transport through the device by means of tunnelling or diffusion. Insufficient hole injection and electron leakage are the common problems during carrier transport in III-nitride based LEDs. In the review by Li et al. [84], methods of improving hole injection were explained and summarized in detail. The reason why electrons have difficulties in being restricted to the active region is that, the p-type layers cannot provide sufficient holes, and the hole mobility is low due to its large effective mass, resulting in asymmetric electron and hole injection [63]. To make matters worse, when emission wavelengths is smaller than 250 nm, the Al compositions of quantum wells and barriers become higher and higher, and the conduction band offset from the active region to the EBL becomes smaller and smaller. The severe electron leakage makes the electron injection efficiency quite low (<20%) [85]. Electron leakage has become one of the most severe problems for the sub-250 nm LEDs. A. Pandey et al. [36] measured the EQE of 245 nm LEDs with different EBL designs, and observed severe efficiency droop at 0.25 A/cm^2, as shown in Figure 6a. Their studies suggested that the observed efficiency droop is largely due to an electrical effect instead of an optical phenomenon.

An effective solution for the severe electron overflow is to introduce a MQB EBL, which can cause multiple reflection effects and increase the effective barrier height. In 1986, Iga et al. [86] theoretically compared the MQB and the traditional single barrier EBL, and the results showed that the thickness modulated MQB EBL increased the effective electron barrier height by two times. In a report by Hirayama et al. [17] in 2010, MQB was introduced into UVC LEDs as the EBL. The EQE and LOP of the 250 nm-LED with MQB EBL reached 1.18% and 4.8 mW, respectively, which was 2.7 times increased compared to that with single barrier EBL. Figure 6b summarizes the wavelength dependent EQE with MQB and single barrier EBLs, proving the significant enhancement of carrier confinement by using MQB

EBL. The influences of EBL thickness and doping on the hole injection and electron blocking effect in sub-250 nm LEDs was also explored. It was shown that EBL without doping could greatly suppress the parasitic emission peaks generated by the surface Mg-related deep level transition, and thickening the EBL could increase the radiative recombination [64]. Other methods to improve hole injection and electron confinement by optimizing EBL have been extensively studied such as wideband gap interlayer and Al composition-graded EBL [87,88]. Concerning that it was still difficult to achieve high quality AlGaN epitaxial layer and high Mg doping concentration with high Al content, Kuo et al. [89] adopted $Al_{0.55}Ga_{0.45}N/Al_{0.75}Ga_{0.25}N$ SLs EBL to replace $Al_{0.85}Ga_{0.15}N$ single barrier EBL. The WPE and LOP were almost the same.

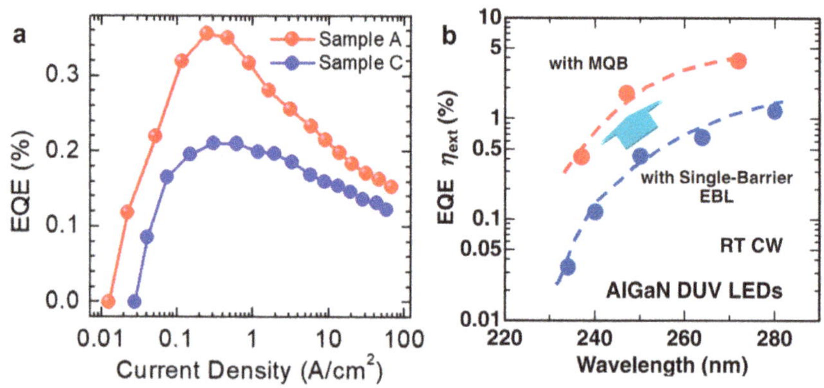

Figure 6. (**a**) Current density dependent EQE under CW bias. Sample A includes an EBL graded from an Al composition of ~95% to ~75%. Sample C is identical to Sample A, except that its EBL is a uniform $Al_{0.75}Ga_{0.25}N$:Mg layer with the same thickness [36]. (**b**) The EQE of DUV LEDs with different wavelength with an MQB and a single barrier EBL [85].

To optimize the quantum barrier (QB) in the active region is also important to improve the carrier confinement and transport efficiency. Velpula et al. [90] inserted a thin AlGaN layer with higher Al content into each QB in the active region, as shown in Figure 7a, and the electron leakage was reduced by approximately 11 times and LOP was increased by approximately 225% at 60 mA compared to the reference LED structure. Lang et al. [91] applied an asymmetric concave QB by inserting an AlGaN layer with lower Al content into the QB, forming a concave region on the energy band, as shown in Figure 7b. This structure can enhance the electron capture ability of QWs, accelerate the hole transport under polarization induced field, and enhance the hole vertical transport. The maximum EQE was 2.32 times higher than that of the reference one. The QB band modulation can also be achieved by inserting a thin layer with high Al content [92], as shown in Figure 7c. By the energy band modulation effect of the delta-AlGaN layer, holes can be accelerated and cross the high barriers with very large kinetic energy, thus enhancing the hole injection into the active region. The maximum EQE is 1.6 times higher than that of the reference one.

There are also many optimizations within the last quantum barrier (LQB) near the EBL, such as the insertion of a thin undoped AlGaN strip layer in the middle of the LQB and Al composition-graded LQB [93–95]. Figure 7d–f exhibits schematic conduction band diagrams of different optimizations within the LQB. It is worth mentioning that composition-graded LQB has two kinds of variations, namely increasing and decreasing [94,95], and specific choice need to be optimized according to the material and structure. In addition, although EBL can suppress electron leakage, the hole injection efficiency is still greatly affected due to the formation of positive polarization surface charge between the LQB and EBL [93,96]. To solve this problem, EBL-free structure was also proposed. Simulation analysis by Shi et al. [97] showed that utilizing Al composition-graded AlGaN insertion layer and EBL-free structure could effectively block electrons and improve hole injection efficiency. The IQE was increased by 64.3% compared to the reference device with EBL.

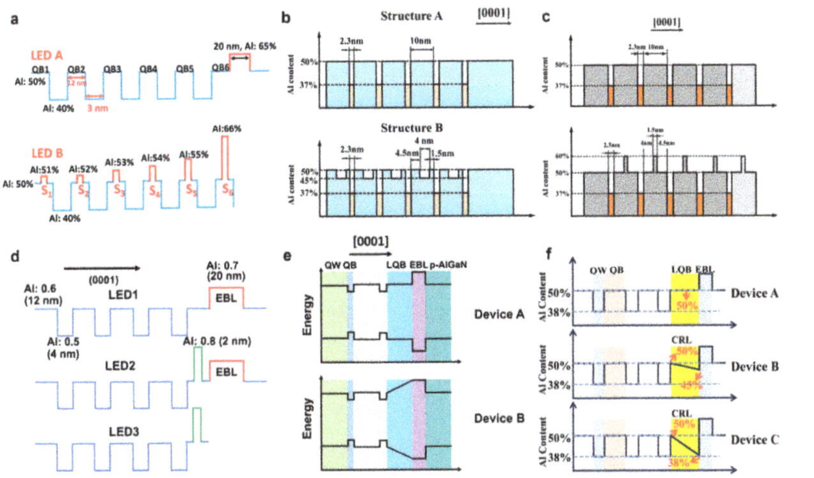

Figure 7. Conduction band diagrams of the active region of the DUV LED (**a**) without (LED A) and with (LED B) inserting thin layers in the QBs [90], (**b**) without (structure A) and with (structure B) asymmetrical concave QBs [91], (**c**) without (upper) and with (lower) delta-accelerating QBs [92], (**d**) without strip-in-a-barrier structure but EBL (LED 1), with strip-in-a-barrier structure and EBL (LED 2), with strip-in-a-barrier structure but EBL (LED 3) [93], (**e**) constant Al content (Device A) and gradually increasing Al content (Device B) LQB, [94] and (**f**) constant Al content (Device A) and gradually decreasing Al content (Device B and C) LQB [95].

Although these energy band and carrier velocity modulation methods within EBL and active region are mainly applied in above-250 nm LEDs, we believe they also have great application potential in sub-250 nm LEDs.

3.4. p-AlGaN with High Al Composition

It is observed that the Mg doping efficiency in AlGaN decreases dramatically with increasing Al mole fraction [98,99]. To achieve high-efficiency p-doped AlGaN has become one of the key issues limiting the development of sub-250 nm LEDs. The low p-doping efficiency of AlGaN is mainly caused by three reasons.

1. The solubility of Mg in GaN and AlN is low. The solubility of Mg replacing Ga or Al atoms in AlGaN is limited by the positive formation enthalpies. The influence of formation enthalpies on the solubility of Mg can be expressed as the following Equation (4)

$$C_{Mg} = N \cdot exp\left(-\frac{\Delta H_f}{kT}\right) \quad (4)$$

   where C_{Mg} is the solubility of Mg atoms, N is the number of lattice points which can be occupied by the doping atoms, ΔH_f is enthalpy change, k is the Boltzmann constant, and T is the temperature. As shown in Figure 8a [100], as the Al content in $Al_xGa_{1-x}N$ increases, both the Mg_{Al} and Mg_{Ga} formation enthalpies monotonically increase, indicating limited Mg solubility;
2. Self-compensation of intrinsic defects [101]. Figure 8b–e show the formation energies as a function of Fermi level for different defects in GaN and AlN [102,103]. The compensation of V_N is an important mechanism resulting in hole concentration decreasing. V_N can provide one, two, or three electrons, and only the charge states of V_N^+ and V_N^{3+} are stable. Since V_N^{3+} has smaller formation energy than V_N^+, V_N^{3+}

is believed to be the main compensating defects in Mg-doped AlGaN alloys [104]. Stampfl et al. [101] found that V_N was more easily formed in AlN than in GaN. The compensation of natural defects requires to be controlled especially in Al-rich AlGaN alloys;

3. The acceptor E_a is very high and gradually increase with Al content, as shown in Figure 8f [105,106]. Mg is generally used as the p-type dopant in AlGaN epilayer. The E_a of Mg in p-GaN films is in the range from 120 to 200 meV, generating a hole concentration of 10^{17}–10^{18} cm^{-3} [107]. However, when the Al content changes from 0 to 100%, the E_a of Mg increases from 160 meV to over 500 meV [99,105]. The hole concentration (p) in wide-bandgap semiconductors can be estimated by the following Formula (5) [108]

$$p = \sqrt{\frac{1}{g} N_A \cdot N_V \cdot exp\left(-\frac{Ea}{2kT}\right)} \quad (5)$$

where g is the acceptor degeneracy (g = 2), N_A is the acceptor concentration, and N_V is the effective state density near the valence band maximum (VBM). It can be seen that the increase of Mg acceptor E_a makes the hole concentration exponentially decrease, leading to the decrease of conductivity.

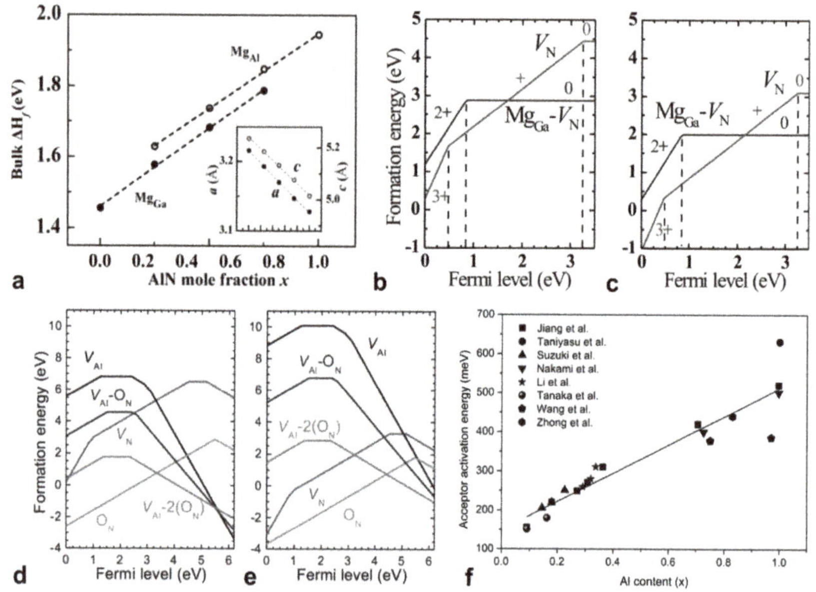

Figure 8. (**a**) Formation enthalpies of Mg_{Ga}/Mg_{Al} in the bulk AlGaN as a function of Al content under N-rich condition [100]. Formation energies as a function of Fermi level for the V_N and the Mg_{Ga}–V_N complex in GaN, under (**b**) N-rich and (**c**) Ga-rich conditions [102]. Formation energies as a function of Fermi level for the V_N, V_{Al}, O_N, $V_{Al}-O_N$ and $V_{Al}-2(O_N)$ in AlN, under (**d**) N-rich and (**e**) Al-rich conditions [103]. (**f**) The activation energy of Mg in $Al_xGa_{1-x}N$ in dependence of Al mole fraction x, data extracted from [105,106].

The surface during epitaxy plays an important role in determining the solubility of the dopant. Tersoff [109] pointed that the elastic interaction of the dopants with the surface relieved the stress caused by atomic size mismatch and lowered the impurity energy owing to the coupling of the impurity stress to the spatially surface stress, thus the solubility of dopants could be significantly improved. Zheng et al. [100] proposed a modified surface engineering technique (MSE) by periodic interruptions under an ultimate V/III ratio condition to enhance Mg effective incorporation without affecting the growth of AlGaN films. Schematic diagram of the Mg incorporation behavior in the AlGaN is shown in Figure 9. For $Al_{0.99}Ga_{0.01}N$, the final doping concentration of Mg can reach 5×10^{19} cm^{-3}.

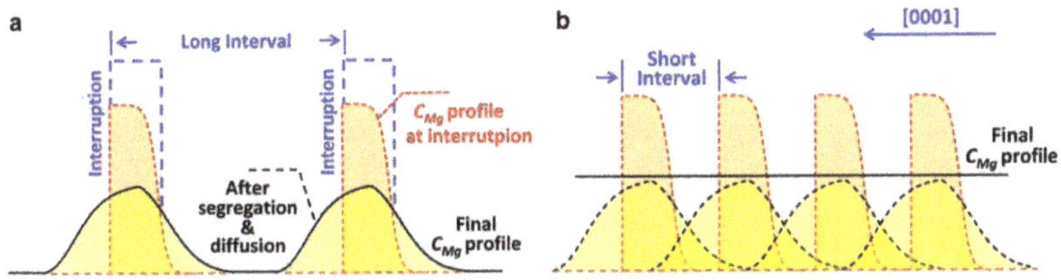

Figure 9. Schematic diagram of the Mg incorporation behavior in the AlGaN grown by the MSE technique. (**a**) When the interruption interval is long, only some peaks distribute locally at the interruptions after Mg segregation and diffusion. (**b**) After optimizing the interruption interval, a high and uniform Mg distribution over the entire AlGaN epilayer was achieved [100].

Self-compensation by V_N complexes can be suppressed by elevating V/III ratio during growth [71,99,104] and defect quasi Fermi level (dQFL) method [110]. Nakarmi et al. [104] increased the V/III ratio up to 5000 in the growth of the p-type $Al_{0.7}Ga_{0.3}N$ with a Mg concentration of 10^{20} cm^{-3} and significantly suppressed the related transition, while T. Kinoshita et al. [99] increased the V/III ratio to 1800 with a Mg concentration of 10^{19} cm^{-3} and achieved a low resistivity of 47 Ω·cm for $Al_{0.7}Ga_{0.3}N$. The discrepancy of the V/III ratio might result from larger formation energy of ${V_N}^{3+}$ and higher crystalline quality under lower Mg concentration [99]. dQFL method utilizes illumination during growth to increase the formation energy of the compensating point defects. A. Klump et al. [110] observed the reduction of V_N-related emission for the dQFL-process samples, and validated the effectiveness of dQFL control process in point defect management for Mg-doped III-nitride films.

To reduce the acceptor E_a for AlGaN with high Al composition, several methods were developed in recent two decades. Under the condition of heavy doping (>2 × 10^{19} cm^{-3}), the hole concentration provided by the impurity band is two orders of magnitude larger than the valence band transport, exhibiting very low acceptor E_a. In addition to forming impurity band, the Mg acceptor E_a can be reduced by Mg pulse or δ-doping [111–118]. δ-doping had been proved to effectively increase the hole concentration without causing a crystal quality droop [119]. The polarization-induced doping [120–122] and superlattice doping [123–125] are two common-used methods to reduce the apparent acceptor E_a. Under biaxial strain, AlGaN layers will generate huge polarization induced electric field (10^5–10^7 V/cm), making the valence and conduction bands tilt, thereby the acceptors ionization probability gets larger. Under polarization-induced electric field, free holes will be swept to the interface, generating two-dimensional hole gas (2DHG) near the interface, and simultaneously generating a depletion layer. If the thickness of the depletion layer is too thick, the vertical transport of holes will be hindered. Thus, the thickness of the thin layer should be limited to 0.6–10 nm [126]. To avoid the free carrier accumulation at the interface, three-dimensional polarization doping was adopted by means of gradually changing Al content, which could generate three-dimensional hole gas (3DHG) [122,127]. SLs doping can reduce the E_a via modulating the band structure based on spontaneous and piezoelectric fields. Kazuaki Ebata et al. [128] achieved high hole concentrations on the order of 10^{18} cm^{-3} and small E_a of 40–67 meV by adopting $AlN/Al_{0.75}Ga_{0.25}N$ SLs with an average Al composition of 0.8. As one kind of SLs doping methods, short-period superlattice (SPSL) is very effective to improve p-doping efficiency [123,129]. Schubert et al. [107] made a theoretical analysis and found that for SPSL (<100 Å) doping, the free carrier concentration could increase by more than ten times when the magnitude of the valance band modulation and the Mg E_a were equal. Wang et al. [130] proposed a desorption-tailoring method and remarkably reduced the Mg E_a to 17.5 meV in p-AlGaN SLs with an average Al composition of 51%. The desorption process, achieved by interrupting the supply of the precursors, was the reverse process of absorption of atoms to the surface. Specified SLs can be grown by adjusting the desorption time and matching the desorption time and epitaxy time. Based on the reduction of activation energy and enhancement of Mg surface-incorporation, the hole concentration in SLs reached 8.1×10^{18} cm^{-3} at room temperature.

Recently, K. Jiang et al. [106] proposed a new method to improve the doping efficiency called "quantum engineering of non-equilibrium p-doping" in which the key points were, firstly, embedding relatively narrow bandgap quantum structure into wide bandgap matrix to produce a new band edge, and secondly, enriching the Mg dopants near the interfaces to improve ionization, as shown in Figure 10. The Mg E_a of AlGaN with Al content of 50~70% was reduced to below 50 meV, and the hole concentration reached higher than 10^{18} cm^{-3} at room temperature.

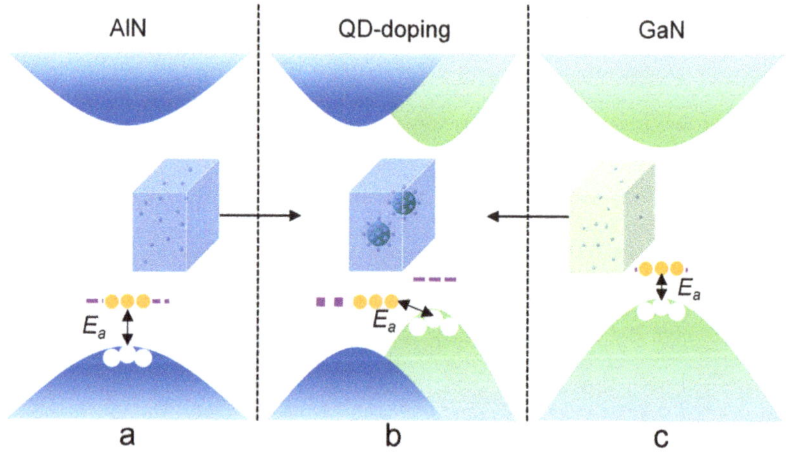

Figure 10. The VBM modulation mode to lower the acceptor E_a. Acceptors are randomly doped in (**a**) AlN and (**c**) GaN. Both have high E_a in this condition. (**b**) GaN QDs are embedded in the AlN matrix and acceptors are doped in the AlN matrix and concentrate near the interface [106].

Although the Mg acceptor E_a in AlGaN with high Al content can be reduced to even lower than that in GaN by various designs, the resistivity is still relatively higher due to significantly reduced hole mobility. Maybe, to increase the hole mobility is the next breakthrough of p-type doping for Al-rich AlGaN.

3.5. *Light Extraction*

According to the current researches, the low LEE may be the most serious problem hindering the UV LED performance, especially for sub-250 nm LEDs. The two main factors restricting the LEE are: (a) the balance between increasing transparency and decreasing driving voltage; (b) the transformation of emission light polarization from TE to transverse magnetic (TM) mode. For instance, to realize 240 nm transparency, the Al content larger than 80% is required within the LED structure. When the emission wavelength is 222 nm, Al mole fraction should be as high as 0.95 to ensure totally transparent, which greatly increases the operation voltage. Hence, a careful trade-off between reducing the optical absorption and increasing the conductivity is needed to achieve relatively high WPE. Furthermore, when the emission wavelength reduces to below 250 nm, the increase of Al mole fraction leads to the rearrangement of valence bands. The main polarization mode of the emission light shifts from TE, which mainly propagates along the (0001) direction that usually vertical to the device surface, to TM mode, which mainly propagates laterally in the c-plane that usually parallel to the device surface [10,131,132]. It was reported that the optical extraction efficiency of TM mode is less than 1/10 that of TE mode [133].

Self light absorption is a severe problem for all UV LEDs and schemes for boost transparency in other UV wavebands can also be applied to sub-250 nm LEDs. The reduction of light absorption can be achieved by using a transparent p-AlGaN contact layer [15,134,135], a highly reflective electrode structure, photonic crystal (PhC) and other reflection structures [13,136–138], etc. Mg-doped $Al_xGa_{1-x}N/Al_yGa_{1-y}N$ SPSL can provide transparent p-type contacts, yet the turn-on voltage is still high due to the high vertical resistivity and low mobility of the free carriers [15]. Muhin et al. [134] investigated the vertical conductivity of $Al_{0.71}Ga_{0.29}N/Al_{0.65}Ga_{0.35}N$:Mg SPSLs and observed that its value was strongly affected by temperature and electric field, to be specific, the vertical conductivity could increase by almost two orders of magnitude when the electric field varied from 0.05 to 0.98 $MV \cdot cm^{-1}$, or the temperature increased from room temperature to 100 °C. Guttmann et al. [135] showed that the average Al mole fraction should be high enough to be transparent for emission wavelength, and a five-fold increase of the on-wafer EQE could be achieved combined with reflective In p-contacts. Kashima et al. [138] improved the EQE of 283 nm LED from 4.8% to 10% by introducing high-reflectance PhC and reflective Ni/Mg p-electrodes on the surface of the contact layer as shown in Figure 11a. Takano et al. [13] reported a 275 nm LED with an EQE of 20.3% by combining various ways to improve the LEE, including the adoption of transparent AlGaN:Mg contact layer and Rh mirror electrode. When reflective p-electrodes were used, EQE could be significantly improved by optimizing the thickness of p-type layers [139]. As shown in Figure 11b, there are two light paths: directly from the active region toward the sapphire (solid arrow) and reflected by the p-electrode toward the sapphire (dashed arrow). When the total optical thickness of the p-type layers changes, the optical path difference between the two paths changes, creating interference and significantly affecting the LOP.

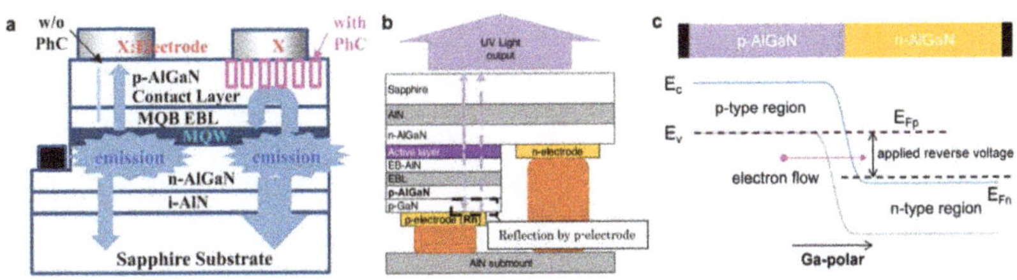

Figure 11. Schematic structure of (**a**) the LEDs with and without a PhC on the p-AlGaN contact layer [138] and (**b**) the flip-chip LED with two light paths [139]. (**c**) Schematic diagram of homogeneous tunnel junction structure and its corresponding energy band as well as carrier transport.

Since the DUV transparent AlGaN:Mg contact layer will lead to the low efficiency of hole injection, it is feasible to utilize the polarization field and tunnel junction. As shown in Figure 11c, the tunnel junction is a p-n junction working at reverse bias, which can achieve high reverse current through interband tunnelling at low reverse bias. In early works, the formation of an effective tunnel junction usually required degenerate p- and n-doping to achieve very narrow depletion regions, so that the carriers could easily tunnel from valance band on one side to conduction band on the other side. Whereas, degenerate p-type doping was almost impossible to achieve in high Al content AlGaN, researchers proposed a tunnel junction based on heterointerface polarization charges, which could eliminate the need for heavy doping [140]. Recently, Mehnke et al. [141] fabricated a tunnel heterojunction LED with an emission wavelength of 232 nm. The LOP and EQE reached to 1.73 mW and 0.35% at 100 mA under pulsed operation, respectively. However, the operation voltage (26 V at 10 mA) was still high compared to the reference traditional LED (6.8 V at 10 mA). The doping of the tunnel junction interface should be optimized to reduce the resistance and operation voltage.

Under the crystal field and spin-orbit coupling, the valence band for wurtzite AlGaN materials contains heavy hole (HH), light hole (LH), and crystal field splitting hole

(CH) sub-bands. The emitting light polarization is determined by the valence sub-band distribution. The topmost CH and HH sub-bands will lead to TM- and TE-dominated emission, respectively. The critical Al mole fraction triggering the transition from TE mode to TM mode depends on the strain state and quantum confinement [142], as shown in Figure 12a. Under compressive strain, the HH and LH sub-bands will be lifted up, while the CH subband will be pulled down, resulting TE mode enhancement. Enhancing the quantum confinement will greatly pull down the CH sub-band due to its light effective mass, even resulting in the intersection to LH and HH and also promoting the TE mode. The QW width and barrier height can affect the quantum confinement, thus controlling the polarization mode. Figure 12b shows that when the QW is relatively thick (>3 nm), the critical Al mole fraction is independent of the QW width. When the QW is relatively thin (<3 nm), the thinner the QW gets, the higher the critical Al mole fraction gets [142]. Figure 12c displays the calculated degree of polarization (DOP, $\frac{I_{TE}-I_{TM}}{I_{TE}+I_{TM}}$) by $k{\cdot}p$ theory of an AlGaN QW with different well and barrier Al contents pseudomorphically grown on AlN bulk substrate with thickness of 1.5 nm [143]. As it can be seen, for a certain well Al content, increasing the barrier Al content can improve the DOP, for a certain barrier Al content, decreasing the well Al content can improve the DOP, and for a certain emission wavelength such as 230 nm, it is necessary to simultaneously increase the barrier Al content and decrease the well Al content.

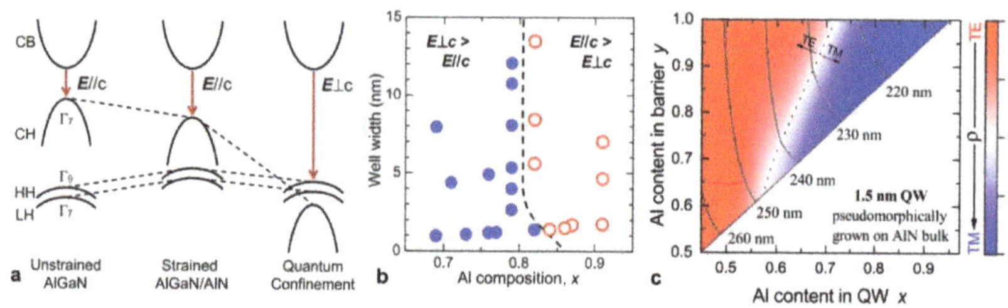

Figure 12. (**a**) Schematic energy band structures for unstrained AlGaN, strained AlGaN on AlN, and AlGaN/AlN QW near the Γ point with high Al content. CB denotes the conduction band. (**b**) Emission polarization mode as function of Al composition and well width [142]. (**c**) DOP of the emission from an AlGaN-based QW with 1.5 nm thickness pseudomorphically grown on AlN bulk calculated by $k{\cdot}p$ theory [143].

Experimentally, when the QW is under strain free or compressive strain states, the TE emission and quantum efficiency will be greatly improved [144–146]. ELOG for reducing TDD can also enhance TE mode emission due to the differences in thermal expansion coefficients. In 2021, Zhang et al. [147] successfully modulated the strain state of AlN template by adopting multiple alternation cycles of low- and high-temperature growth. Hence, the in-plane compressive strain of the QWs had increased. The DOP was enhanced from 41.5% to 61.9%. When the GaN quantum well width is reduced to a few atom monolayers (MLs), strong quantum confinement will appear, which can increase the effective transition energy of GaN by several eV. Some groups [23,24,148,149] have carried out related studies on AlN-delta-GaN QW LEDs and achieved lower wavelength emission. The advantage of this method is that in addition to increasing TE emission, the spontaneous emission rate is also significantly increased by suppressing the QCSE. At present, researches on AlN-delta-GaN QW LEDs are still in the early stage, further and deeper researches need to be carried out and its feasibility requires to be confirmed.

4. Conclusions and Prospect

Nowadays, the technology roadmap of AlGaN based sub-250 nm LEDs is basically similar to the conventional DUV LEDs. The performance collapse mainly results from the problem accumulation with Al composition increasing of each layer in the device structure.

First of all, improving the material crystal quality is the key issue to promote the sub-250 nm LED performance. Before the AlN single-crystal substrate fabrication and homogeneous epitaxy technologies mature, the AlN/Sapphire template is still the best choice. HTA is a promising method, which can provide cost-effective, stable, high-quality, and compressive strain AlN/Sapphire, benefiting the IQE and LEE.

The contact and sheet resistivity of the n-AlGaN layer contribute part of the driving voltage. When Al mole fraction is larger than 0.8, shallow donor Si gradually changes into a deeper donor, and finally becomes a stable charge capture center, leading to a significant increase in resistivity. Controlling the formation of *DX* centers is a critical technical challenge for high Al composition AlGaN. Optimization of n-layers, including modulating growth conditions and searching for suitable contact metals, need to be further explored. For carrier confinement and transport, although a large number of energy band engineering studies have been carried out to enhance carrier confinement, electron overflow is still one of the most severe problems in sub-250 nm LEDs due to the physic limitations. The MQB EBL is a promising method to increase the effective electron barrier.

Resulting from the increase E_a of Mg dopants in AlGaN, p-AlGaN layers cannot offer sufficient free carriers and enough high conductance. Different methods, such as δ-doping, quantum engineering doping, polarization induced doping, and superlattice doping, etc., have been proposed to improve the doping efficiency, while it is still facing daunting challenges. In addition, how to increase the hole mobility may be the next problem to be solved. Serious light absorption and emission mode transition problem greatly reduces the LEE in this wavelength range. Adopting tunnel junction may be a quite hopeful method to achieve fully transparent LED.

Author Contributions: Conceptualization, K.J., X.S. and D.L.; methodology, K.J. and C.Z.; writing—original draft preparation, C.Z.; writing—review and editing, K.J.; project administration, K.J., X.S. and D.L.; funding acquisition, K.J., X.S. and D.L. All authors have read and agreed to the published version of the manuscript.

Funding: This research was funded by National Key R&D Program of China (grant number 2022YFB3605103); National Natural Science Foundation of China (grant numbers 62004196, U22A2084, 62121005, and 61827813); Youth Innovation Promotion Association of Chinese Academy of Sciences; and Youth Talent Promotion Project of the Chinese Institute of Electronics (grant number 2020QNRC001).

Data Availability Statement: Not applicable.

Conflicts of Interest: The authors declare no conflict of interest.

References

1. Zollner, C.J.; DenBaars, S.P.; Speck, J.S.; Nakamura, S. Germicidal ultraviolet LEDs: A review of applications and semiconductor technologies. *Semicond. Sci. Technol.* **2021**, *36*, 123001. [CrossRef]
2. Mehnke, F.; Guttmann, M.; Enslin, J.; Kuhn, C.; Reich, C.; Jordan, J.; Kapanke, S.; Knauer, A.; Lapeyrade, M.; Zeimer, U.; et al. Gas Sensing of Nitrogen Oxide Utilizing Spectrally Pure Deep UV LEDs. *IEEE J. Sel. Top. Quantum Electron.* **2017**, *23*, 29–36. [CrossRef]
3. Mellqvist, J.M.; Rosén, A. DOAS for flue gas monitoring—I. Temperature effects in the U.V./visible absorption spectra of NO, NO_2, SO_2 and NH_3. *J. Quant. Spectrosc. Radiat. Transf.* **1996**, *56*, 187–208. [CrossRef]
4. Hodgkinson, J.; Tatam, R.P. Optical gas sensing: A review. *Meas. Sci. Technol.* **2012**, *24*, 012004. [CrossRef]
5. Xu, Z.; Sadler, B.M. Ultraviolet Communications: Potential and State-Of-The-Art. *IEEE Commun. Mag.* **2008**, *46*, 67–73.
6. Sun, K.X.; Allard, B.; Buchman, S.; Williams, S.; Byer, R.L. LED deep UV source for charge management of gravitational reference sensors. *Class. Quantum Gravity* **2006**, *23*, S141–S150. [CrossRef]
7. Narita, K.; Asano, K.; Morimoto, Y.; Igarashi, T.; Hamblin, M.R.; Dai, T.; Nakane, A. Disinfection and healing effects of 222-nm UVC light on methicillin-resistant Staphylococcus aureus infection in mouse wounds. *J. Photochem. Photobiol. B Biol.* **2018**, *178*, 10–18. [CrossRef]
8. Zwicker, P.; Schleusener, J.; Lohan, S.B.; Busch, L.; Sicher, C.; Einfeldt, S.; Kneissl, M.; Kühl, A.A.; Keck, C.M.; Witzel, C.; et al. Application of 233 nm far-UVC LEDs for eradication of MRSA and MSSA and risk assessment on skin models. *Sci. Rep.* **2022**, *12*, 2587. [CrossRef]

9. Guttmann, M.; Lobo-Ploch, N.; Gundlach, H.; Mehnke, F.; Sulmoni, L.; Wernicke, T.; Cho, H.K.; Hilbrich, K.; Külberg, A.; Friedler, M.; et al. Spectrally pure far-UVC emission from AlGaN-based LEDs with dielectric band pass filters. *J. Phys. D Appl. Phys.* **2022**, *55*, 205105. [CrossRef]
10. Guttmann, M.; Mehnke, F.; Belde, B.; Wolf, F.; Reich, C.; Sulmoni, L.; Wernicke, T.; Kneissl, M. Optical light polarization and light extraction efficiency of AlGaN-based LEDs emitting between 264 and 220 nm. *Jpn. J. Appl. Phys.* **2019**, *58*, SCCB20. [CrossRef]
11. Mehnke, F.; Sulmoni, L.; Guttmann, M.; Wernicke, T.; Kneissl, M. Influence of light absorption on the performance characteristics of UV LEDs with emission between 239 and 217 nm. *Appl. Phys. Express* **2019**, *12*, 012008. [CrossRef]
12. Kneissl, M.; Rass, J. *III-Nitride Ultraviolet Emitters: Technology and Applications*, 1st ed.; Springer: Berlin/Heidelberg, Germany, 2016; pp. 90–92.
13. Takano, T.; Mino, T.; Sakai, J.; Noguchi, N.; Tsubaki, K.; Hirayama, H. Deep-ultraviolet light-emitting diodes with external quantum efficiency higher than 20% at 275 nm achieved by improving light-extraction efficiency. *Appl. Phys. Express* **2017**, *10*, 031002. [CrossRef]
14. Inoue, S.i.; Tamari, N.; Taniguchi, M. 150 mW deep-ultraviolet light-emitting diodes with large-area AlN nanophotonic light-extraction structure emitting at 265 nm. *Appl. Phys. Lett.* **2017**, *110*, 141106. [CrossRef]
15. Taniyasu, Y.; Kasu, M.; Makimoto, T. An Aluminium Nitride Light-Emitting Diode with a Wavelength of 210 Nanometres. *Nature* **2006**, *441*, 325–328. [CrossRef] [PubMed]
16. Hirayama, H.; Noguchi, N.; Fujikawa, S.; Norimatsu, J.; Kamata, N.; Takano, T.; Tsubaki, K. 222–282 nm AlGaN and InAlGaN-based deep-UV LEDs fabricated on high-quality AlN on sapphire. *Phys. Status Solidi Appl. Res.* **2009**, *206*, 1176–1182. [CrossRef]
17. Hirayama, H.; Tsukada, Y.; Maeda, T.; Kamata, N. Marked Enhancement in the Efficiency of Deep-Ultraviolet AlGaN Light-Emitting Diodes by Using a Multiquantum-Barrier Electron Blocking Layer. *Appl. Phys. Express* **2010**, *3*, 031002. [CrossRef]
18. Lobo-Ploch, N.; Mehnke, F.; Sulmoni, L.; Cho, H.K.; Guttmann, M.; Glaab, J.; Hilbrich, K.; Wernicke, T.; Einfeldt, S.; Kneissl, M. Milliwatt power 233nm AlGaN-based deep UV-LEDs on sapphire substrates. *Appl. Phys. Lett.* **2020**, *117*, 111102. [CrossRef]
19. Jo, M.; Itokazu, Y.; Hirayama, H. Milliwatt-power far-UVC AlGaN LEDs on sapphire substrates. *Appl. Phys. Lett.* **2022**, *120*, 211105. [CrossRef]
20. Taniyasu, Y.; Kasu, M. Surface 210 nm light emission from an AlN p–n junction light-emitting diode enhanced by A-plane growth orientation. *Appl. Phys. Lett.* **2010**, *96*, 221110. [CrossRef]
21. Liu, D.; Cho, S.J.; Park, J.; Seo, J.H.; Dalmau, R.; Zhao, D.; Kim, K.; Gong, J.; Kim, M.; Lee, I.K.; et al. 229nm UV LEDs on aluminum nitride single crystal substrates using p-type silicon for increased hole injection. *Appl. Phys. Lett.* **2018**, *112*, 081101. [CrossRef]
22. Liu, D.; Cho, S.J.; Park, J.; Gong, J.; Seo, J.H.; Dalmau, R.; Zhao, D.; Kim, K.; Kim, M.; Kalapala, A.R.K.; et al. 226nm AlGaN/AlN UV LEDs using p-type Si for hole injection and UV reflection. *Appl. Phys. Lett.* **2018**, *113*, 011111. [CrossRef]
23. Liu, C.; Ooi, Y.K.; Islam, S.M.; Xing, H.G.; Jena, D.; Zhang, J. 234nm and 246nm AlN-Delta-GaN quantum well deep ultraviolet light-emitting diodes. *Appl. Phys. Lett.* **2018**, *112*, 011101. [CrossRef]
24. Islam, S.M.; Protasenko, V.; Lee, K.; Rouvimov, S.; Verma, J.; Xing, H.G.; Jena, D. Deep-UV emission at 219nm from ultrathin MBE GaN/AlN quantum heterostructures. *Appl. Phys. Lett.* **2017**, *111*, 091104. [CrossRef]
25. Nishida, T.; Makimoto, T.; Saito, H.; Ban, T. AlGaN-based ultraviolet light-emitting diodes grown on bulk AlN substrates. *Appl. Phys. Lett.* **2004**, *84*, 1002–1003. [CrossRef]
26. Yoshikawa, A.; Hasegawa, R.; Morishita, T.; Nagase, K.; Yamada, S.; Grandusky, J.; Mann, J.; Miller, A.; Schowalter, L.J. Improve efficiency and long lifetime UVC LEDs with wavelengths between 230 and 237 nm. *Appl. Phys. Express* **2020**, *13*, 022001. [CrossRef]
27. Yin, X.; Zhao, S. High Internal Quantum Efficiency AlGaN Epilayer Grown by Molecular Beam Epitaxy on Si Substrate. *ECS J. Solid State Sci. Technol.* **2021**, *10*, 076001. [CrossRef]
28. Zhang, Q.; Parimoo, H.; Martel, E.; Zhao, S. Vertical semiconductor deep ultraviolet light emitting diodes on a nanowire-assisted aluminum nitride buffer layer. *Sci. Rep.* **2022**, *12*, 7230. [CrossRef]
29. Zhao, S.; Connie, A.; Tavakoli Dastjerdi, H.; Kong, X.; Wang, Q.; Djavid, M.; Sadaf, S.; Liu, X.; Shih, I.; Guo, H.; et al. Aluminum nitride nanowire light emitting diodes: Breaking the fundamental bottleneck of deep ultraviolet light sources. *Sci. Rep.* **2015**, *5*, 8332. [CrossRef]
30. Amano, H.; Collazo, R.; Santi, C.D.; Einfeldt, S.; Funato, M.; Glaab, J.; Hagedorn, S.; Hirano, A.; Hirayama, H.; Ishii, R.; et al. The 2020 UV emitter roadmap. *J. Phys. D Appl. Phys.* **2020**, *53*, 503001. [CrossRef]
31. Ruschel, J.; Glaab, J.; Susilo, N.; Hagedorn, S.; Walde, S.; Ziffer, E.; Cho, H.K.; Ploch, N.L.; Wernicke, T.; Weyers, M.; et al. Reliability of UVC LEDs fabricated on AlN/sapphire templates with different threading dislocation densities. *Appl. Phys. Lett.* **2020**, *117*, 241104. [CrossRef]
32. Glaab, J.; Haefke, J.; Ruschel, J.; Brendel, M.; Rass, J.; Kolbe, T.; Knauer, A.; Weyers, M.; Einfeldt, S.; Guttmann, M.; et al. Degradation effects of the active region in UV-C light-emitting diodes. *J. Appl. Phys.* **2018**, *123*, 104502. [CrossRef]
33. Meneghini, M.; Pavesi, M.; Trivellin, N.; Gaska, R.; Zanoni, E.; Meneghesso, G. Reliability of Deep-UV Light-Emitting Diodes. *IEEE Trans. Device Mater. Reliab.* **2008**, *8*, 248–254. [CrossRef]
34. Glaab, J.; Ruschel, J.; Lobo Ploch, N.; Cho, H.; Mehnke, F.; Sulmoni, L.; Guttmann, M.; Wernicke, T.; Weyers, M.; Einfeldt, S.; et al. Impact of operation parameters on the degradation of 233 nm AlGaN-based far-UVC LEDs. *J. Appl. Phys.* **2022**, *131*, 014501. [CrossRef]

35. Moe, C.G.; Sugiyama, S.; Kasai, J.; Grandusky, J.R.; Schowalter, L.J. AlGaN Light-Emitting Diodes on AlN Substrates Emitting at 230 nm. *Phys. Status Solidi Appl. Res.* **2018**, *215*, 1700660. [CrossRef]
36. Pandey, A.; Gim, J.; Hovden, R.; Mi, Z. Electron overflow of AlGaN deep ultraviolet light emitting diodes. *Appl. Phys. Lett.* **2021**, *118*, 241109. [CrossRef]
37. Mehnke, F.; Kuhn, C.; Guttmann, M.; Reich, C.; Kolbe, T.; Kueller, V.; Knauer, A.; Lapeyrade, M.; Einfeldt, S.; Rass, J.; et al. Efficient charge carrier injection into sub-250nm AlGaN multiple quantum well light emitting diodes. *Appl. Phys. Lett.* **2014**, *105*, 051113. [CrossRef]
38. Piprek, J. Efficiency droop in nitride-based light-emitting diodes. *Phys. Status Solidi* **2010**, *207*, 2217–2225. [CrossRef]
39. Ban, K.; Yamamoto, J.I.; Takeda, K.; Ide, K.; Iwaya, M.; Takeuchi, T.; Kamiyama, S.; Akasaki, I.; Amano, H. Internal Quantum Efficiency of Whole-Composition-Range AlGaN Multiquantum Wells. *Appl. Phys. Express* **2011**, *4*, 052101. [CrossRef]
40. Hasan, S.; Mamun, A.; Hussain, K.; Gaevski, M.; Ahmad, I.; Khan, A. Growth evolution of high-quality MOCVD aluminum nitride using nitrogen as carrier gas on the sapphire substrate. *J. Mater. Res.* **2021**, *36*, 4360–4369. [CrossRef]
41. Sun, X.; Li, D.; Chen, Y.; Song, H.; Jiang, H.; Li, Z.; Miao, G.; Zhiwei, Z. In situ observation of two-step growth of AlN on sapphire using high-temperature metal–organic chemical vapour deposition. *CrystEngComm* **2013**, *15*, 6066–6073. [CrossRef]
42. Zhang, Y.; Yang, J.; Zhao, D.; Liang, F.; Chen, P.; Liu, Z. High-quality AlN growth on flat sapphire at relatively low temperature by crystal island shape control method. *Appl. Surf. Sci.* **2022**, *606*, 154919. [CrossRef]
43. Zhang, Y.; Yang, J.; Zhao, D.; Liang, F.; Chen, P.; Liu, Z. Adjustment of Al atom migration ability and its effect on the surface morphology of AlN grown on sapphire by metal–organic chemical vapor deposition. *Semicond. Sci. Technol.* **2021**, *36*, 105010. [CrossRef]
44. Zhang, L.; Xu, F.; Wang, J.; He, C.; Guo, W.; Wang, M.; Sheng, B.; Lu, L.; Qin, Z.; Wang, X.Q.; et al. High-quality AlN epitaxy on nano-patterned sapphire substrates prepared by nano-imprint lithography. *Sci. Rep.* **2016**, *6*, 1–8. [CrossRef]
45. Lee, D.; Lee, J.W.; Jang, J.; Shin, I.S.; Jin, L.; Park, J.H.; Kim, J.; Lee, J.; Noh, H.S.; Kim, Y.I.; et al. Improved performance of AlGaN-based deep ultraviolet light-emitting diodes with nano-patterned AlN/sapphire substrates. *Appl. Phys. Lett.* **2017**, *110*, 191103. [CrossRef]
46. Okada, N.; Saito, T.; Fujikawa, S.; Maeda, N.; Jo, M.; Hirayama, H.; Tadatomo, K. Investigation of off-cut angle of sapphire for epitaxial lateral overgrowth of AlN and fabrication of high-quality AlN template. *J. Cryst. Growth* **2022**, *588*, 126640. [CrossRef]
47. Imura, M.; Nakano, K.; Narita, G.; Fujimoto, N.; Okada, N.; Balakrishnan, K.; Iwaya, M.; Kamiyama, S.; Amano, H.; Akasaki, I.; et al. Epitaxial lateral overgrowth of AlN on trench-patterned AlN layers. *J. Cryst. Growth* **2007**, *298*, 257–260. [CrossRef]
48. Imura, M.; Nakano, K.; Kitano, T.; Fujimoto, N.; Narita, G.; Okada, N.; Balakrishnan, K.; Iwaya, M.; Kamiyama, S.; Amano, H.; et al. Microstructure of epitaxial lateral overgrown AlN on trench-patterned AlN template by high-temperature metal-organic vapor phase epitaxy. *Appl. Phys. Lett.* **2006**, *89*, 221901. [CrossRef]
49. Nakano, K.; Imura, M.; Narita, G.; Kitano, T.; Hirose, Y.; Fujimoto, N.; Okada, N.; Kawashima, T.; Iida, K.; Balakrishnan, K.; et al. Epitaxial lateral overgrowth of AlN layers on patterned sapphire substrates. *Phys. Status Solidi* **2006**, *203*, 1632–1635. [CrossRef]
50. Liu, B.; Xu, F.; Wang, J.; Lang, J.; Zhang, N.; Fang, X.; Wang, L.; Yang, X.; Kang, X.; Wang, X.; et al. High quality AlN with uniform in-plane strain on nano-patterned AlN templates achieved by preset strain modulation. *Jpn. J. Appl. Phys.* **2021**, *60*, 120903. [CrossRef]
51. Shojiki, K.; Ishii, R.; Uesugi, K.; Funato, M.; Kawakami, Y.; Miyake, H. Impact of face-to-face annealed sputtered AlN on the optical properties of AlGaN multiple quantum wells. *AIP Adv.* **2019**, *9*, 125342. [CrossRef]
52. Ben, J.; Shi, Z.; Zang, H.; Sun, X.; Liu, X.; Lü, W.; Li, D. The formation mechanism of voids in physical vapor deposited AlN epilayer during high temperature annealing. *Appl. Phys. Lett.* **2020**, *116*, 251601. [CrossRef]
53. Xiao, S.; Suzuki, R.; Miyake, H.; Harada, S.; Ujihara, T. Improvement mechanism of sputtered AlN films by high-temperature annealing. *J. Cryst. Growth* **2018**, *502*, 41–44. [CrossRef]
54. Liu, S.; Yuan, Y.; Sheng, S.; Wang, T.; Zhang, J.; Huang, L.; Zhang, X.; Kang, J.; Luo, W.; Li, Y.; et al. Four-inch high quality crack-free AlN layer grown on a high-temperature annealed AlN template by MOCVD. *J. Semicond.* **2021**, *42*, 122804. [CrossRef]
55. Miyake, H.; Nishio, G.; Suzuki, S.; Hiramatsu, K.; Fukuyama, H.; Kaur, J.; Kuwano, N. Annealing of an AlN buffer layer in N2–CO for growth of a high-quality AlN film on sapphire. *Appl. Phys. Express* **2016**, *9*, 025501. [CrossRef]
56. Wang, M.X.; Xu, F.J.; Xie, N.; Sun, Y.H.; Liu, B.Y.; Ge, W.K.; Kang, X.N.; Qin, Z.X.; Yang, X.L.; Wang, X.Q.; et al. High-temperature annealing induced evolution of strain in AlN epitaxial films grown on sapphire substrates. *Appl. Phys. Lett.* **2019**, *114*, 112105. [CrossRef]
57. Koshelev, O.A.; Nechaev, D.V.; Brunkov, P.N.; Ivanov, S.V.; Jmerik, V.N. Stress control in thick AlN/c-Al2O3 templates grown by plasma-assisted molecular beam epitaxy. *Semicond. Sci. Technol.* **2021**, *36*, 035007. [CrossRef]
58. Hirayama, H.; Yatabe, T.; Noguchi, N.; Ohashi, T.; Kamata, N. 231–261 nm AlGaN deep-ultraviolet light-emitting diodes fabricated on AlN multilayer buffers grown by ammonia pulse-flow method on sapphire. *Appl. Phys. Lett.* **2007**, *91*, 071901. [CrossRef]
59. Hirayama, H.; Noguchi, N.; Yatabe, T.; Kamata, N. 227 nm AlGaN Light-Emitting Diode with 0.15 mW Output Power Realized using a Thin Quantum Well and AlN Buffer with Reduced Threading Dislocation Density. *Appl. Phys. Express* **2008**, *1*, 051101. [CrossRef]
60. Li, F.; Wang, L.; Yao, W.; Meng, Y.; Yang, S.; Wang, Z. Analysis of growth rate and crystal quality of AlN epilayers by flow-modulated metal organic chemical vapor deposition. *Superlattices Microstruct.* **2020**, *137*, 106336. [CrossRef]

61. Banal, R.G.; Funato, M.; Kawakami, Y. Initial nucleation of AlN grown directly on sapphire substrates by metal-organic vapor phase epitaxy. *Appl. Phys. Lett.* **2008**, *92*, 241905. [CrossRef]
62. Chen, S.; Li, Y.; Ding, Y.; Li, S.; Zhang, M.; Wu, Z.; Fang, Y.; Dai, J.; Chen, C. Defect Reduction in AlN Epilayers Grown by MOCVD via Intermediate-Temperature Interlayers. *J. Electron. Mater.* **2015**, *44*, 217–221. [CrossRef]
63. Chen, Y.; Zhang, Z.; Jiang, H.; Li, Z.; Miao, G.; Song, H. Optimized growth of AlN templates for back-illuminated AlGaN-based solar-blind ultraviolet photodetectors by MOCVD. *J. Mater. Chem. C* **2018**, *6*, 1201–1206. [CrossRef]
64. He, C.; Zhao, W.; Wu, H.; Liu, N.; Zhang, S.; Li, J.; Jia, C.; Zhang, K.; He, L.; Chen, Z.; et al. Fast growth of crack-free thick AlN film on sputtered AlN/sapphire by introducing high-density nano-voids. *J. Phys. D Appl. Phys.* **2020**, *53*, 405303. [CrossRef]
65. Li, D.B.; Aoki, M.; Miyake, H.; Hiramatsu, K. Improved surface morphology of flow-modulated MOVPE grown AlN on sapphire using thin medium-temperature AlN buffer layer. *Appl. Surf. Sci.* **2007**, *253*, 9395–9399. [CrossRef]
66. Wang, H.M.; Zhang, J.P.; Chen, C.Q.; Fareed, Q.; Yang, J.W.; Khan, M.A. AlN/AlGaN superlattices as dislocation filter for low-threading-dislocation thick AlGaN layers on sapphire. *Appl. Phys. Lett.* **2002**, *81*, 604–606. [CrossRef]
67. Kim, J.; Pyeon, J.; Jeon, M.; Nam, O. Growth and characterization of high quality AlN using combined structure of low temperature buffer and superlattices for applications in the deep ultraviolet. *Jpn. J. Appl. Phys.* **2015**, *54*, 081001. [CrossRef]
68. Qi, Y.; Wang, Y.; Pang, Z.; Dou, Z.; Wei, T.; Gao, P.; Zhang, S.; Xu, X.; Chang, Z.; Deng, B.; et al. Fast Growth of Strain-Free AlN on Graphene-Buffered Sapphire. *J. Am. Chem. Soc.* **2018**, *140*, 11935–11941. [CrossRef]
69. Dalmau, R.; Moody, B.; Schlesser, R.; Mita, S.; Xie, J.; Feneberg, M.; Neuschl, B.; Thonke, K.; Collazo, R.; Rice, A.; et al. Growth and Characterization of AlN and AlGaN Epitaxial Films on AlN Single Crystal Substrates. *J. Electrochem. Soc.* **2011**, *158*, H530–H535. [CrossRef]
70. Reddy, P.; Khachariya, D.; Mecouch, W.; Breckenridge, M.; Bagheri, P.; Guan, Y.; Kim, J.; Pavlidis, S.; Kirste, R.; Mita, S.; et al. Study on avalanche breakdown and Poole–Frenkel emission in Al-rich AlGaN grown on single crystal AlN. *Appl. Phys. Lett.* **2021**, *119*, 182104. [CrossRef]
71. Bagheri, P.; Klump, A.; Washiyama, S.; Breckenridge, M.; Kim, J.; Guan, Y.; Khachariya, D.; Quiñones-García, C.; Sarkar, B.; Rathkanthiwar, S.; et al. Doping and compensation in heavily Mg doped Al-rich AlGaN films. *Appl. Phys. Lett.* **2022**, *120*, 082102. [CrossRef]
72. Collazo, R.; Mita, S.; Xie, J.; Rice, A.; Tweedie, J.; Dalmau, R.; Sitar, Z. Progress on n-type doping of AlGaN alloys on AlN single crystal substrates for UV optoelectronic applications. *Phys. Status Solidi* **2011**, *8*, 2031–2033. [CrossRef]
73. Mehnke, F.; Trinh, X.; Pingel, H.; Wernicke, T.; Janzén, E.; Son, N.; Kneissl, M. Electronic properties of Si-doped $Al_xGa_{1-x}N$ with aluminum mole fractions above 80%. *J. Appl. Phys.* **2016**, *120*, 145702. [CrossRef]
74. Bryan, I.; Bryan, Z.; Washiyama, S.; Reddy, P.; Gaddy, B.; Sarkar, B.; Breckenridge, M.; Guo, Q.; Bobea, M.; Tweedie, J.; et al. Doping and compensation in Al-rich AlGaN grown on single crystal AlN and sapphire by MOCVD. *Appl. Phys. Lett.* **2018**, *112*, 062102. [CrossRef]
75. Lang, D.V.; Logan, R.A.; Jaros, M. Trapping characteristics and a donor-complex (DX) model for the persistent-photoconductivity trapping center in Te-doped $Al_xGa_{1-x}As$. *Phys. Rev. B* **1979**, *19*, 1015–1030. [CrossRef]
76. Breckenridge, M.H.; Bagheri, P.; Guo, Q.; Sarkar, B.; Khachariya, D.; Pavlidis, S.; Tweedie, J.; Kirste, R.; Mita, S.; Reddy, P.; et al. High n-type conductivity and carrier concentration in Si-implanted homoepitaxial AlN. *Appl. Phys. Lett.* **2021**, *118*, 112104. [CrossRef]
77. Washiyama, S.; Reddy, P.; Sarkar, B.; Breckenridge, M.H.; Guo, Q.; Bagheri, P.; Klump, A.; Kirste, R.; Tweedie, J.; Mita, S.; et al. The role of chemical potential in compensation control in Si:AlGaN. *J. Appl. Phys.* **2020**, *127*, 105702. [CrossRef]
78. Almogbel, A.S.; Zollner, C.J.; Saifaddin, B.K.; Iza, M.; Wang, J.; Yao, Y.; Wang, M.; Foronda, H.; Prozheev, I.; Tuomisto, F.; et al. Growth of highly conductive Al-rich AlGaN:Si with low group-III vacancy concentration. *AIP Adv.* **2021**, *11*, 095119. [CrossRef]
79. Hayden Breckenridge, M.; Guo, Q.; Klump, A.; Sarkar, B.; Guan, Y.; Tweedie, J.; Kirste, R.; Mita, S.; Reddy, P.; Collazo, R.; et al. Shallow Si donor in ion-implanted homoepitaxial AlN. *Appl. Phys. Lett.* **2020**, *116*, 172103. [CrossRef]
80. Moses, P.; Miao, M.; Yan, Q.; Van de Walle, C. Hybrid functional investigations of band gaps and band alignments for AlN, GaN, InN, and InGaN. *J. Chem. Phys.* **2011**, *134*, 084703. [CrossRef]
81. Sulmoni, L.; Mehnke, F.; Mogilatenko, A.; Guttmann, M.; Wernicke, T.; Kneissl, M. Electrical properties and microstructure formation of V/Al-based n-contacts on high Al mole fraction n-AlGaN layers. *Photonics Res.* **2020**, *8*, 1381–1387. [CrossRef]
82. Schweitz, K.; Wang, P.; Mohney, S.; Gotthold, D. V/Al/Pt/Au Ohmic contact to n-AlGaN/GaN heterostructures. *Appl. Phys. Lett.* **2002**, *80*, 1954–1956. [CrossRef]
83. Nagata, N.; Senga, T.; Iwaya, M.; Takeuchi, T.; Kamiyama, S.; Akasaki, I. Reduction of contact resistance in V-based electrode for high AlN molar fraction n-type AlGaN by using thin SiNx intermediate layer. *Phys. Status Solidi C* **2017**, *14*, 1600243.
84. Li, L.; Zhang, Y.; Xu, S.; Bi, W.; Zhang, Z.H.; Kuo, H.C. On the Hole Injection for III-Nitride Based Deep Ultraviolet Light-Emitting Diodes. *Materials* **2017**, *10*, 1221. [CrossRef]
85. Hirayama, H.; Maeda, N.; Fujikawa, S.; Toyoda, S.; Kamata, N. Recent progress and future prospects of AlGaN-based high-efficiency deep-ultraviolet light-emitting diodes. *Jpn. J. Appl. Phys.* **2014**, *53*, 100209. [CrossRef]
86. Iga, K.; Uenohara, H.; Koyama, F. Electron reflectance of multiquantum barrier (MQB). *Electron. Lett.* **1986**, *22*, 1008–1010. [CrossRef]
87. Sumiya, S.; Zhu, Y.; Zhang, J.; Kosaka, K.; Miyoshi, M.; Shibata, T.; Tanaka, M.; Egawa, T. AlGaN-Based Deep Ultraviolet Light-Emitting Diodes Grown on Epitaxial AlN/Sapphire Templates. *Jpn. J. Appl. Phys.* **2008**, *47*, 43–46. [CrossRef]

88. Chen, S.; Tian, W.; Wu, Z.; Fang, Y.; Dai, J.; Chen, C. Advantages of AlGaN-Based 310-nm UV Light-Emitting Diodes With Al Content Graded AlGaN Electron Blocking Layers. *Photonics J. IEEE* **2013**, *5*, 8200309.
89. Kuo, Y.K.; Chen, F.M.; Chang, J.Y.; Huang, M.F.; Liou, B.T.; Shih, Y.H. Design and Optimization of Electron-Blocking Layer in Deep Ultraviolet Light-Emitting Diodes. *IEEE J. Quantum Electron.* **2020**, *56*, 1–6. [CrossRef]
90. Velpula, R.T.; Jain, B.; Velpula, S.; Nguyen, H.D.; Nguyen, H.P.T. High-performance electron-blocking-layer-free deep ultraviolet light-emitting diodes implementing a strip-in-a-barrier structure. *Opt. Lett.* **2020**, *45*, 5125–5128. [CrossRef]
91. Lang, J.; Xu, F.J.; Sun, Y.H.; Zhang, N.; Wang, J.M.; Liu, B.Y.; Wang, L.B.; Xie, N.; Fang, X.Z.; Kang, X.N.; et al. Carrier Velocity Modulation by Asymmetrical Concave Quantum Barriers to Improve the Performance of AlGaN-Based Deep Ultraviolet Light Emitting Diodes. *IEEE Photonics J.* **2021**, *13*, 1–8. [CrossRef]
92. Lang, J.; Xu, F.J.; Ge, W.K.; Liu, B.Y.; Zhang, N.; Sun, Y.H.; Wang, M.X.; Xie, N.; Fang, X.Z.; Kang, X.N.; et al. High performance of AlGaN deep-ultraviolet light emitting diodes due to improved vertical carrier transport by delta-accelerating quantum barriers. *Appl. Phys. Lett.* **2019**, *114*, 172105. [CrossRef]
93. Velpula, R.T.; Jain, B.; Bui, H.Q.T.; Shakiba, F.M.; Jude, J.; Tumuna, M.; Nguyen, H.D.; Lenka, T.R.; Nguyen, H.P.T. Improving carrier transport in AlGaN deep-ultraviolet light-emitting diodes using a strip-in-a-barrier structure. *Appl. Opt.* **2020**, *59*, 5276. [CrossRef] [PubMed]
94. Wu, J.; Li, P.; Zhou, X.; Wu, J.; Hao, Y. Increasing the Carrier Injection Efficiency of GaN-Based Ultraviolet Light-Emitting Diodes by Double Al Composition Gradient Last Quantum Barrier and p-Type Hole Supply Layer. *IEEE Photonics J.* **2021**, *13*, 1–8. [CrossRef]
95. He, L.; Zhao, W.; Zhang, K.; He, C.; Wu, H.; Liu, X.; Luo, X.; Li, S.; Chen, Z. Marked enhancement in the efficiency of deep ultraviolet light-emitting diodes by using a $Al_xGa_{1-x}N$ carrier reservoir layer. *Appl. Phys. Express* **2019**, *12*, 062013. [CrossRef]
96. Chu, C.; Tian, K.; Che, J.; Shao, H.; Kou, J.; Zhang, Y.; Li, Y.; Wang, M.; Zhu, Y.; Zhang, Z.H. On the origin of enhanced hole injection for AlGaN-based deep ultraviolet light-emitting diodes with AlN insertion layer in p-electron blocking layer. *Opt. Express* **2019**, *27*, A620. [CrossRef] [PubMed]
97. Shi, L.; Du, P.; Guoyi, T.; Liu, Z.; Luo, W.; Liu, S.; Zhou, S. High efficiency electron-blocking-layer-free deep ultraviolet LEDs with graded Al-content AlGaN insertion layer. *Superlattices Microstruct.* **2021**, *158*, 107020. [CrossRef]
98. Bremser, M.D.; Perry, W.G.; Zheleva, T.; Edwards, N.V.; Nam, O.H.; Parikh, N.; Aspnes, D.E.; Davis, R.F. Growth, doping and characterization of $Al_xGa_{1-x}N$ thin film alloys on 6H-SiC(0001) substrates. *Diam. Relat. Mater.* **1997**, *6*, 196–201. [CrossRef]
99. Kinoshita, T.; Obata, T.; Yanagi, H.; Inoue, S.i. High p-type conduction in high-Al content Mg-doped AlGaN. *Appl. Phys. Lett.* **2013**, *102*, 012105. [CrossRef]
100. Zheng, T.; Lin, W.; Cai, D.; Weihuang, Y.; Jiang, W.; Chen, H.; Li, J.; Li, S.; Kang, J. High Mg effective incorporation in Al-rich $Al_xGa_{1-x}N$ by periodic repetition of ultimate V/III ratio conditions. *Nanoscale Res. Lett.* **2014**, *9*, 40. [CrossRef]
101. Stampfl, C.; Neugebauer, J.; Van de Walle, C.G. Doping of AlxGa1xN alloys. *Mater. Sci. Eng. B* **1999**, *59*, 253–257. [CrossRef]
102. Yan, Q.; Janotti, A.; Scheffler, M.; Van de Walle, C.G. Role of nitrogen vacancies in the luminescence of Mg-doped GaN. *Appl. Phys. Lett.* **2012**, *100*, 142110. [CrossRef]
103. Yan, Q.; Janotti, A.; Scheffler, M.; Van de Walle, C.G. Origins of optical absorption and emission lines in AlN. *Appl. Phys. Lett.* **2014**, *105*, 111104. [CrossRef]
104. Nakarmi, M.L.; Nepal, N.; Lin, J.Y.; Jiang, H.X. Photoluminescence studies of impurity transitions in Mg-doped AlGaN alloys. *Appl. Phys. Lett.* **2009**, *94*, 091903. [CrossRef]
105. Liang, Y.H.; Towe, E. Progress in efficient doping of high aluminum-containing group III-nitrides. *Appl. Phys. Rev.* **2018**, *5*, 011107. [CrossRef]
106. Jiang, K.; Sun, X.; Shi, Z.; Zang, H.; Ben, J.; Deng, H.X.; Li, D. Quantum engineering of non-equilibrium efficient p-doping in ultra-wide band-gap nitrides. *Light Sci. Appl.* **2021**, *10*, 69. [CrossRef]
107. Bryan, Z.; Hoffmann, M.; Tweedie, J.; Kirste, R.; Callsen, G.; Bryan, I.; Rice, A.; Bobea, M.; Mita, S.; Xie, J.; et al. Fermi Level Control of Point Defects During Growth of Mg-Doped GaN. *J. Electron. Mater.* **2012**, *42*, 815–819. [CrossRef]
108. Schubert, E.F.; Grieshaber, W.; Goepfert, I.D. Enhancement of deep acceptor activation in semiconductors by superlattice doping. *Appl. Phys. Lett.* **1996**, *69*, 3737–3739. [CrossRef]
109. Tersoff, J. Enhanced Solubility of Impurities and Enhanced Diffusion near Crystal Surfaces. *Phys. Rev. Lett.* **1995**, *74*, 5080–5083. [CrossRef]
110. Klump, A.; Hoffmann, M.P.; Kaess, F.; Tweedie, J.; Reddy, P.; Kirste, R.; Sitar, Z.; Collazo, R. Control of passivation and compensation in Mg-doped GaN by defect quasi Fermi level control. *J. Appl. Phys.* **2020**, *127*, 045702. [CrossRef]
111. Wu, R.; Shen, L.; Yang, M.; Sha, Z.; Cai, Y.; Feng, Y.; Huang, Z.; Wu, Q. Enhancing hole concentration in AlN by Mg:O codoping: Ab initio study. *Phys. Rev. B* **2008**, *77*, 073203. [CrossRef]
112. Aoyagi, Y.; Takeuchi, M.; Iwai, S.; Hirayama, H. High hole carrier concentration realized by alternative co-doping technique in metal organic chemical vapor deposition. *Appl. Phys. Lett.* **2011**, *99*, 112110. [CrossRef]
113. Aoyagi, Y.; Takeuchi, M.; Iwai, S.; Hirayama, H. Formation of AlGaN and GaN epitaxial layer with high p-carrier concentration by pulse supply of source gases. *AIP Adv.* **2012**, *2*, 012177. [CrossRef]
114. Al tahtamouni, T.M.; Lin, J.Y.; Jiang, H.X. Effects of Mg-doped AlN/AlGaN superlattices on properties of p-GaN contact layer and performance of deep ultraviolet light emitting diodes. *AIP Adv.* **2014**, *4*, 047122. [CrossRef]

115. Nakarmi, M.L.; Kim, K.H.; Li, J.; Lin, J.Y.; Jiang, H.X. Enhanced p-type conduction in GaN and AlGaN by Mg–doping. *Appl. Phys. Lett.* **2003**, *82*, 3041–3043. [CrossRef]
116. Chen, Y.; Wu, H.; Han, E.; Yue, G.; Chen, Z.; Wu, Z.; Wang, G.; Jiang, H. High hole concentration in p-type AlGaN by indium-surfactant-assisted Mg-delta doping. *Appl. Phys. Lett.* **2015**, *106*, 162102. [CrossRef]
117. Kim, K.; Li, J.; Jin, S.; Lin, J.Y.; Jiang, H. III-nitride ultraviolet light-emitting diodes with delta doping. *Appl. Phys. Lett.* **2003**, *83*, 566–568. [CrossRef]
118. Bayram, C.; Pau, J.; McClintock, R.; Razeghi, M. Performance enhancement of GaN ultraviolet avalanche photodiodes with p-type -doping. *Appl. Phys. Lett.* **2008**, *92*, 241103. [CrossRef]
119. Li, T.; Simbrunner, C.; Wegscheider, M.; Navarro-Quezada, A.; Quast, M.; Schmidegg, K.; Bonanni, A. GaN:-Mg grown by MOVPE: Structural properties and their effect on the electronic and optical behavior. *J. Cryst. Growth* **2008**, *310*, 13–21. [CrossRef]
120. Ambacher, O.; Smart, J.; Shealy, J.R.; Weimann, N.G.; Chu, K.; Murphy, M.; Schaff, W.J.; Eastman, L.F.; Dimitrov, R.; Wittmer, L.; et al. Two-dimensional electron gases induced by spontaneous and piezoelectric polarization charges in N- and Ga-face AlGaN/GaN heterostructures. *J. Appl. Phys.* **1999**, *85*, 3222–3233. [CrossRef]
121. Li, S.; Ware, M.; Wu, J.; Minor, P.; Wang, Z.; Wu, Z.; Jiang, Y.; Salamo, G.J. Polarization induced pn-junction without dopant in graded AlGaN coherently strained on GaN. *Appl. Phys. Lett.* **2012**, *101*, 122103. [CrossRef]
122. Simon, J.; Protasenko, V.; Lian, C.; Xing, H.; Jena, D. Polarization-Induced Hole Doping in Wide–Band-Gap Uniaxial Semiconductor Heterostructures. *Science* **2010**, *327*, 60–64. [CrossRef]
123. Xi, Y.; Chen, K.; Mont, F.; Kim, J.K.; Schubert, E.; Liu, W.; Li, X.; Smart, J. Comparative study of n-type AlGaN grown on sapphire by using a superlattice layer and a low-temperature AlN interlayer. *J. Cryst. Growth* **2007**, *299*, 59–62. [CrossRef]
124. Kozodoy, P.; Hansen, M.; DenBaars, S.P.; Mishra, U.K. Enhanced Mg doping efficiency in Al0.2Ga0.8N/GaN superlattices. *Appl. Phys. Lett.* **1999**, *74*, 3681–3683. [CrossRef]
125. Kumakura, K.; Makimoto, T.; Kobayashi, N. Enhanced Hole Generation in Mg-Doped AlGaN/GaN Superlattices Due to Piezoelectric Field. *Jpn. J. Appl. Phys.* **2000**, *39*, 2428–2430. [CrossRef]
126. Zhang, J.; Gao, Y.; Zhou, L.; Gil, Y.U.; Kim, K.M. Surface hole gas enabled transparent deep ultraviolet light-emitting diode. *Semicond. Sci. Technol.* **2018**, *33*, 07LT01. [CrossRef]
127. Zheng, T.; Lin, W.; Liu, R.; Cai, D.; Li, J.; Li, S.; Kang, J. Improved p-type conductivity in Al-rich AlGaN using multidimensional Mg-doped superlattices. *Sci. Rep.* **2016**, *6*, 21897. [CrossRef] [PubMed]
128. Ebata, K.; Nishinaka, J.; Taniyasu, Y.; Kumakura, K. High hole concentration in Mg-doped AlN/AlGaN superlattices with high Al content. *Jpn. J. Appl. Phys.* **2018**, *57*, 04FH09. [CrossRef]
129. Allerman, A.; Crawford, M.; Miller, M.; Lee, S.R. Growth and characterization of Mg-doped AlGaN–AlN short-period superlattices for deep-UV optoelectronic devices. *J. Cryst. Growth* **2010**, *312*, 756–761. [CrossRef]
130. Wang, J.; Wang, M.; Xu, F.; Liu, B.; Lang, J.; Zhang, N.; Kang, X.; Qin, Z.; Yang, X.; Wang, X.Q.; et al. Sub-nanometer ultrathin epitaxy of AlGaN and its application in efficient doping. *Light Sci. Appl.* **2022**, *11*, 71. [CrossRef]
131. Nam, K.B.; Li, J.; Nakarmi, M.L.; Lin, J.Y.; Jiang, H.X. Unique optical properties of AlGaN alloys and related ultraviolet emitters. *Appl. Phys. Lett.* **2004**, *84*, 5264–5266. [CrossRef]
132. Kolbe, T.; Knauer, A.; Chua, C.; Yang, Z.; Einfeldt, S.; Vogt, P.; Johnson, N.M.; Weyers, M.; Kneissl, M. Optical polarization characteristics of ultraviolet (In)(Al)GaN multiple quantum well light emitting diodes. *Appl. Phys. Lett.* **2010**, *97*, 171105. [CrossRef]
133. Ryu, H.Y.; Choi, I.G.; Choi, H.S.; Shim, J.I. Investigation of Light Extraction Efficiency in AlGaN Deep-Ultraviolet Light-Emitting Diodes. *Appl. Phys. Express* **2013**, *6*, 062101. [CrossRef]
134. Muhin, A.; Guttmann, M.; Kuhn, C.; Mickein, E.; Aparici, J.R.; Ziffer, E.; Susilo, N.; Sulmoni, L.; Wernicke, T.; Kneissl, M. Vertical conductivity and Poole–Frenkel-ionization of Mg acceptorsin AlGaN short-period superlattices with high Al mole fraction. *Appl. Phys. Lett.* **2020**, *117*, 252101. [CrossRef]
135. Guttmann, M.; Susilo, A.; Sulmoni, L.; Susilo, N.; Ziffer, E.; Wernicke, T.; Kneissl, M. Light extraction efficiency and internal quantum efficiency of fully UVC-transparent AlGaN based LEDs. *J. Phys. D: Appl. Phys.* **2021**, *54*, 335101. [CrossRef]
136. Khan, M.A.; Maeda, N.; Jo, M.; Akamatsu, Y.; Tanabe, R.; Yamada, Y.; Hirayama, H. 13 mW operation of a 295–310 nm AlGaN UV-B LED with a p-AlGaN transparent contact layer for real world applications. *J. Mater. Chem. C* **2018**, *7*, 143–152. [CrossRef]
137. Nakashima, T.; Takeda, K.; Shinzato, H.; Iwaya, M.; Kamiyama, S.; Takeuchi, T.; Akasaki, I.; Amano, H. Combination of Indium–Tin Oxide and SiO2/AlN Dielectric Multilayer Reflective Electrodes for Ultraviolet-Light-Emitting Diodes. *Jpn. J. Appl. Phys.* **2013**, *52*, 08JG07. [CrossRef]
138. Kashima, Y.; Maeda, N.; Matsuura, E.; Jo, M.; Iwai, T.; Morita, T.; Kokubo, M.; Tashiro, T.; Kamimura, R.; Osada, Y.; et al. High external quantum efficiency (10%) AlGaN-based deep-ultraviolet light-emitting diodes achieved by using highly reflective photonic crystal on p-AlGaN contact layer. *Appl. Phys. Express* **2017**, *11*, 012101. [CrossRef]
139. Matsukura, Y.; Inazu, T.; Pernot, C.; Shibata, N.; Kushimoto, M.; Deki, M.; Honda, Y.; Amano, H. Improving light output power of AlGaN-based deep-ultraviolet light-emitting diodes by optimizing the optical thickness of p-layers. *Appl. Phys. Express* **2021**, *14*, 084004. [CrossRef]
140. Schubert, M.F. Interband tunnel junctions for wurtzite III-nitride semiconductors based on heterointerface polarization charges. *Phys. Rev. B* **2010**, *81*, 035303. [CrossRef]

141. Mehnke, F.; Kuhn, C.; Guttmann, M.; Sulmoni, L.; Montag, V.; Glaab, J.; Wernicke, T.; Kneissl, M. Electrical and optical characteristics of highly transparent MOVPE-grown AlGaN-based tunnel heterojunction LEDs emitting at 232 nm. *Photon. Res.* **2021**, *9*, 1117–1123. [CrossRef]
142. Banal, R.G.; Funato, M.; Kawakami, Y. Optical anisotropy in [0001]-oriented $Al_xGa_{1-x}N/AlN$ quantum wells ($x > 0.69$). *Phys. Rev. B* **2009**, *79*, 121308. [CrossRef]
143. Reich, C.; Guttmann, M.; Feneberg, M.; Wernicke, T.; Mehnke, F.; Kuhn, C.; Rass, J.; Lapeyrade, M.; Einfeldt, S.; Knauer, A.; et al. Strongly transverse-electric-polarized emission from deep ultraviolet AlGaN quantum well light emitting diodes. *Appl. Phys. Lett.* **2015**, *107*, 142101. [CrossRef]
144. Zheng, J.; Li, J.; Zhong, Z.; Lin, W.; Chen, L.; Li, K.; Wang, X.; Chou, C.; Li, S.; Kang, J. Effect of electrical injection-induced stress on interband transitions in high Al content AlGaN MQWs. *RSC Adv.* **2017**, *7*, 55157–55162. [CrossRef]
145. Bryan, Z.; Bryan, I.; Mita, S.; Tweedie, J.; Sitar, Z.; Collazo, R. Strain dependence on polarization properties of AlGaN and AlGaN-based ultraviolet lasers grown on AlN substrates. *Appl. Phys. Lett.* **2015**, *106*, 232101. [CrossRef]
146. Long, H.; Wang, S.; Dai, J.; Wu, F.; Zhang, J.; Chen, J.; Liang, R.; Feng, Z.; Chen, C. Internal strain induced significant enhancement of deep ultraviolet light extraction efficiency for AlGaN multiple quantum wells grown by MOCVD. *Opt. Express* **2018**, *26*, 680. [CrossRef]
147. Zhang, S.; Zhang, Y.; Tang, N.; Wang, W.; Chen, X.; Fu, L.; He, C.; Lv, Y.; Feng, Z.; Xu, F.; et al. Compressive strain induced enhancement of transverse-electric polarized ultraviolet light emission for AlGaN quantum wells. *Superlattices Microstruct.* **2020**, *150*, 106749. [CrossRef]
148. Bayerl, D.; Islam, S.; Jones, C.M.; Protasenko, V.; Jena, D.; Kioupakis, E. Deep ultraviolet emission from ultra-thin GaN/AlN heterostructures. *Appl. Phys. Lett.* **2016**, *109*, 241102. [CrossRef]
149. Gao, N.; Chen, J.; Feng, X.; Lu, S.; Lin, W.; Li, J.; Chen, H.; Huang, K.; Kang, J. Strain engineering of digitally alloyed AlN/GaN nanorods for far-UVC emission as short as 220 nm. *Opt. Mater. Express* **2021**, *11*, 1282. [CrossRef]

crystals

MDPI

Review

Reliability, Applications and Challenges of GaN HEMT Technology for Modern Power Devices: A Review

Naeemul Islam [1], Mohamed Fauzi Packeer Mohamed [1,*], Muhammad Firdaus Akbar Jalaludin Khan [1], Shaili Falina [2,3,*], Hiroshi Kawarada [3,4] and Mohd Syamsul [4,5,*]

1 School of Electrical and Electronic Engineering, Universiti Sains Malaysia, Nibong Tebal 14300, Malaysia
2 Collaborative Microelectronic Design Excellence Center (CEDEC), Universiti Sains Malaysia, Sains@USM, Bayan Lepas 11900, Malaysia
3 Faculty of Science and Engineering, Waseda University, Tokyo 169-8555, Japan
4 The Kagami Memorial Laboratory for Materials Science and Technology, Waseda University, Tokyo 169-0051, Japan
5 Institute of Nano Optoelectronics Research and Technology (INOR), Universiti Sains Malaysia, Sains@USM, Bayan Lepas 11900, Malaysia
* Correspondence: fauzi.packeer@usm.my (M.F.P.M.); shailifalina@usm.my (S.F.); nasyriq@usm.my (M.S.)

Abstract: A new generation of high-efficiency power devices is being developed using wide bandgap (WBG) semiconductors, like GaN and SiC, which are emerging as attractive alternatives to silicon. The recent interest in GaN has been piqued by its excellent material characteristics, including its high critical electric field, high saturation velocity, high electron mobility, and outstanding thermal stability. Therefore, the superior performance is represented by GaN-based high electron mobility transistor (HEMT) devices. They can perform at higher currents, voltages, temperatures, and frequencies, making them suitable devices for the next generation of high-efficiency power converter applications, including electric vehicles, phone chargers, renewable energy, and data centers. Thus, this review article will provide a basic overview of the various technological and scientific elements of the current GaN HEMTs technology. First, the present advancements in the GaN market and its primary application areas are briefly summarized. After that, the GaN is compared with other devices, and the GaN HEMT device's operational material properties with different heterostructures are discussed. Then, the normally-off GaN HEMT technology with their different types are considered, especially on the recessed gate metal insulator semiconductor high electron mobility transistor (MISHEMT) and p-GaN. Hereafter, this review also discusses the reliability concerns of the GaN HEMT which are caused by trap effects like a drain, gate lag, and current collapse with numerous types of degradation. Eventually, the breakdown voltage of the GaN HEMT with some challenges has been studied.

Keywords: GaN HEMT; normally off; reliability; challenges; power devices; semiconductor devices; wide bandgap

Citation: Islam, N.; Mohamed, M.F.P.; Khan, M.F.A.J.; Falina, S.; Kawarada, H.; Syamsul, M. Reliability, Applications and Challenges of GaN HEMT Technology for Modern Power Devices: A Review. *Crystals* **2022**, *12*, 1581. https://doi.org/10.3390/cryst12111581

Academic Editors: Peng Chen, Zhizhong Chen and Dmitri Donetski

Received: 29 September 2022
Accepted: 2 November 2022
Published: 7 November 2022

1. Introduction

Electrical power has been a fundamental driver of humanity's progress and is indispensable in our daily life. Power electronics are critical components of electrical power usage because they control, convert, and manage electric currents, voltages, or powers. They are extensively utilized in consumer items, energy harvesting, and usage, such as switching power supplies, power converters, power inverters, motor drives, etc. Figure 1 shows the different fields of operation of power devices with the required voltage and current ratings [1,2]. As depicted in Figure 2, a typical power electronic system includes a power source, a filtering mechanism, a power conditioner, a load, and a control circuit. A power conditioner is composed of a series of semiconductor devices working in a switched mode, in which the devices are turned between the "OFF" and "ON" states under the direction of the gate driver [1]. When turned "off", an ideal switch should have no leakage current and a full voltage drop across it. If switched "on", it should have a zero voltage drop across it (Figure 3). In actuality, the power devices exhibit a resistance during the on-state (R_{ON}) and a leakage current during the off-state. Under the common biasing conditions, the voltage across it is restricted by the off-state drain to source the breakdown voltage (BV_{DSS}). Thus, to maximize the energy efficiency and minimize conversion losses, the most crucial properties of power devices to accomplish are a low R_{ON} and a high BV_{DSS}.

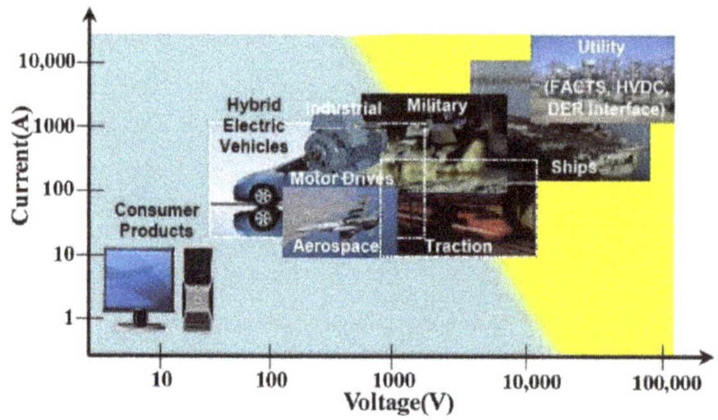

Figure 1. Applications of power electronics in power management and control. Reproduced with permission [2]. Copyright 2019, the author(s). Published by Elsevier.

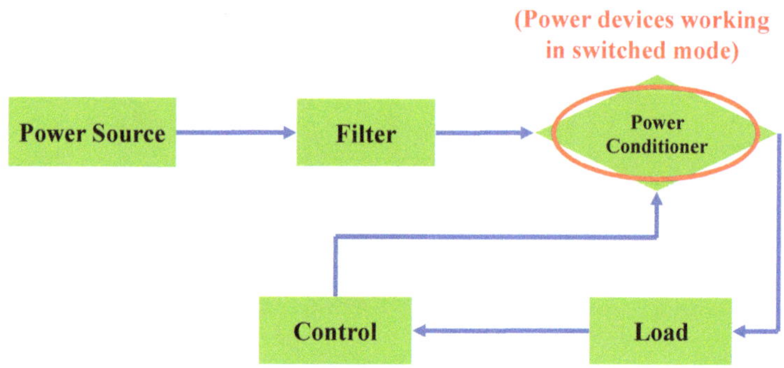

Figure 2. Components of a power electronic system and power semiconductor devices in power conditioners.

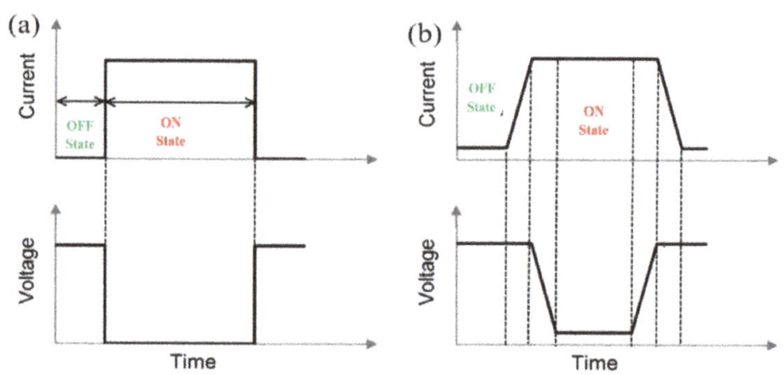

Figure 3. The current and voltage across (**a**) an ideal power switching device and (**b**) a power switching device.

Moreover, Gallium Nitride (GaN) has risen as an attractive material for fabricating semiconductor devices over the last decade. Its excellent performance characteristics, like its wide band gap, high electron mobility, high breakdown field, low noise, high saturation velocity, and low thermal impedance, are essential for modern power device technology [3].

This review article will provide a basic overview of the various technological and scientific elements of the current GaN HEMT technology, where we summarize the recent and future potential market strategy of GaN HEMT power devices. Additionally, the previous and present data of the most extensively used commercially device, the normally-off GaN HEMT, are amalgamated, focusing explicitly on the insulator's role in the recessed MISHEMT region and the metal on the p-GaN performances, which makes this review article more comprehensive compared to other review articles. Moreover, this review also discusses the reliability concerns of the GaN HEMT which are caused by trap effects like a drain, gate lag, and current collapse with numerous types of degradation.

1.1. Market Strategy of GaN HEMT Power Devices

GaN-based devices are being reported in various market-driven sectors, including RF power devices, photonics, high-frequency communications, control, and high-power conversion, which assists commerce's and corporations in meeting the constantly expanding need for more outstanding metrics. The inherent robustness of the device enables it to compete effectively in the current markets for tiny, rugged, and highly dependable electronics, such as in vehicle, military, and space applications. GaN engineering appears promising for achieving the critical performance requirements and the possibility of sustained economic benefits upon its maturity.

Figures 4 and 5 show that according to Yole Développement (Yole), the power GaN market will double by 2020, surpassing the USD one billion mark in 2026. As an additional validation of this remarkable GaN market development, Yole projects that the telecom and datacom and automotive sectors will contribute to the total growth in the mid- to long-term, owing to GaNs rise in fast chargers. In the consumer market, GaN had a good year in 2020, owing to the technology adoption by various businesses, including Xiaomi, Lenovo, Samsung, Realme, Dell, LG, and other Chinese aftermarket firms. Yole anticipates that the GaN consumer power supply industry will be the primary driver, growing from USD 29 million in 2020 to about USD 672 million in 2026 at a compound annual growth rate (CAGR) of 69%. In the telecom and datacom sectors, where more efficient, smaller power supplies must comply with the stricter energy consumption standards, data centers and telecom operators are already interested in GaN devices.

Following Eltek, Delta, and BelPower's recent small volume adoption of the GaN-based power supply, it was predicted that there would be a more significant market for GaN in 2020, with a market value of USD 9.1 million and a CAGR of 71% from 2020 to 2026, reaching more than USD 223 million in 2026. The automotive and mobility markets are also paying close attention to GaN because of significant incentives for vehicle electrification and a need to increase the driving range via A system efficiency improvement. AEC-qualified manufacturers include EPC, Transphorm, GaN Systems, and Texas Instruments. STMicroelectronics, a major IDM, also pursues GaN for EVs via partnerships and acquisitions. In 2022, GaN will likely infiltrate at limited levels, primarily via OEM and Tier-1 samples. Yole anticipates that the automobile and mobility industry will exceed a value of USD 155 million in 2026 [4].

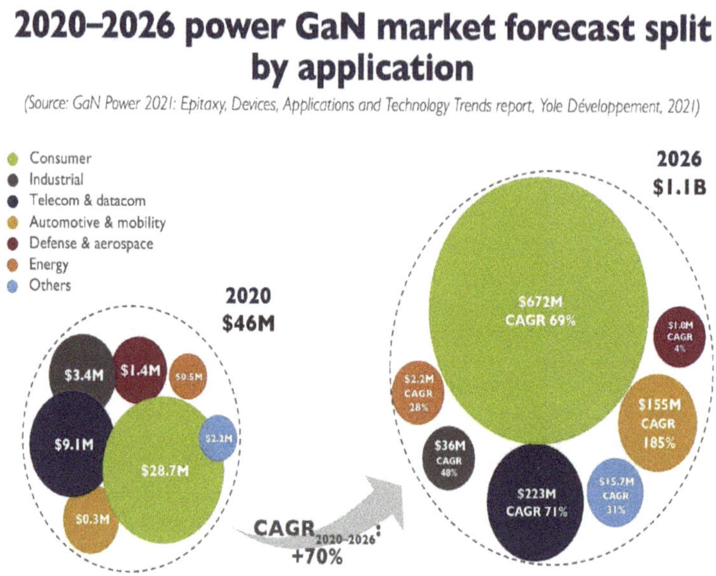

Figure 4. GaN power market forecast split by application prediction by Yole. Reproduced with permission [4]. Copyright 2021, published by Yole Développement.

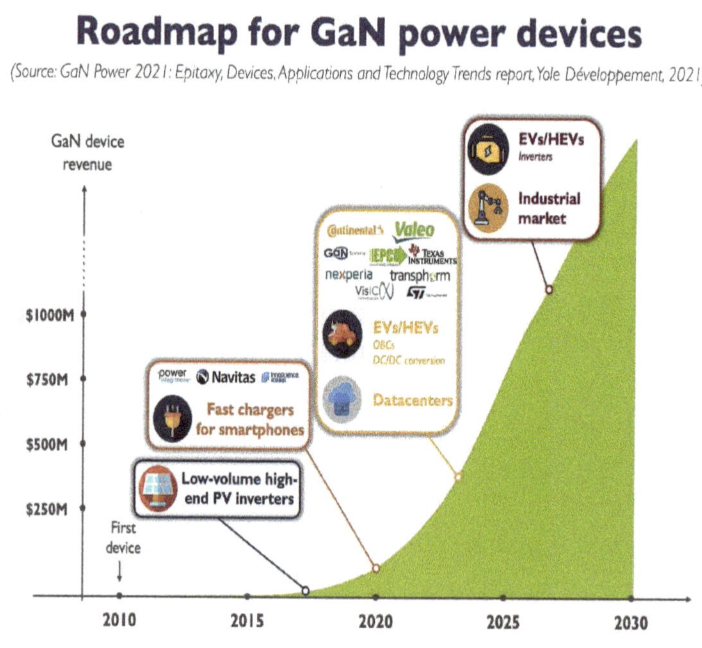

Figure 5. Roadmap for GaN power devices prediction by Yole. Reproduced with permission [4]. Copyright 2021, published by Yole Développement.

While GaN continues to gain traction in the general consumer market, telecom, datacom, automotive, and mobility businesses will benefit from the "economy of scale impact" and price erosion. Indeed, Yole anticipates that a GaN adoption will increase in these areas, where reliability and affordability are critical, beginning in 2023–2024. In the long run, after GaN has shown its durability and high current capability at a lower price, it may be able to reach the more difficult EV/HEV inverter market and the conservative industrial industry, creating an extraordinary high-volume potential for GaN. Indeed, Nexperia and VisIC are developing GaN-based xEV inverter systems to compete with SiC and Si. Furthermore, as various industry experts previously said, GaN is more suitable for low-to-medium voltage applications (200–600 V), including a sizable share of the consumer electronic market (e.g., phone chargers, audio amplifiers, and computer power supplies). GaN is the most excellent contender to replace conventional Si devices in this voltage range. The 600–900 V voltage range is strategically important since it includes converters for electric cars (EVs), hybrid EVs, and inverters for sustainable energy (e.g., photovoltaic). GaN devices are projected to compete or coexist alongside SiC devices in this voltage range. Recent industry analyst studies [5] show that the GaN power device market is expected to grow dramatically, surpassing a value of USD 700 million in 2025, as illustrated in Figure 6.

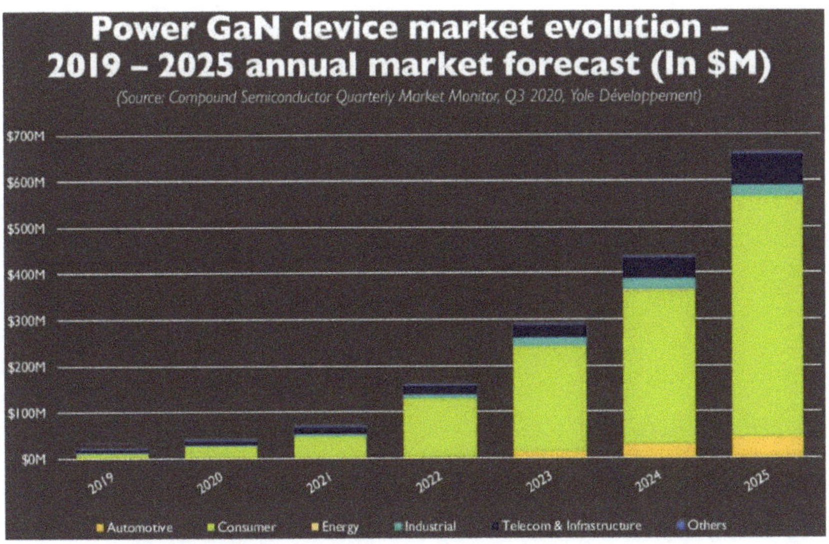

Figure 6. GaN Power device market size prediction by Yole. Reproduced with permission [5]. Copyright 2020, published by Yole Développement.

1.2. Market Strategy of GaN HEMT RF Devices

The gallium nitride (GaN) radio-frequency (RF) device market is rising at a compound annual growth rate (CAGR) of 18% from a USD 891 million value in 2020 to a more than USD 2.4 billion value by 2026, forecasts the market analyst firm Yole Développement [6]. It has been forecasted that the market will be dominated by defense and 5G telecom infrastructure applications, representing 49% and 41% of the entire market by 2026, respectively. In particular, the GaN-based macro/micro-cell sector will represent more than 95% of the GaN telecom infrastructure market in 2026, as reflected in Figure 7.

In the RF GaN industry, everything started with GaN-on-SiC (gallium nitride on silicon carbide) technology. Launched more than 20 years ago, GaN-on-SiC is now a serious rival to silicon LDMOS and gallium arsenide (GaAs) in RF power applications. Dominated by GaN-on-SiC technology, a vertical supply chain integration has been preferred in defense and 5G telecom applications. Therefore, GaN-on-SiC technology is still preferred in terms of its high-power density and thermal conductivity. In addition to its deep penetration in the military radar, GaN-on-SiC has also been the choice of telecom OEMs such as Huawei, Nokia, and Samsung for 5G massive MIMO infrastructures. Due to their high bandwidth and efficiency, GaN-on-SiC devices continuously take a share from LDMOS of the 5G market and are starting to benefit from the 6″ wafer platform transition. In this context, the GaN-on-SiC device market is growing at a 17% CAGR from USD 342 million in 2020 to USD 2.222 billion in 2026. However, as a key challenger, GaN-on-Si is still in the game, promising cost-efficient and scalable solutions. The recent entry of foundries and synergy with the emerging power electronics GaN-on-Si industry can also help GaN-on-Si RF to gain momentum in the longer term, says Yole. Driven by handsets but also defense and 5G telecom infrastructure applications, the GaN-on-Si device market is growing at a CAGR of 86% from less than USD 5 million in 2020 to USD 173 million in 2026 [6], as illustrated in Figure 8.

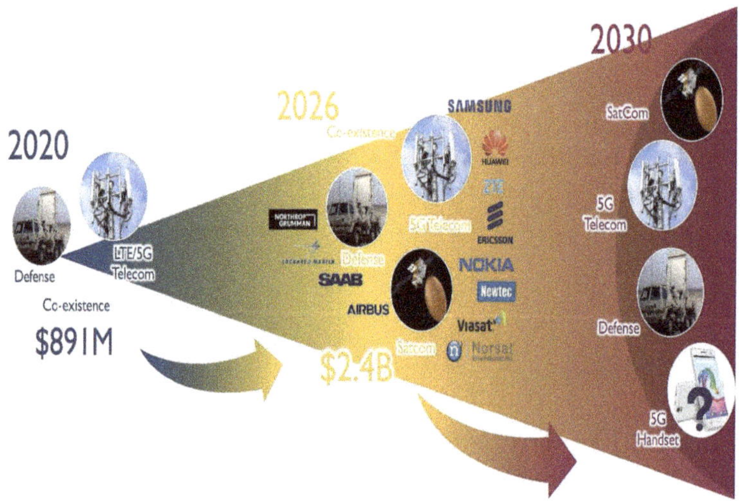

Figure 7. GaN Power RF market evaluation by Yole. Reproduced with permission [6]. Copyright 2021, published by Yole Développement.

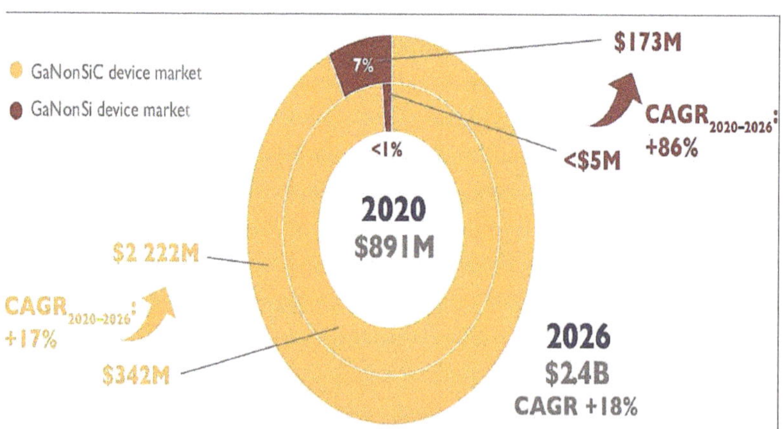

Figure 8. GaN RF application breakdown by substrate markets as compiled in 2021 by Yole. Reproduced with permission [6]. Copyright 2021, published by Yole Développement.

2. Wide Bandgap Semiconductors for Power Devices

Due to the many constraints of silicon-based power devices and the rising interest in increasing performances, new semiconductor materials for next-generation power electronic devices are being developed. Because of their better electrical properties, wide bandgap (WBG) semiconductors like GaN, β-Ga_2O_3, and SiC have emerged as market leaders in power electronic applications. Since the material's small bandgap is responsible for many of the constraints for Si-based devices, wide bandgap power devices have a better performance with a larger blocking voltage, dependability, and efficiency with less thermal constraint [7–9]. Thus, GaN, β-Ga_2O_3, and SiC have a bandgap which is two to four times that of Silicon. A bigger energy bandgap makes it more challenging to break the bonding and generate free charge carriers for a current conduction, which results in a less conductive material and transistors with lower leakage currents and a more excellent stability at high temperatures. On the contrary, β-Ga_2O_3 devices have a substantially greater breakdown voltage compared to Si devices when their on-resistances are comparable. Moreover, GaN devices have a larger saturation on-current than Si devices due to the higher saturated electron velocity of GaN. Additionally, the carrier drift velocity directly connects with the switching speed of semiconductor devices, and this device has a quicker recovery time and a lower reverse recovery current. On the other hand, SiC presents a thermal conductivity three times higher than Si and GaN and 16 times higher than β-Ga_2O_3. The characteristics of Si are compared to those of WBG semiconductors such as SiC, β-Ga_2O_3, and GaN in Figure 9 [10–12].

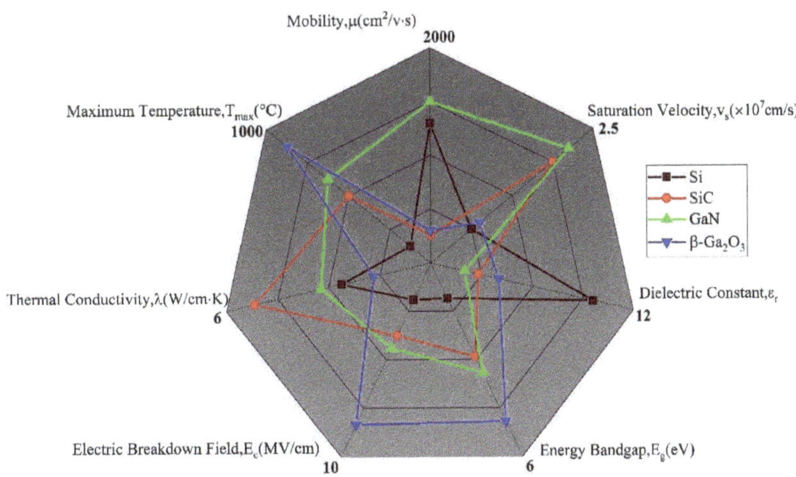

Figure 9. Comparison of wide bandgap materials with Si.

However, numerous figures of merit (FOM) have been proposed to compare the merit of several semiconductor materials for a provided application. A higher figure of merit number indicates a better performance. Hence, Johnson's figure of merit (JFOM) provides an indication of the material suitability for high-power applications at high frequencies, which can be estimated by using Equation (1) [13]:

$$\text{JFOM} = \left(\frac{E_c V_{sat}}{2\pi}\right)^2 \tag{1}$$

Here, V_{sat} denotes the saturation electron velocity, and E_c denotes the critical electric field. On the contrary, to analyze the optimum results of field effect transistors in low-frequency power switching applications, Baliga's figure of merit (BFOM) was devised where conductive loss predominates, as shown in Equation (2) [14]:

$$BFOM = \varepsilon_r \mu E_g{}^3 \tag{2}$$

Here, ε_r denotes the semiconductor's dielectric constant, and μ denotes the electron mobility. For a high frequency, the Baliga high-frequency FOM (BHFFOM) in Equation (3) evaluates the devices where the switching losses are the most significant [14]:

$$BHFFOM = \frac{\mu E_c{}^2 V_g{}^{\frac{1}{2}}}{2V_{BR}{}^{\frac{3}{2}}} \tag{3}$$

Here, the gate drive voltage is V_g, while the breakdown voltage is V_{BR}. By using the device's on-state specific resistance, R_{on}, and the critical electric field, E_c, the latter is coupled to an experimental figure of merit on Equation (4). Eventually, to account for the high power, high temperature, and high-frequency performance simultaneously, the combined FOM (CFOM) was devised, as shown in Equation (5) [15]:

$$FOM = \frac{V_{BR}{}^2}{R_{on}} = \varepsilon_r \mu \frac{E_c{}^3}{4} \tag{4}$$

$$CFOM = \chi \varepsilon_r \mu V_{sat} E_c{}^2 \tag{5}$$

Table 1 shows the multiple figures of merit for the β-Ga_2O_3 potential high-frequency and high-power performance compared to other competing semiconductor materials. According to this table, β-Ga_2O_3 is a great contender for high-frequency power applications. In JFOM, the β-Ga_2O_3 is more remarkable than approximately 1.5 times that of GaN, 10 times that of SiC, and about a thousand times that of Si. Consequently, β-Ga_2O_3 outperforms Si, GaN, and SiC in terms of its high-frequency/high-power performance. Similarly, in BFOM, β-Ga_2O_3 is more remarkable than around four times that of GaN, ten times that of SiC, and a thousand times that of Si [16]. Moreover, wide-bandgap semiconductors, regardless of their FOM, provide better performance features for high-power, high-frequency applications. Thus, due to the high values of the Baliga FOM and the breakdown field, β-Ga_2O_3 is a promising candidate for the next generation of high-power devices, including HEMT, Schottky barrier diodes (SBDs), and field-effect transistors (FETs).

Table 1. Figures of merit of GaN and competing semiconductor materials in power electronics [12].

Material	Johnson's Figure of Merit, (JFOM)	Keyes Figure of Merit, (KFOM)	Baliga's Figure of Merit, (BFOM)	Baliga's High Frequency Figure of Merit, (BHFFOM)	Combined Figure of Merit, (CFOM)
Si	1.0	1.0	1.0	1.0	1.0
SiC	277.8	3.6	317.1	46.3	248.6
GaN	1089.0	1.8	846.0	100.8	353.8
β-Ga_2O_3	2844.4	0.2	3214.1	142.2	37.0

2.1. *Lattice Structure and Polarization of GaN-Based Semiconductors*

Wurtzite hexagonal close-packed (HCP) (α-phase), rock-salt, and cubic zincblende (β-phase) are examples of III-N-based semiconductors, mainly GaN-based semiconductors [17,18]. Figure 10 depicts the wurtzite crystal structure, which is particularly important for electronics and optoelectronics applications today. The hexagonal unit cell of wurtzite comprises two intersecting hexagonal close packed (HCP) sub-lattices. As illustrated in Figure 10, it is defined by two lattice parameters, the height of the cell c_0, and the length of a side of the hexagonal base a_0, in an ideal ratio $\frac{c_0}{a_0} = \sqrt{\frac{8}{3}} \approx 1.633$. Each sub-lattice is made up of one kind of atom. In relation to one another, the internal cell parameter $u_0 = 3/8$ moves them along the c-axis. Table 2 presents the structural and polarization properties of III-Ns.

Table 2. Structural and polarization parameters of III-N wurtzite semiconductors [17].

Parameter	GaN	AlN	InN
a_0(A°)	3.197	3.108	3.580
c_0/a_0	1.6297	1.6033	1.6180
$\epsilon_1 = (u_0 - u_{ideal}) \times 10^{-3}$	1.9	6.4	3.7

These ideal values are absent in the typical nitride compounds GaN, InN, and AlN and their associated alloys. The c_0/a_0 ratio deviates from the ideal lattice's 1.633 as the non-ideality of the lattice grows [19].

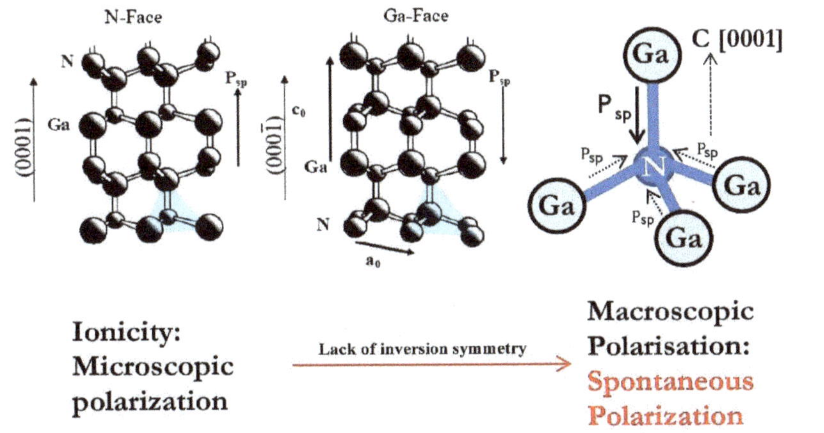

Figure 10. The alignment of P_{SP} and the tetrahedral bond structure of Ga-face and N-face heterostructures. Reproduced with permission [20]. Copyright 2019, Elsevier.

Because of the very high electronegativity of the N atom, which is 3.04 on Pauling's scale, and for Aluminum (1.61), Gallium (1.81), and Indium (1.78) atoms [19], consequently, the data for u_0 and c_0/a_0 deviate from the optimum values given earlier [21]. Group III-N semiconductors have a particularly high polarization due to the crystal structure's intrinsic electronic charge redistribution, which means a spontaneous polarization, P_{sp}, [19]. The directions <0001> and <000$\bar{1}$> are not comparable due to a lack of inversion symmetry shown along its c-axis in the wurtzite structure. Hence, cation-face, i.e., Ga-face and anion-face, i.e., N-face, are the alternative polarities of GaN. The polarization field points change their position in the cation-face, i.e., Ga-face, surface towards the substrate. On the contrary, the polarization field points move in the opposite direction in the anion-face, i.e., N-face.

Because of the lack of this inversion symmetry, when the group III-N semiconductors' lattice is imposed by stress in the <0001> direction, the ideal lattice values of the crystal structure's c_0 and a_0 will shift to adapt to the stress. As a result, the strength of the polarization would shift. In the strained group III-N crystals, this extra polarization is known as piezoelectric polarization, P_{pz} [22]. For instance, the in-plane constant of the lattice a_0 would drop, and the vertical constant of the lattice c_0 would rise if the crystal of the nitride is subjected towards a biaxial compressive stress. Due to the piezoelectric and spontaneous polarizations working in opposing directions, the ratio of c_0/a_0 will climb to 1.633 of the perfect lattices, and the overall polarization strength will be reduced in the crystal. Moreover, when the tensile stress is imposed on the crystal, the spontaneous and piezoelectric polarization work simultaneously. As a result, the total polarization will rise. The piezoelectric coefficients e_{33} and e_{13} are evaluated for the piezoelectric polarization, P_{pz} [23,24], as shown in Equation (6),

$$P_{pz} = e_{33}\varepsilon_3 + e_{13}(\varepsilon_1 + \varepsilon_2) \tag{6}$$

Here, a_0 and c_0 are the lattice parameter equilibrium values, $\varepsilon_3 = (c - c_0)/c_0$ is the strain along the c axis, and $\varepsilon_1 = \varepsilon_2 = (a - a_0)/a_0$ is the isotropic in-plane strain. The various strains in the lattice are connected in Equation (7). The elastic constants denoted by c_{13} and c_{33}. Combining Equations (1) and (2) provides the corresponding formula in Equation (8):

$$\varepsilon_3 = -2\frac{c_{13}}{c_{33}}\varepsilon_1 \tag{7}$$

$$P_{pz} = 2\frac{a - a_0}{a_0}\left[e_{13} - e_{33}\frac{c_{13}}{c_{33}}\right] \tag{8}$$

The equation $\left(e_{13} - e_{33}\frac{c_{13}}{c_{33}}\right)$ will be negative if e_{33}, c_{13}, and c_{33} are positive, and the piezoelectric coefficient e_{13} is negative in wurtzite III-nitrides [25]. Thus, it is claimed that piezoelectric and spontaneous polarization are parallel under tensile stress layers, but they are anti-parallel under compressive stress layers. Polarization and gradients induce fixed sheet charges at the surfaces and interfaces of the AlGaN/GaN heterostructures. As a result, there are high electric fields within the heterostructure. The electric field in nitrides may approach 3×10^6 V/cm, which enhances the hole or electron formation at the AlGaN/GaN contacts. This is called polarization-induced doping, which is the root of the two-dimensional electrons gas (2DEG) (discussed more in the below segment).

Figure 11a shows that growing a tiny epi AlGaN layer on top of a thicker GaN layer generates an AlGaN/GaN heterostructure. A 2DEG is generated at the interface between the AlGaN barrier and the GaN channel because of the P_{sp} and P_{pz}, which may be a conduction channel consisting of exceptionally high mobility electrons even without any n-type doping, which is ideal for transistors. Both GaN and AlGaN have a significant P_{sp}. When the material is strained, a P_{pz} component develops in GaN, AlN, or AlGaN. The piezoelectric polarization arises in the thin layer of AlGaN because it is beneath the tensile strain. In contrast, the layer of GaN is considerably thicker and nearly relaxed in an AlGaN/GaN heterostructure. P_{sp} and P_{pz} are parallel when the AlGaN layer is under a tensile strain (Figure 11a) [23]. So, to acquire the AlGaN characteristics and determine the actual polarization-induced charge densities, we apply Equations (9) through to (14) [26].

Similarly, 2DEG is generated at the interface between the AlN barrier and the GaN channel for the reason of the P_{sp} and P_{pz}, as illustrated in Figure 11b. However, due to the large difference in the spontaneous and piezoelectric polarizations between the GaN and AlN layers, the 2DEG, which forms near the AlN/GaN interface, can reach over 3×10^{13} cm^{-2} for an extremely thin AlN barrier layer thickness (d < 5 nm), along with a high mobility (>1000 cm^2/V·s) and a very low sheet resistance (R_{sh} < 150 Ω/□) [27]. On the contrary, due to the high aluminum content and the associated degree of bond polarity, a spontaneous polarization in in the $In_{0.18}Al_{0.82}N$/GaN heterostructures generates two-dimensional electrons gas (2DEG) densities which are competitive with AlGaN/GaN structures without the need for a strain-induced piezoelectric component. For InAlN alloys, the strain state can go from tensile when the In % is lower than ~18%, to compressive when the In % is higher than ~18%, as shown in Figure 11c ((1), (2)), respectively; when the InAlN is the lattice matched to the GaN buffer, the In % is around 18%. Moreover, the implementation of the InAlN barriers with the In % around 18% allows for the strain to be controlled in the InAlN/GaN heterostructure, and a lattice-matched $In_{0.18}Al_{0.82}N$ layer to the GaN sheet charge density is almost three times higher than for the conventionally grown $Al_{0.25}Ga_{0.75}N$ barriers on the GaN channels [28].

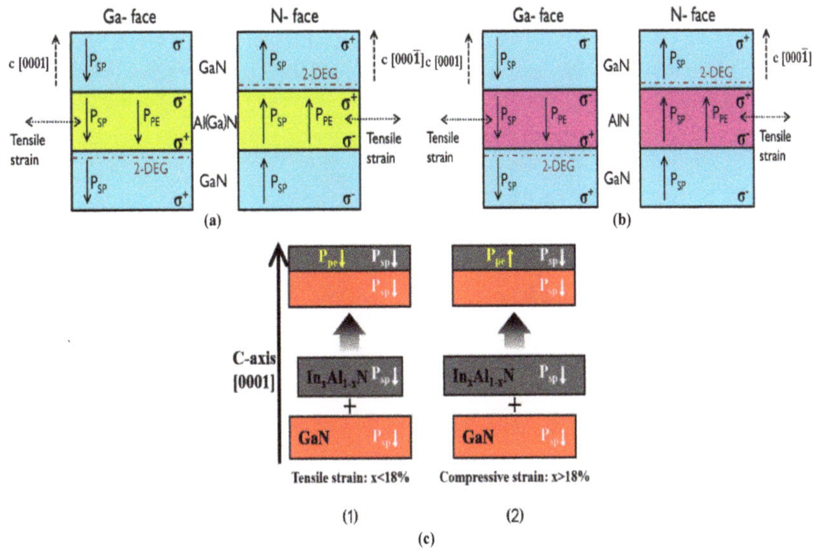

Figure 11. Polarization charges in an (**a**) AlGaN/GaN, (**b**) AlN/GaN, and (**c**) InAlN/GaN heterostructure. Reproduced with permission [20]. Copyright 2019, Elsevier.

Spontaneous polarization can be observed in Equation (9). As an alternative, for a precise interpolation, a bowing parameter b might be utilized in Equation (10) [29]. The constants of the lattice for $Al_xGa_{1-x}N$ are in Equations (11) and (12) where x = Al mole fraction. The elastic constants are presented in Equations (13) and (14).

$$P_{sp}(x) = (-0.052x - 0.029)\ \frac{c}{m^2} \tag{9}$$

$$P_{sp}(x) = ((xP_{sp}(AlN) + (1-x)P_{sp}(GaN) + bx(1-x)) \tag{10}$$

$$a_0\ (x) = (-0.077x + 3.189) \times 10^{-10}\ m \tag{11}$$

$$c_0\ (x) = (-0.203x + 5.189) \times 10^{-10}\ m \tag{12}$$

$$c_{13}\ (x) = (5x + 103)\ GPa \tag{13}$$

$$c_{33}\ (x) = (-32x + 405)\ GPa \tag{14}$$

The above equations indicate that if the Al content in the AlGaN layer rises, the overall polarization will soar. A polarization charge density (ρp) corresponds to a polarization gradient (P) in space ($\rho p = \nabla P$). Therefore, at the AlGaN/GaN abrupt contact, a fixed polarization charge would be induced. In the case of AlGaN/GaN, fixed positive induced polarization charges form at the interface. Even without n-type doping, the positive polarization-induced charges at the interface of AlGaN/GaN attract the free electrons. Because of this, in the AlGaN/GaN interface, 2DEG is generated. The carrier density of 2DEG may also be estimated analytically by using polarization [30].

The 2DEG creation is explained in a few Figures. Thus, in the undoped AlGaN/GaN heterojunction, the 2DEG creation may be described on the AlGaN surface by introducing donor states. An the isolated AlGaN material energy band gap is depicted in Figure 12 with the presence of surface donor states. The isolated AlGaN material is believed to be free of compressive/tensile forces. The Fermi level is attained if the density of the AlGaN material is enough. So, the electrons are advanced to the conduction band and drawn to the other side by polarization-induced electric fields after the Fermi level exceeds the donor state-level E_S. The Fermi level causes the dropping of the electrons into the GaN side when the AlGaN and GaN materials come into contact. The 2DEG results from a buildup of electrons at the interface, as displayed in Figure 13. Thus, combined with the ionized surface donor, it produces an electric field that points between the interface and the surface in the AlGaN layer to minimize the polarization field [31].

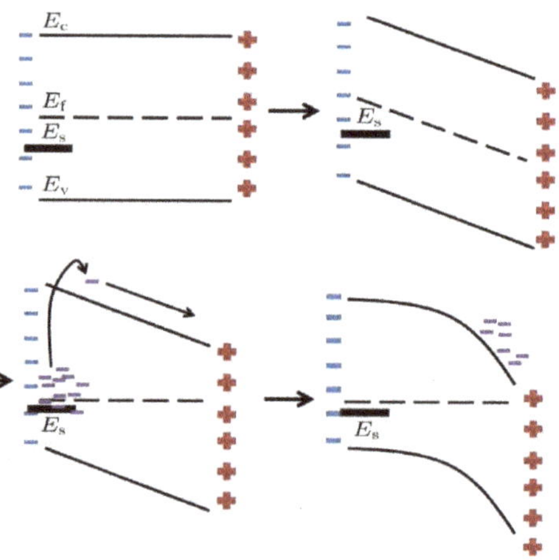

Figure 12. An undoped AlGaN material with E_S surface donor states has an energy band. Electrons are advanced into the conduction band and gravitate toward the electric field created by positive polarization. Reproduced with permission from Chin. Phys. B. [31].

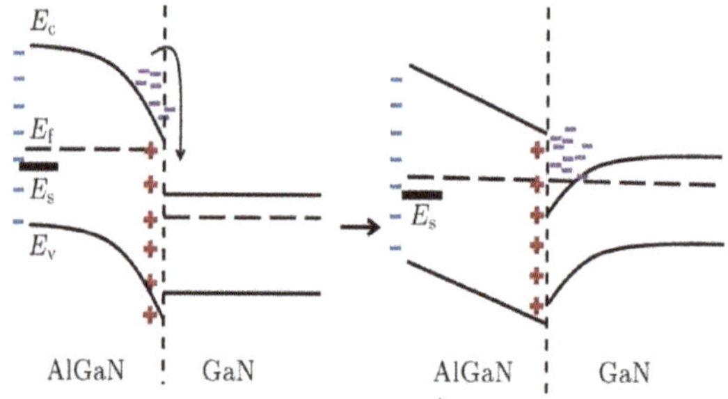

Figure 13. In the 2DEG quantum well, the buildup of electrons is shown in the energy band of an undoped AlGaN/GaN heterojunction. Reproduced with permission from Chin. Phys. B. [31].

As noted previously, the creation of 2DEG occurs naturally at the interface when a tiny strained AlGaN layer is developed on top of a thicker relaxed GaN layer. Due to the confinement of high mobility electrons, this 2DEG exhibits very conductively in the quantum well. Because of the reduced surface scattering, the electron mobility rises between the unstrained GaN and 2DEG areas at roughly 1000 $cm^2/V{\cdot}s$ and 1500–2000 $cm^2/V{\cdot}s$, respectively. For HEMTs, a transistor channel with a higher mobility and electron concentration seems appropriate. Figure 14 depicts a basic AlGaN/GaN HEMT design. The substrates are mostly SiC, silicon, sapphire, or GaN, and a buffer layer is formed to relieve the strain induced by a foreign substrate's lattice mismatch. The 2DEG creates a native channel between the source and drain of the device.. In most early AlGaN/GaN transistors, a Schottky gate electrode is formed by the deposition of Ni/Au or Pt metal on top of the AlGaN layer [32]. The GaN HEMT, with the 2DEG present at the interface of the AlGaN/GaN, was first shown in early 1993 [33]. The channel is regulated by supplying a gate voltage (V_{GS}) and, therefore, a vertical electrical field to deplete or augment the channel, making it less conductive in the off-state or more conductive in the on-state. Later, researchers from all over the globe have progressively reported high-performance GaN HEMTs for high-power and RF applications [34,35].

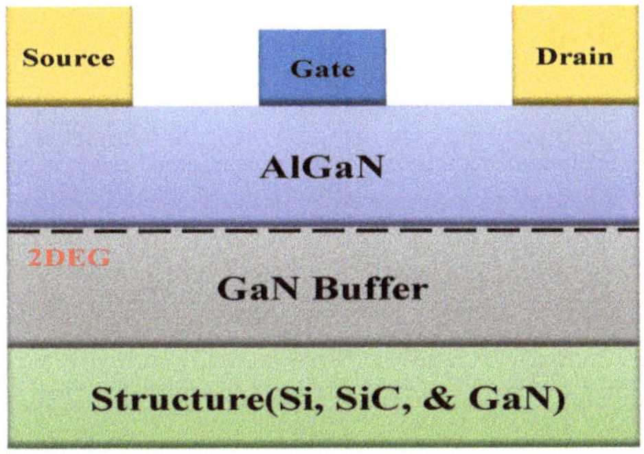

Figure 14. AlGaN/GaN HEMT layout and 2DEG generated in AlGaN/GaN contact are shown by the dashed line.

As said previously, when no bias voltage is applied from the gate to the source at the AlGaN/GaN interface, i.e., $V_{GS} = 0$ V, the 2DEG channel exists. This kind of HEMT is known as a normally-on HEMT or a depletion-mode (D-mode) HEMT. It signifies that a current may freely flow between the source and the drain. It also indicates that the AlGaN/GaN HEMT's threshold voltage (V_{TH}) is less than 0 V, requiring a V_{GS} to deplete the 2DEG and switch off the HEMT, as shown in Figure 15. Meanwhile, this feature is also one of the primary problems for the state-of-the-art AlGaN/GaN HEMTs utilized in higher current and power applications [36]. It has several practical limitations, including a high consumption, possible electrical safety risk, intrinsic fail-safe functioning, and complex circuit layout [32,36]. As a result, a normally-off operation, or an enhance-mode (E-mode) HEMT, is needed due to a high and positive threshold voltage (V_{TH}).

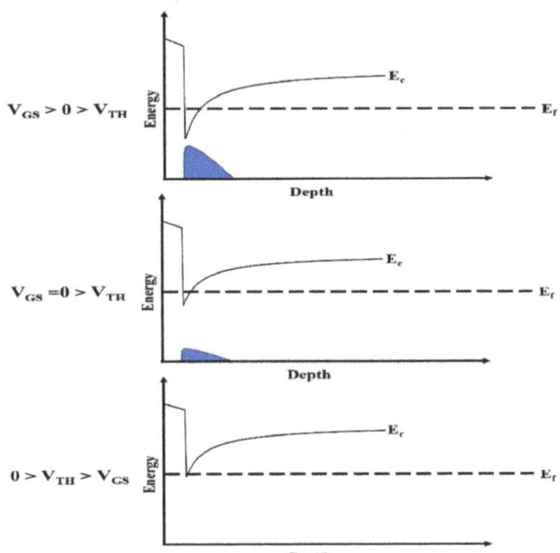

Figure 15. The band diagrams for the three locations of the 2DEGs $V_{GS} > 0 > V_{TH}$, $V_{GS} = 0 > V_{TH}$ and $0 > V_{TH} > V_{GS}$.

2.2. Lateral and Vertical GaN Power Transistors

Power transistors based on GaN are considered for two types of design: lateral and vertical structures. The lateral structure of the GaN HEMT is illustrated in Figure 16, where 2DEG formed at the AlGaN/GaN heterostructure interface. A high electron mobility (2000 cm^2/Vs) and electron velocity (1.3×10^7 cm/s saturation velocity and 2.5×10^7 cm/s peak velocity) characterize the AlGaN/GaN structure. Furthermore, due to a significant spontaneous and piezoelectric polarization, the GaN HEMT structure exhibits a density of sheet carriers over 1×10^{13} cm^{-2} in III-nitride materials. GaN HEMTs have a low on-resistance because of the device's carrier density and high electron mobility. This structure is discussed deeply in the normally-off GaN HEMT section.

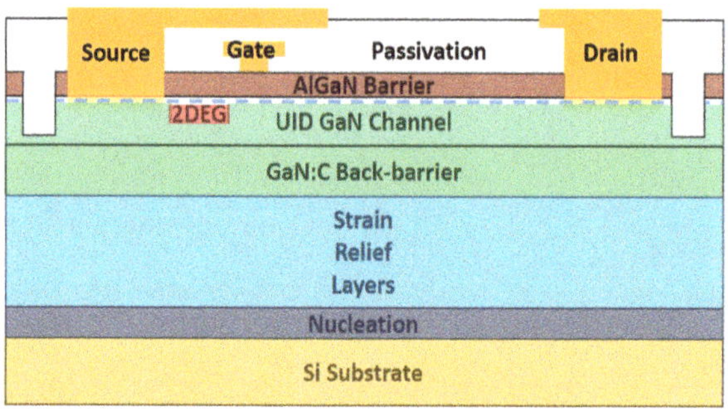

Figure 16. Typical lateral AlGaN/GaN HEMT structure. Reproduce from IOP science under the terms of the Creative Commons Attribution 3.0 license [36].

Similarly, the vertical layout is beneficial for obtaining a high breakdown voltage and low on-resistance characteristics, as displayed in the current aperture vertical electron transistor (CAVET) in Figure 17 [36]. Here, the drain at the bottom, the gate, and the source are on the top of the structure. The gate controls the current via an aperture among the current blocking layers (CBLs) into the drain, where vast amounts of material flow, which is commonly produced by an isolation implantation or P-type doping of the GaN layer. The AlGaN/GaN layer's horizontal high mobility electron channel is combined with a thick GaN drift region to obtain a low R_{ON} and a high breakdown voltage. Compared to the lateral designs, vertical devices into the bulk material of the device maintain the blocking voltage in the vertical direction, resulting in a lower chip area for a particular operation current.

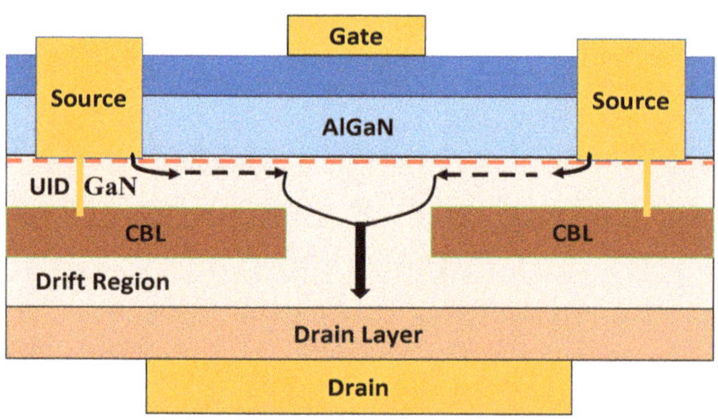

Figure 17. Vertical CAVET AlGaN/GaN HEMT structure. Reproduce from IOP science under the terms of the Creative Commons Attribution 3.0 license [36].

3. Normally on and off GaN HEMT Power Device Structure

3.1. GaN HEMT Technology (Normally on)

When the AlGaN/GaN HEMT structure is normally on, it is known as the depletion-mode (D-mode) structure. Figure 18 illustrates one of the examples of it. Where the buffer layer (1–5 μm) is deposited on the substrate for compensating the lattice mismatch stress, after that, to form a heterojunction, the GaN (UID) layer, AlN layer (0.7–1.2 nm), and $Al_xGaN_{1-x}N$ barrier (15–30 nm) layer is deposited. The thickness and Al molar fraction x of the $Al_xGaN_{1-x}N$ (usually 0.15 to 0.4) varies to maintain the number of acceptable electric charges under the relaxation's critical thickness [37]. The heterojunction's energy bands are bent downward to produce a quantum well with a sharp, where electrons of a high intensity are confined at the interface of AlGaN/GaN due to a bandgap offset and the polarization impact of AlGaN/GaN [38]. Then, the 2DEG channel generates the high-density current from the Ohmic-contact source to the drain [39]. Consequently, a Schottky gate is needed for the pinching off the 2DEG channel, which is normally on. When sufficient negative voltages are applied to the devise gate, the height of the Schottky barrier increases, allowing for the conduction band to pass through the AlGaN barrier's gate region; thereby, the HEMT is turned off. Eventually, for the device surface protection, passivating layers (often SiO_2 or SiN_x) are required [40–42].

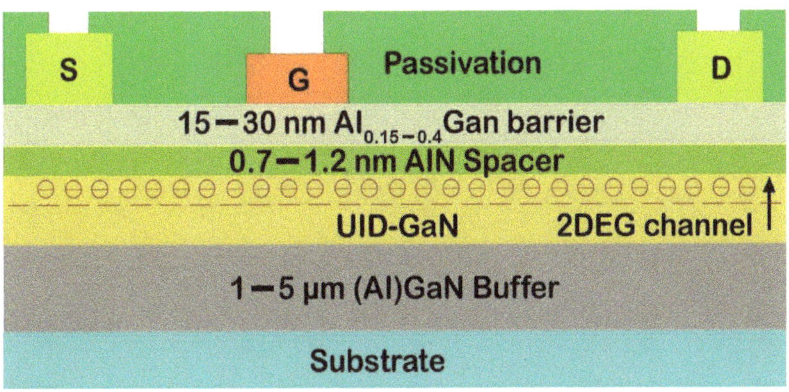

Figure 18. Device layout for a GaN HEMT structure that is normally on.

3.2. GaN HEMT Technology (Normally off)

In the field of power conversion, a negative bias is required to turn off the devices, which is the major issue for the normally-on devices. On the contrary, a normally off operation is widely wanted for safety reasons. Consequently, academia and industry (e.g., Gan Systems, Infineon, Panasonic, STMicroelectronics, and On Semiconductors) are working to create and market dependable normally-off HEMTs [43]. Thus, the threshold voltage V_{TH} in AlGaN/GaN HEMT is determined by numerous variables relating to the heterojunction and gate metal characteristics, shown in Equation (15) [44]:

$$V_{TH}(x) = \varphi_B(x) - \Delta E_C(x) - \frac{\sigma(x)}{\varepsilon_0 \varepsilon_{AlGaN}(x)} t - \frac{qN_D}{2\varepsilon_0 \varepsilon_{AlGaN}(x)} (t)^2 \quad (15)$$

where between the gate metal and the AlGaN barrier layer, $\varphi B(x)$ denotes the Schottky barrier height, $\sigma(x)$ denotes the polarization charge at the interface of the AlGaN/GaN, $\Delta E_C(x)$ denotes the conduction band discontinuity at the interface of the AlGaN/GaN, $\varepsilon_{AlGaN}(x)$ denotes the AlGaN layer permittivity, ε_0 denotes the permittivity vacuum, t denotes the AlGaN thickness, N_D denotes the doping, q denotes the electric charge, and x denotes the Al content in the barrier layer. As ns increases at a zero bias, more significant polarization discrepancies between the AlGaN and GaN and thicker AlGaN barriers result in a more negative V_{TH}. Hence, it is clear from the equation that the AlGaN barrier layer or modifying Schottky barrier height relies on the thickness and Al content to tune the V_{TH}. Therefore, various approaches have been suggested for obtaining the normally-off GaN HEMTs: the (a) cascode configuration, (b) fluorine implantation and thin/ultrathin-barrier, (c) p-GaN Gate, and (d) recessed gate.

3.2.1. HEMTs with Cascode Configuration (Normally off)

Figure 19 reflects a cascode setup of high-voltage normally-on GaN HEMTs and low-voltage normally-off Si MOSFETs [45,46]. When the V_{TH} of a Si MOSFET less than the positive gate-to-drain voltage is applied to the system, the GaN HEMT's gate-source voltage is 0 V; as a result, the opening of a GaN HEMT is normally on. By contrast, deactivating the Si MOSFET results in the GaN HEMT having a significant negative gate to source voltage, thus, the 2DEG channel is pinching off. Evidently, the cascode setup generates a positive V_{TH} for regulating the GaN HEMT switch that is normally on. Additionally, normally-off GaN HEMTs with the cascode design are commercially available at 600 V [43,45]. However, the parasitic effects of the packing and the high-temperature stability of the Si MOSFETs remain significant disadvantages [47,48]. If the GaN HEMT has a relatively high on-resistance compared to the Si MOSFET, the "cascode" configuration performs well at that time. In fact, since the on-resistance increases with the rated breakdown voltage, the "cascode" approach is advantageous when the normally on GaN HEMT is one of a high-voltage, and the Si MOSFET is low-voltage [49]. As an illustration, the Si MOSFET will provide just a 3% on-resistance to a 600 V GaN HEMT "cascode". On the contrary, for a lower targeted breakdown, the on-resistance of the GaN HEMT decreases; thus, the Si MOSFETs contribution becomes significant. Hence, the "cascode" approach is practically advantageous for applications above 200 V [49].

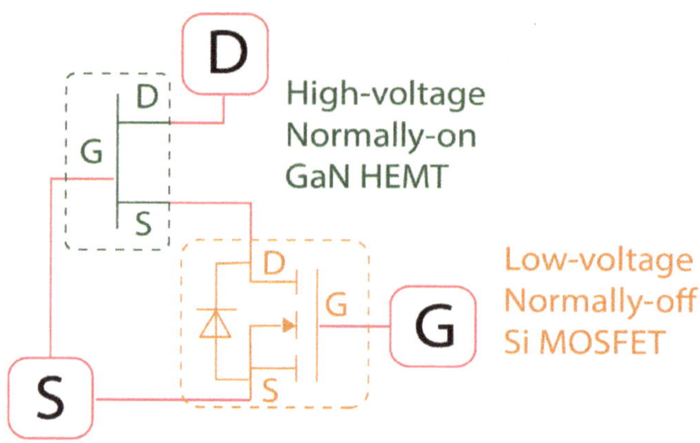

Figure 19. Normally-off HEMTs with cascode configuration structure.

3.2.2. HEMTs with Fluorine Implantation and Thin/Ultrathin Barrier (Normally off)

Figure 20a shows an alternative method of performing the normally-off operation to avoid the dry etching, which could cause damage to the device operation. Due to their considerable electron-negativity, fluorine given by the ion implantation and in the AlGaN layers may easily boost the AlGaN barrier electron potential and empty the 2DEG channel [50]. Likewise, by adjusting the height of the AlGaN conduction band, the 2DEG can be depleted by a GaN HEMT design with a thin/ultrathin barrier [51,52], as displayed in Figure 20b. When the AlGaN barrier thickness falls below a critical value, the V_{TH} shifts to a negative bias in the recessed-gate configuration. However, in various application issues, like for fluorine gates, the injection method must be repeatable, and for thin barrier HEMT designs, the output current is low, restricting these two devices [51,53].

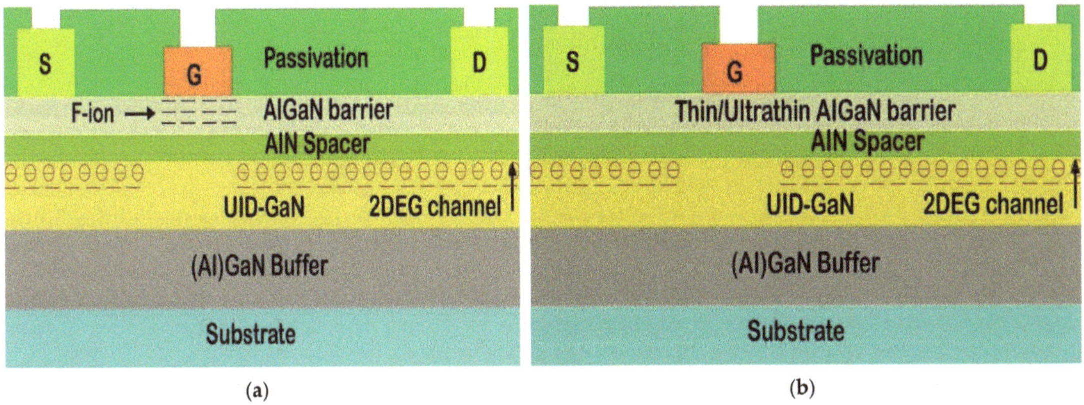

Figure 20. (**a**) Normally-off HEMTs with fluorine implantation structure.; (**b**) normally-off HEMTs with thin/ultrathin-barrier structure.

3.2.3. HEMTs with p-GaN Gate (Normally off)

A p-GaN (or p-AlGaN) layer is applied to the AlGaN/GaN heterostructure at the gate region. As illustrated in Figure 21, it is probably the most promising technique to accomplish a normally-off operation [54–56]. In fact, it is presently the only commercially available normally-off GaN HEMT technology. GaN HEMTs of this sort have received much attention from the scientific community and the industry. The p-GaN gate HEMTs working principle is generally represented in Figure 22. The AlGaN conduction band is elevated by the p-GaN cap layer on the AlGaN, resulting in 2DEG depletion. Hence, it is possible to obtain the device's normally-off mode [57]. Earlier, Uemoto et al. [55] proposed a GaN HEMT with a *p*-AlGaN gate that is normally off, with a V_{TH} of 1.0 V and a BV_{DSS} of 800 V.

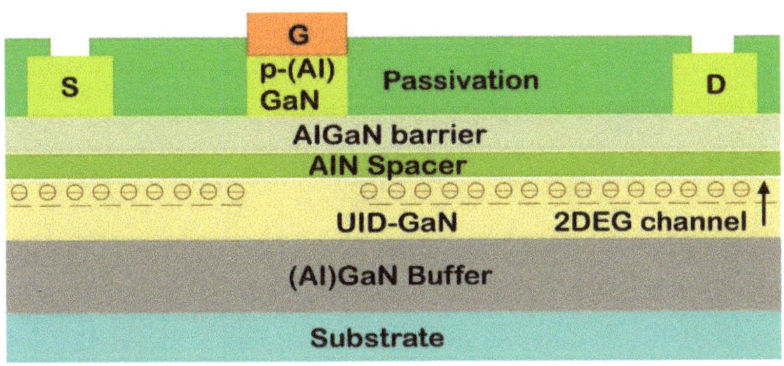

Figure 21. Normally-off HEMTs with p- GaN (or p-AlGaN) layout.

The characteristics of the AlGaN/GaN heterostructure should be specified appropriately to enable an effective depletion of the 2DEG and a high V_{TH} ($V_{TH} > 0$) [48–60]. The layer thickness of the AlGaN barrier is usually between 10 and 15 nm, whereas the content of Al is between 15 and 25%. For the effective depletion of 2DEG, a high amount of doping in the p-GaN layer (>10^{18} cm^{-3}) is frequently needed [61]. In this regard, increasing the Mg electrical activation is one of the most critical factors in improving the threshold voltage V_{TH} of the p-GaN layer for a specific Mg concentration. Adequate growth parameters for the p-GaN layer and annealing conditions may achieve the latter goal [62,63]. On the contrary, because of the high energy of ionization (in the range from 150 to 200 m eV) of Mg as a p-type dopant, i.e., an acceptor, obtaining a high activation of Mg in p-GaN is problematic [64,65]. In most cases, an acceptor concentration of 2-–5 × 10^{19} cm^{-3} is employed, roughly two orders of magnitude more than the concentration of holes. A greater Mg concentration in p-GaN might cause the layer's crystalline quality to deteriorate, reducing the electrically active acceptors [66,67]. As a result, in the p-GaN gate HEMTs, getting a high V_{TH} is difficult. Earlier, a Pd-based Ohmic gate was implemented on the cap layer of p-AlGaN, which improved the device's hole injection and current capabilities. Therefore, the transistor was also known as a "gate injection transistor" (GIT) [55]. The GaN–GIT structure was modified to a hybrid drain-embedded GIT (HD-GIT) by Kaneko et al. [68] The Panasonic group contains a p-doped GaN area adjacent to the ohmic drain. The injected holes discharge the trapped electrons near the drain edge in the off-state from the p-GaN layer, which helps prevent the current collapse. Compared to the standard GIT structure, this new modification boosts the breakdown voltage to 850 V with a slight increase in the device's leakage current and the on-resistance of 2.6 mΩ·cm^2.

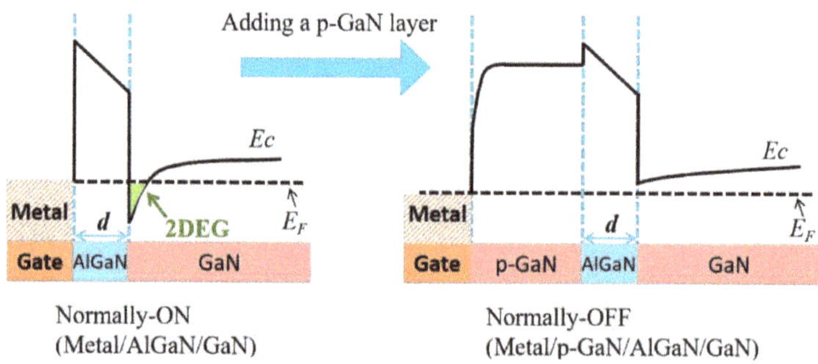

Figure 22. Metal/AlGaN/GaN energy band diagram with and without a p-GaN gate under equilibrium. Reproduced with permission [57]. Copyright 2017, Elsevier.

In contrast, the metal/p-GaN/AlGaN/GaN system TCAD simulations indicate that on the p-GaN, a Schottky metal gate must have a higher V_{TH} and less leakage than an Ohmic gate [55,63,64]. In the on-state, Meneghini et al. [60] demonstrated that a Schottky gate based on WSiN to the p-GaN rather than a typical Ni/Au Ohmic contact could improve the gate voltage swing of a transistor and lower the gate leakage current by roughly four orders of magnitude. Generally, the continuous power consumption of GaN HEMTs is due to the gate leakage and heat generation in the gate driver. The lack of a considerable gate side current injection due to the significant Schottky barrier is thus essential for HEMTs with a p-GaN gate for a decreased power consumption. Because of this, on the p-GaN, the Schottky gate is preferable over the Ohmic gate solution nowadays. Furthermore, several data from the literature are shown in Table 3, where some metals have been used in the Schottky gate contact to the p-GaN. Subsequently, because of its thermal and chemical durability and processing compatibility, adopting a TiN gate is now a good choice [69]. Eventually, several device and reliability difficulties still need to be investigated further, such as the low V_{TH}, high gate leakage currents, the poor gate BV, and the influence of fabrication techniques on the device's performance [57,70].

Table 3. Survey of literature data on different Schottky Contact.

Matel Gate	Doping n_{dop} and p-GaN Thickness T	V_{TH} (V)	R_{ON} (Ω·mm)	Ref
Ti $_{(40\ nm)}$/Au $_{(100\ nm)}$	$n_{dop} = 4 \times 10^{19}$ cm^{-3}, T = 100 nm	1.82	11.9	[71]
Ti $_{(25\ nm)}$/Au $_{(120\ nm)}$	$n_{dop} = 1 \times 10^{18}$ cm^{-3}, T = 70 nm	3	13.9	[72]
Ni/Au	$n_{dop} = 3.5 \times 10^{18}$ cm^{-3}, T = 50 nm	1.5	8	[73]
Ti $_{(50\ nm)}$/Au $_{(150\ nm)}$	N/A, T = 70 nm	1.2	N/A	[74]
Ni $_{(20\ nm)}$/Au $_{(200\ nm)}$	$n_{dop} = 5 \times 10^{19}$ cm^{-3}, T = 100 nm	1.5	N/A	[75]
Ni $_{(25\ nm)}$/Au $_{(200\ nm)}$	$n_{dop} = 1 \times 10^{18}$ cm^{-3}, T = 60 nm	1.7–2.1	5.65 & 5.05	[76]
Ni/Au	$n_{dop} = 3 \times 10^{19}$ cm^{-3}, T = 85 nm	1.5	29.5	[77]
Ni $_{(25\ nm)}$/Au $_{(120\ nm)}$	N/A, T = 100 nm	1.5	20	[78]
Ni/Au	$n_{dop} = 4 \times 10^{19}$ cm^{-3}, T = 70 nm	1.02	15.4	[79]
Ni $_{(15\ nm)}$/Au $_{(280\ nm)}$	$n_{dop} = 4 \times 10^{19}$ cm^{-3}, T = 70 nm	2.2	43.6	[80]
Ti/Au	$n_{dop} = 4 \times 10^{19}$ cm^{-3}, T = 85 nm	2.1	21	[81]
TiN	$n_{dop} = 3 \times 10^{19}$ cm^{-3}, T = 80 nm	1.6	17.8	[82]
W	$n_{dop} = 1 \times 10^{19}$ cm^{-3}, T = 75 nm	2.1	49	[83]
TiN	N/A, T = 60 nm	2.1	15	[62]
Pd	$n_{dop} = 3 \times 10^{19}$ cm $^{-3}$, T = 100 nm	1.7	8.5	[84]
Mo $_{(100\ nm)}$/Ni $_{(20\ nm)}$	$n_{dop} = 3 \times 10^{19}$ cm $^{-3}$, T = 80 nm	1.08	10.7	[85]
W	$n_{dop} = 1 \times 10^{19}$ cm $^{-3}$, T = 75 nm	1.6	26.5	[86]
Ti $_{(25\ nm)}$/Au $_{(120\ nm)}$	N/A, T = 75 nm	3.2 & 1.8	16	[87]
Ti $_{(45\ nm)}$/Au $_{(200\ nm)}$	$n_{dop} = 2 \times 10^{18}$ cm^{-3} & 2×10^{19} cm^{-3}, T = 50 nm	1.30 & 1.45	9.66 & 9.51	[88]
Ti/Au/Ni	$n_{dop} = 4 \times 10^{17}$ cm^{-3}, T = 50 nm	1.5	N/A	[89]

3.2.4. HEMTs with Recessed Gate (Normally off)

This method reduces the layer thickness of the AlGaN barrier under the gate via a plasma etching, which is the last option offered for achieving an HEMT (normally off) [90]. Because the leakage current of the gate is caused by a tunneling phenomenon that is very sensitive to the barrier layer thickness and uniformity, this approach necessitates a perfect control of the AlGaN etching process. Furthermore, etching-induced damage might increase the leakage current of the gate and non-uniformity effects in V_{TH} [91]. Additionally, there are two types of gates recessed available such as the slight MIS-HEMT of the recessed-gate design underneath the gate dielectric, a thin AlGaN barrier, and the complete MIS-FET of the recessed-gate design, known as a MOS-HFET hybrid, as shown in Figure 23a,b [92,93].

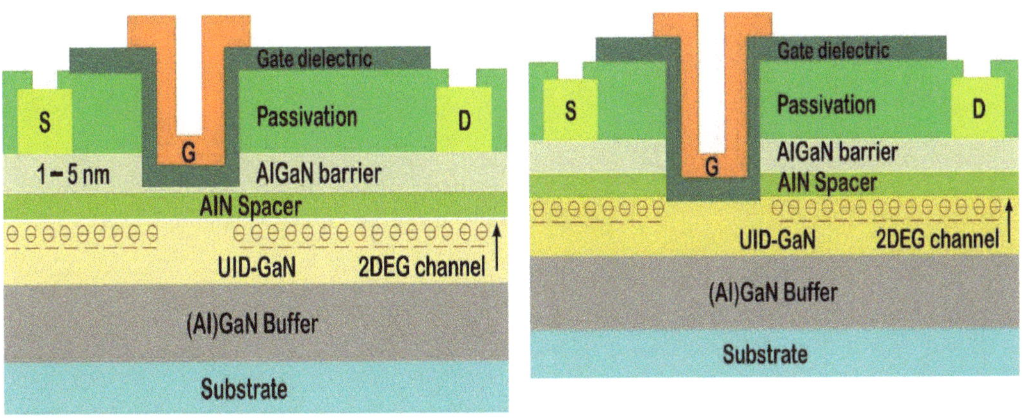

(a) (b)

Figure 23. (**a**) Normally-off HEMTs with slight MIS-HEMT of recessed-gate design; (**b**) normally-off HEMTs with complete MIS-FET of recessed-gate design.

Standard MIS-FETs may obtain an E-mode performance by entirely eliminating the AlGaN barrier layer underneath the gate, resulting in the device turning off at a zero gate voltage. Whenever the positive voltage of the gate exceeds V_{TH}, an accumulation layer of an electron forms at the interface of the gates, which operates the devices' conductive channel and is switched on. It has benefited from a high threshold voltage (V_{TH}) but suffers from a decrease in the channel mobility (μ_{FE}) due to rough surfaces in the recessed region and electrically active faults in the MIS structure. In addition, a poor channel mobility will raise the device on-resistance (R_{ON}) and system power consumption. Therefore, a slight MIS-HEMT of the recessed-gate has been offered to improve the device's channel mobility (μ_{FE}) and decline its on-resistance (R_{ON}), where the thin layer of the AlGaN barrier will hinder the 2DEG channel from the MIS interface. As a result, the channel mobility (μ_{FE}) increased and reduced the on-resistance (R_{ON}) [94]. Thus, the channel mobility (μ_{FE}) was defined as Equation (16):

$$\mu_{FE} = \frac{L_g}{WC_{ox}V_{DS}}\left(\frac{\partial I_{DS}}{\partial V_{GS}}\right) \tag{16}$$

Here, L_g and W donate the channel length and width, while C_{ox} donates the gate insulator capacitance per unit area. Moreover, Fiorenza et al. [95] have evaluated in MISHEMTs of the recessed gate the field-effect mobility by using gate insulator SiO_2 with several variables (the roughness of the surface, the temperature, the field of electric, the quality of the dielectric, and so on). This study demonstrated in the insulator/GaN system the necessity of reducing the interface state density to enhance the field effect's mobility and decrease the precise on-resistance. Eventually, these two recessed gates improved the threshold voltage, device performance, channel mobility, the reliability of the device, and overcame the gate leakage issue [96–98]. Additionally, the gate dielectric for the gate recessed depends on several characteristics, such as the permittivity, the (Al) GaN band offset, the bandgap, and the insulator chemical stability [38,94,95]. Especially for power-switching devices, a wide bandgap material is also significant. The (Al) GaN band offsets are required to reduce the gate leakage currents appropriately, even while operating the forward gate bias. On the contrary, a high permittivity value is advantageous for obtaining high transconductances [43]. In the case of MIS-HEMTs, the capacitive contribution of the gate dielectric is minimized by a high permittivity dielectric, facilitating a stronger connection between the 2DEG channel and the gate and therefore maintaining high transconductances, which is especially important for RF devices. In addition, compared to Schottky gate HEMTs, the high permittivity materials may limit the changing of the values of the threshold voltage from positive to negative of normally-on MIS-HEMTs, which helps lower the static power usage and improve the performance and the energy efficiency of the device [99].

In GaN-based insulated-gate transistors, Figure 24 demonstrates the bandgap-permittivity connection for nitride compounds and the insulators for different gate dielectrics [38,95,96]. Figure 25a illustrates that Robertson et al. [100] calculated the insulators on the GaN band offsets; they were the first to forecast the band alignment of the GaN and the insulators using charge neutrality levels (ECNL) calculations. Figure 25b shows the dielectrics on the $Al_{0.3}Ga_{0.7}N$ band offsets recently obtained by Reddy et al. [101] using the same approach. As a gate dielectric, SiO_2 on AlGaN/GaN MIS-HEMTs was first implemented by Khan et al. [102] and they noticed that it enhances the gate voltage swing capabilities and controls the leakage currents in the gates for its attractive characteristic-like large band offset to (Al)GaN, the chemical stability, and the large bandgap.

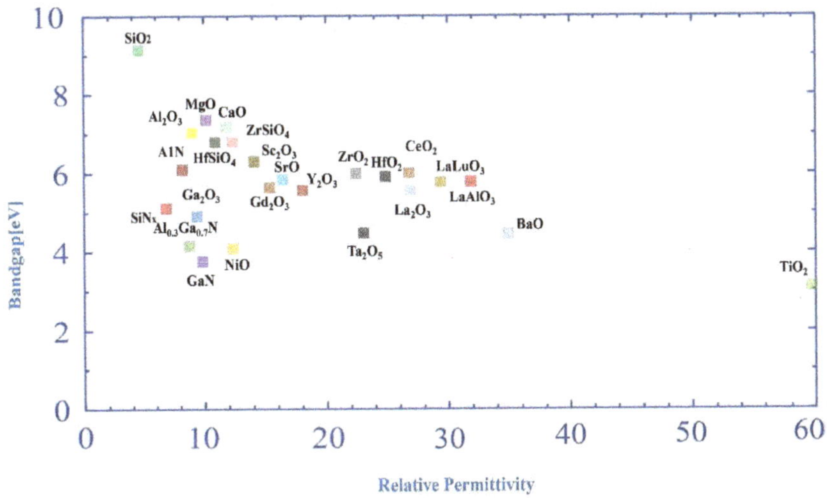

Figure 24. For primary insulators and GaN compounds, the energy bandgap (eV) vs. permittivity is shown. Reprinted with permission [42]. Copyright 2017, the author(s). Published by Elsevier.

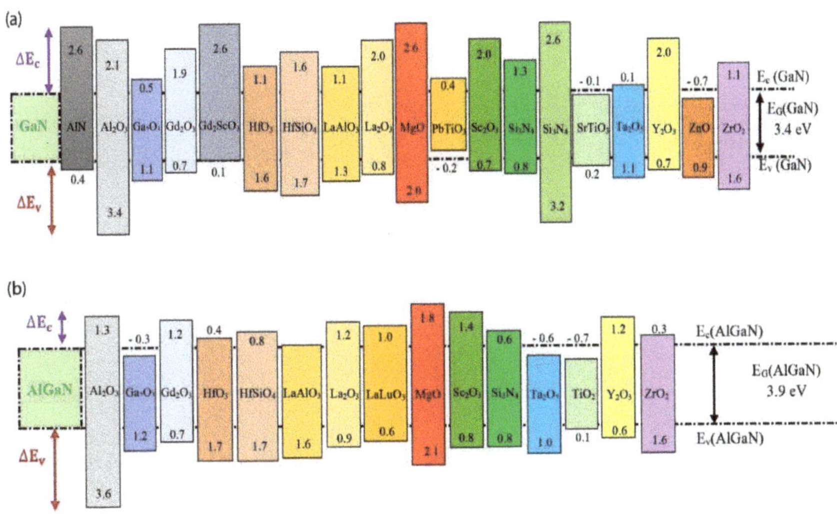

Figure 25. Various dielectric materials regarding (**a**) GaN [100] and (**b**) $Al_{0.3}Ga_{0.7}N$ [101] have different valence and conduction band offsets (ΔE_V and ΔE_C).

In contrast, for MIS gate designs utilizing dielectrics such as Ga_2O_3, SiNx, because of its small conduction band offsets, resulted in large gate leakage currents [103–105]. Likewise, the minor AlN and (Al)GaN lattice mismatch has been shown in a few investigations, whereby it is appropriate for the gate insulator and passivation layer [106,107]. Moreover, numerous high-permittivity dielectrics have been used in the MIS gate architectures of GaN HEMTs [108], including HfO_2, ZrO_2, Ta_2O_5, La_2O_3, CeO_2, and TiO_2 [109–111]. Thus, Table 4 represents the data gathered from the literature study of normally-off HEMTs with a recessed gate where several types of gate insulators (SiN, SiO_2, Al_2O_3, and processing (PECVD, LPCVD)) are shown.

Table 4. Survey literature data on difference types of gate insulators and processes.

Gate Insulator	Insulator Processing	μ_{FE} ($cm^2 \cdot V^{-1} \cdot S^{-1}$)	V_{TH}(V)	R_{ON} (Ω·mm)	On/off Ratio	G_m (mS/mm)	Ref
$SiN_{(30\ nm)}$/ $SiO_{2\ (3\ nm)}$	SiO_2 ALD + post annealing at 890 °C in N_2 + SiN LPCVD at 665 °C	116	2.4	15.9 Ω·mm at V_{GS} = 12 V	6 × 10^8	52	[112]
$SiN_{(100\ nm\ and\ 300\ nm)}$/ $ZrO_{2(23\ nm)}$	SiN PECVD + ZrO_2 ALD at 200 °C	850	2.19	9.2 Ω·mm at V_{GS} = 8 V	~10^9	135	[113]
$SiO_{2\ (50\ nm)}$	PECVD + post annealing at 850 °C in N_2	110	0.7	N/A	N/A	N/A	[95]
$Al_2O_{3\ (25\ nm)}$	ALD + TMA and ozone	N/A	5	12.9 Ω·mm at V_{GS} = 19.7 V	N/A	44	[114]
$HfO_{2\ (13\ nm)}$	ALD+ ozone	1482	3.1	6.0 Ω·mm at V_{GS} = 8 V	10^5	N/A	[115]
$Al_2O_{3\ (15\ nm)}$	ALD+ ozone	1991	2.6	5.5 Ω·mm at V_{GS} = 8 V	10^5	N/A	[115]
SiN $_{(2\ nm)}$/ SiN $_{(15\ nm)}$	LT- PECVD at 850 °C + HT-LPCVD at 780 °C	160	2.37	13.2 Ω·mm at V_{GS} = 15 V	N/A	17	[116]
SiN $_{(15\ nm)}$	HT-LPCVD at 780 °C	38	1.28	20 Ω·mm at V_{GS} = 15 V	N/A	2	[116]
$TiO_{2(3.4\ nm)}$	Thermal oxidation	1270	4.2	N/A	2.3 × 10^8	N/A	[110]
$Al_2O_{3\ (25\ nm)}$	ALD + post annealing at 830 °C in N_2	2033	2.5	6.8 Ω·mm at V_{GS} = 12 V	10^8	92	[117]
SiN $_{(20\ nm)}$	LPCVD on photo-electrochemical recess	49	0.8	26 Ω·mm at V_{GS} = 15 V	N/A	2	[118]
SiN $_{(15\ nm)}$	SiN LPCVD at 780 °C	141	1.3	12 Ω·mm at V_{GS} = 15 V	N/A	13	[118]
SiN $_{(30\ nm)}$/ AIN $_{(5\ nm)}$	AIN PEALD + SiN LPCVD	198.80	6.28	11.62 Ω·mm at V_{GS} = 15 V	10^8	85.75	[119]
$HfO_{2\ (20\ nm)}$	ALD + post annealing at 600 °C in N_2	N/A	2.5	3.8 Ω·mm at V_{GS} = 6.5 V	2.1×10^9	81.38	[120]
$Al_2O_{3\ (30\ nm)}$	ALD + post annealing at 850 °C in N_2	1602	2.6	10.2 Ω·mm at V_{GS} = 10 V	N/A	58	[121]
AIN $_{(7\ nm)}$/ SiN $_{(7\ nm)}$	MOCVD	180	1.2	N/A	(5−−6) × 10^8	60	[122]
$Al_2O_{3\ (4\ nm)}$	ALD + post annealing at 900 °C in O_2	1450	1.55	7.1 Ω·mm at V_{GS} = 12 V	N/A	54	[123]
SiNx $_{(20\ nm)}$/ SiON $_{(10\ nm)}$	SiON PECVD at 350 °C in SiH_4, NH_3, N_2O, and N_2 atmospheres + SiNx LPCVD at 780 °C with an ammonia flow of 280 sccm, a SiH_2Cl_2 flow of 70 sccm.	1793	0.81	31.2 Ω·mm at V_{GS} = 12 V	N/A	N/A	[124]
$Al_2O_{3\ (18\ nm)}$	ALD + post annealing at 400 °C in N_2	65	7.6	19.5 Ω·mm at V_{GS} = 14 V	N/A	N/A	[125]
$Al_2O_{3\ (30\ nm)}$	ALD + post annealing at 850 °C in N_2	1670	0.53	24.4 Ω·mm at V_{GS} = 10 V	~10^8	42	[126]
$Al_2O_{3\ (5\ nm)}$/ SiN $_{(7\ nm)}$	SiN LPCVD at 780 °C	122	1.7	12.9 Ω·mm at V_{GS} = 18 V	~10^8	53	[127]
$NiO_{x(100\ nm)}$	Sputtering	N/A	0.45	9.42 Ω·mm at V_{GS} = 4 V	10^8	~75	[128]
$SiO_{2\ (20\ nm)}$	ALD + post annealing at 780 °C	1700	1.56	7.4 Ω·mm at V_{GS} = 6 V	10^{10}	190	[129]
$Al_2O_{3\ (20\ nm)}$	ALD + post annealing at 900 °C in N_2	245	5	N/A at V_{GS} = 12 V	10^{10}	19	[130]
$Si_3N_{4\ (10\ nm)}$/ $TiO_{2\ (20\ nm)}$	SiN PECVD + TiO_2 ALD	1200	1.81	43.81 Ω·mm at V_{GS} = 10 V	N/A	112	[131]

Subsequently, the AlGaN thickness is another important factor for the threshold voltage (V_{TH}). Medjdoub et al. [132] had shown the relation between the AlGaN thickness with a threshold voltage. From Figure 26a, notice that if the AlGaN thickness (t_{RA}) reduces, the V_{TH} increases. Similarly, other authors, Saito et al. [133], presented the relationship between the V_{TH} and the specific on-resistance ($R_{on}A$) by calculating the equation, as displayed in Figure 26b. On the contrary, the on-resistance ($R_{ON}A$) will be increased if the 2DEG mobility is decreased. Thus, from Figure 26b, the equation for calculating the specific on-resistance ($R_{on}A$) is denoted by Equation (17),

$$R_{on}A = \left(R_{gs} + R_{ch} + R_{gd} + 2R_{con}\right) \times \left(L_{gs} + L_{ch} + L_{gd} + 2L_{con}\right) \tag{17}$$

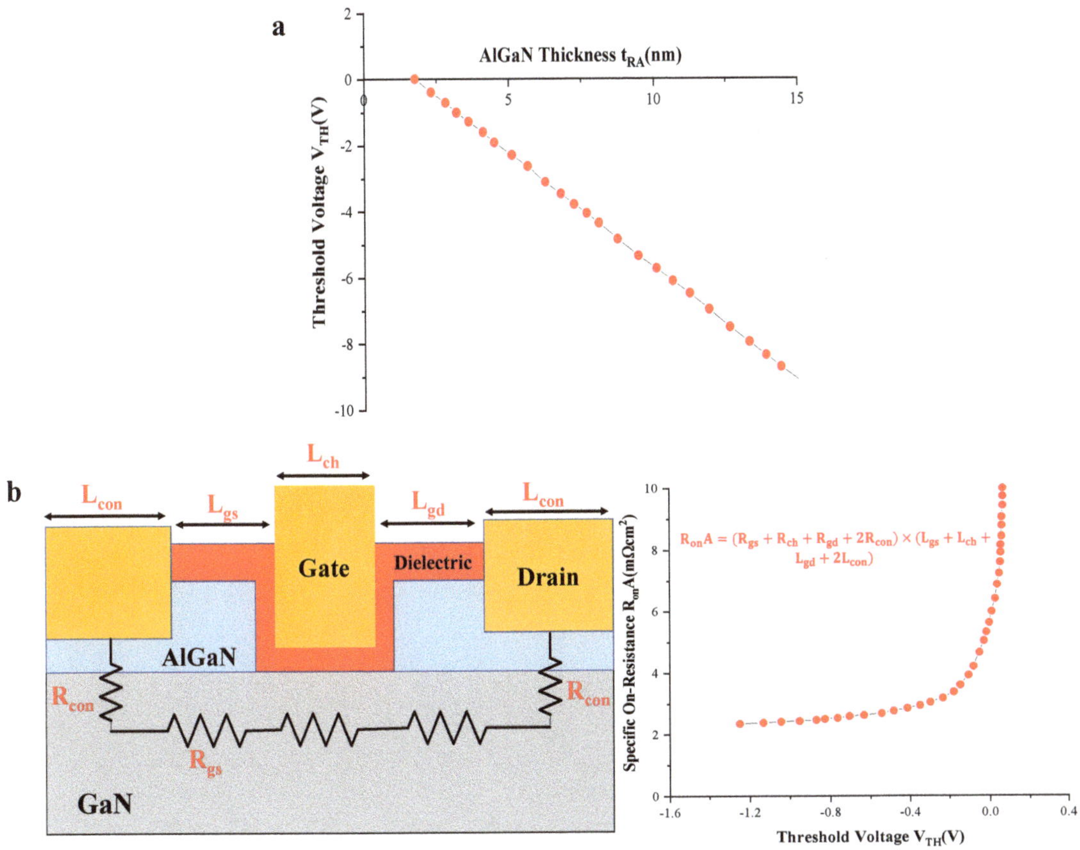

Figure 26. (**a**) The AlGaN layer thickness and gate threshold voltage relation. (**b**) In the recessed-gate GaN HEMT, the calculation of specific on-resistance and the relation between the threshold voltage and the specific on-resistance. Reprinted under the open access policy from Bentham Open [132].

Additionally, the field plate has used a device structure to reduce the current collapse. For GaN-based devices, current collapse is a well-known phenomenon that results in a trade-off between the breakdown voltage and the on-resistance by inserting the source-gate field plate; the electric field at the gate-drain edge is redistributed [134–136], as shown in Figure 16. The current collapse is discussed in the below segment.

4. Reliability of GaN HEMTs

4.1. Degradation Creation

While GaN HEMTs have improved remarkably, several material-related challenges must be resolved before they are revealed as a reliable and sustainable technology. AlGaN/GaN HEMTs' actual device performance has not yet reached the theoretically predicted levels. Even on unaged GaN HEMTs, parasitic effects could be noticed that degrade the performance metrics but do not significantly impact the reliability. Nevertheless, over time or under aging stresses, the degradation processes that accelerate with time impair not only the electrical response but also the device's resilience and dependability, resulting in shorter lifetimes and failure. The GaN industry's principal aim is to improve the dependability of its products while also reducing the parasitic effects. Further degradation difficulties with GaN devices include thermal mismatches, heteroepitaxy, and polarization effects with the substrate, all of which are inherent to the material. Compared to other conventional technologies, AlGaN/GaN HEMTs can work at an extremely high bias, electric fields, and temperatures. Figure 27 demonstrates the various types of degradation affecting AlGaN/GaN HEMTs [137,138].

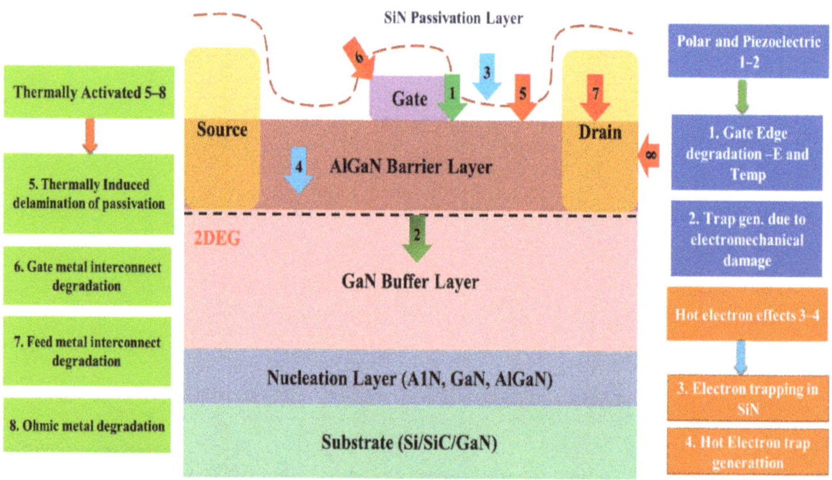

Figure 27. Several degrading problems affect AlGaN/GAN HEMTs with their origin. Reprinted with permission [138]. Copyright 2010, Cambridge University Press, and the European Microwave Association.

It is important to note that five to eight issues are associated with thermally induced degradation processes that have already been reported in other semiconductor systems (SiC, GaAs, Si, etc.). As a result, these failure mechanisms are simpler to explain and exclusive to the metallization scheme which has been adopted. Both three and four mechanisms are connected to the introduction of the hot electrons; it is a usual problem that field-effect transistors of a high voltage face. Other semiconductor devices have been proven to suffer from a deterioration, caused by hot electrons (Si, GaAs, InP, etc.). On the other hand, because of this semiconductor material's piezoelectric and polar properties, one to two issues are specific for GaN devices; consequently, those processes have not been fully defined or investigated. As a result, the GaN HEMT reliability analysis is a unique issue that needs to be devoted, as well as an in-depth physics-based study. Furthermore, high Schottky-gate leakage currents are also one of the main issues for degrading the device performance and reliability. Thus, the reduction in BV_{GD} raises the noise figure (NF) and decreases the power-added efficiency (PAE) due to the high gate leakage current [139]. In addition, there are numerous mechanisms of gate leakages, including the emission of a thermionic field [140], the emission of thermionic [141], the dislocation-assisted tunnelling [142], the trap assisted tunnelling [143], the Fowler–Nordheim tunnelling [144], the defect hoping [145], the space charged limited current [146], and the Frenkel–Poole emission [147].

4.2. Reliability Issues

These days, researchers focus on enhancing the AlGaN/GaN HEMTs reliability device because of the trapping effect during the AlGaN/GaN HEMT operation. One of the major obstacles causing performance degradation is the trapping effects in AlGaN/GaN HEMTs during operation, causing current dispersion between DC to RF or otherwise, DC to pulsed $I_D - V_D$ characteristics. Trapping effects like the frequency dispersion of transconductance, the sensitivity of the light, the drain lag and gate lag transients, and the limited microwave power output have all been detected [148].

As shown in Figure 28, there are various trapping mechanisms usually seen in AlGaN/GaN HEMTs, such as:

- Under a high negative VGS, barrier traps and a gate electron are injected onto the surface;
- Under a high on-state VDS, the barrier/buffer traps are injected using hot electrons;
- During material growth, the electrons captured by the deep layers are induced.

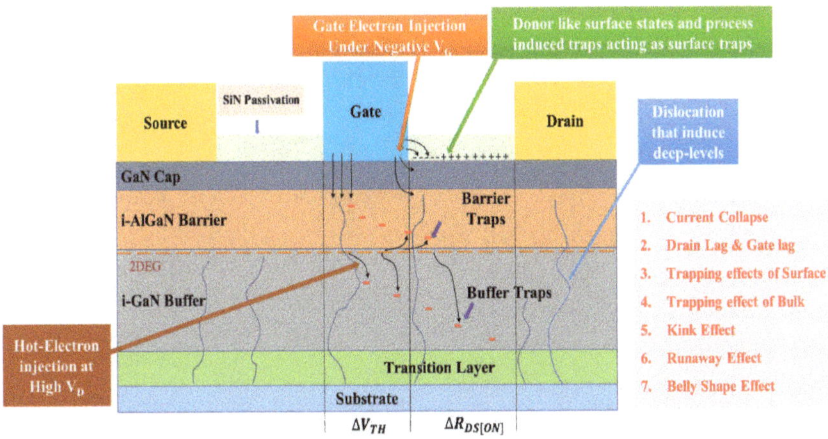

Figure 28. The AlGaN/GaN heterostructure is shown schematically, with several trapping methods and their effects on device performance. Reprinted with permission [149]. Under the terms of the Creative Commons Attribution 4.0 license.

4.2.1. Current Collapse

In order to maintain the charge neutrality, trapping electrons reduces in the 2DEG channel the density of the sheet carriers, resulting in a decreased current density in the drain. This decrease in the drain current is referred to as the "current collapse". M. A. Khan et al. [150] released the first article on AlGaN/GaN HFETs current collapse in 1994. The "drain-lag" and "gate-lag" conditions of the quiescent bias (Q-bias) are the main parameters used to characterize the current collapse phenomena [148]. In Figure 29, the characteristics curve of I_D–V_D demonstrates the current collapse's effect on the output power of HEMT. The solid and dotted lines indicate the device's I_D–V_D characteristics curve before and after the current collapse phenomena. The current collapse causes the knee voltage (V_{knee}) to grow as the maximum current decreases (I_{Dmax}). It reduces the maximum output power (P_{max}) that may be achieved, which is provided by the given equation:

$$P_{max} = \frac{(\Delta V_D \times \Delta I_D)}{8} \tag{18}$$

The dissimilarity between the V_{knee} and the breakdown voltage (BV_{GD}) equals the highest possible voltage swing (ΔV_D). I_{Dmax} specifies the maximum current swing. P_{max} is depicted in Figure 29 by the shaded region, and the current collapse phenomena have narrowed this operating window. Thus, the main reason for the current collapse is caused by the reduction in positive charges of polarization induced from the heterojunction transistor surface, followed by the loss of an equivalent number of electrons in the 2DEG [151]. $\Delta R_{DS[ON]}$ characterizes the current collapse, as illustrated in Figure 29, because the electrons are trapped in the gate-drain access area by the surface, buffer, or barrier traps. Similarly, under the gate are the areas of electron trapping (see Figure 28), resulting in a threshold voltage shift (ΔV_{TH}), which characterizes a current collapse; therefore, the pulsing I_D–V_D characteristics are caused by a drop in the drain current in the device's saturation zone, as seen in Figure 29.

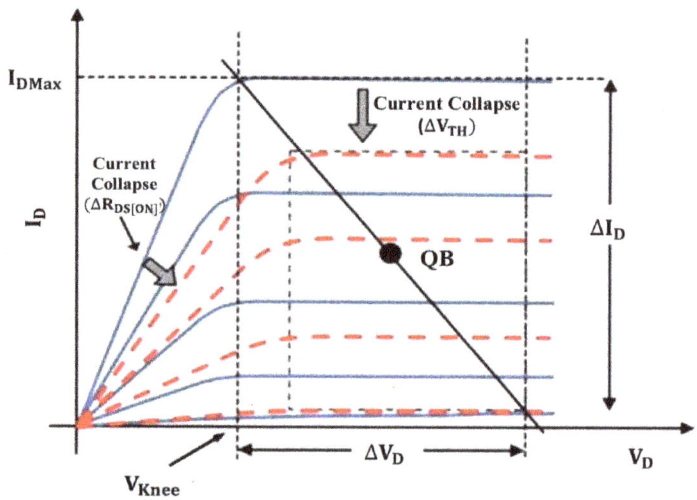

Figure 29. The I_D–V_D characteristics curve of HEMT before (solid lines) and after (graded lines) current collapse is shown in this diagram. Reprinted with permission [152]. Copyright 2003, the American Vacuum Society.

4.2.2. Drain and Gate Lag

The gate lag refers to the drain current's transient response to the gate voltage, and a drain lag is the transient response of the drain current to the drain voltage, which remains constant when the drain or gate voltage matches. According to Figure 30, Zhang et al. [153] investigated the magnitudes of collapse between the saturation and the knee voltages for the various quiescent biases. Hence, the collapse considerably reduces the performance near the knee voltage. From curve (a), when V_{DS} = 4.5 V, the collapse percentage was 4.2 and 9.8, respectively, with the Al/SiN passivation layer. On the contrary, the collapse rate increased by a 5.5 and 13.8 percentage, with the SiN passivation layer at the V_{DS} = 5 V, approximately, on the curve (b).

However, the effects of acceptor- and donor-like traps on the HEMT device were analyzed by Tirado et al. [154] The accumulating hole near the surface counteracts the acceptor-like traps, and the device is insensitive to the acceptor-like traps. Therefore, the surface donor-like traps are the primary cause of the gate lag [155]. Figure 31a presents that in addition to the dipole charge owing to polarization-induced charges, the charge due to 2DEG, the charge due to the donor-like trap ionization, and the charge due to the holes accumulating close to the AlGaN surface, the filling and emptying of the donor-like traps in the surface can impact the density of the 2DEG. When a new device is strained, the electrons that directly tunnel through the Schottky gate contact can occupy the donor-like traps and turn them neutral [156]. The 2DEG density is dropped with the reducing donor-like traps due to the charge neutrality criteria, which causes an initial fall in the current in Figure 31b [154], where the drain current instantly rises as the on-state remains stable for a while. After that, until saturation, its soars again. At the final stage, the drain current change is referred to as the delay of the current, known as a gate lag.

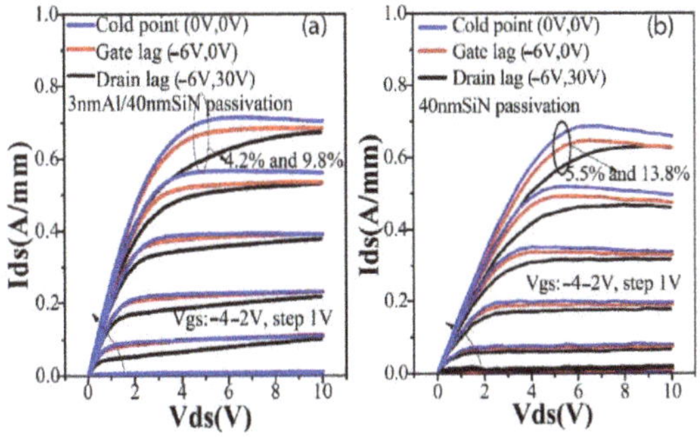

Figure 30. The gate lag and drain lag of (**a**) Al/SiN passivated and (**b**) SiN passivated AlGaN/GaN HEMT devices were measured at different passivation layers. Reproduced with permission [153]. Copyright 2019, the author (s), published by AIP Publishing.

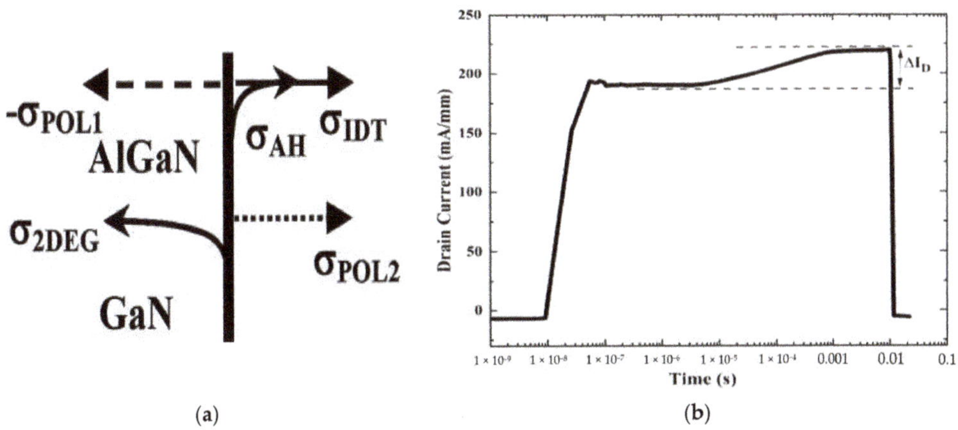

Figure 31. (**a**) Diagram of an HEMT device's space charge element. The two-dimensional electron gas is referred to as σ_{2DEG}, the dipole charge induced by polarization is denoted by $\pm\, \sigma_{POL1,2}$, Ionized donor-like traps are referred to as σ_{IDT}, and σ_{AH} symbolizes the accumulation of holes near the AlGaN surface. (**b**) AlGaN/GaN HEMT gate lag phenomenon's transient waveform of the drain current. Reproduced with permission from IOP Publishing [154].

When measured at the turn-on pulsing mode, the observed drain current tends to reduce with time, as shown in Figure 32a. When evaluated at the turn-off pulsing mode, Figure 32b indicates a decline in the drain current when the voltage at the drain returns to 0 V. These two varieties of the drain current curves are frequently referred to as the drain lag [157]. Due to the trapping of the electron in the channel, this current reduction occurred. Some electrons might tunnel through the AlGaN barrier while the device is in the on-state and trapped by deep-level traps. Eventually, to overcome these issues, the passivation layer and field plate had to be used on the device, as presented in Figure 16.

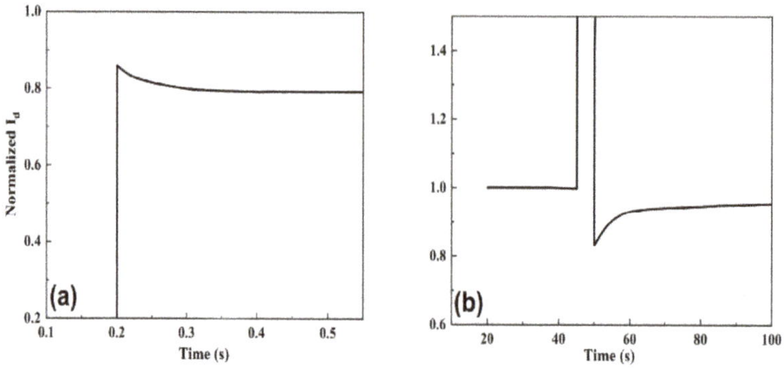

Figure 32. For AlGaN/GaN HEMTs, the transient waveform of the drain current from the drain lag phenomena. (**a**) turn-on pulsing mode (**b**) turn-off pulsing mode. Reproduced with permission [157]. Copyright 2009, the author (s), published by AIP Publishing.

The gate lag measurements compare the prompt and steady-state I_D at a constant low field drain voltage (avoid self-heating contributions) while the gate bias is modulated from a quiescent point, usually below the pinch-off voltage ~$V_G = -6$ V, to an open channel condition such as $V_G = 0$ V. At the pinch-off, a high amount of channel electrons are trapped in bulk or surface states, which do not immediately feed back into the channel instantaneously when the gate turns on. With the large-signal recovery time of constants up to the order of seconds, the responsible traps must have a significantly large activation energy and/or be fed by a "slow" conducting mechanism (like hopping) to explain the long time constants. The surface states are believed to dominate in the gate lag since the passivation significantly improves the current response.

For the drain lag measurements, the gate voltage is kept constant at around 0 V, while the drain voltage is shifted from a low equilibrium value of ~10–100 mV to a higher value of 15–30 V. The buffer traps are reported to be dominant here since the devices with higher buffer layer conductivities display lower drain lag ratios. High-drain biases inject electrons into the buffer, where they stay trapped. The passivation has a minimal effect, and the hot electrons can contribute significantly. The recovery time is often in minutes, and the reduction in the knee voltages and $I_{D,max}$ is amplified. The threshold shifts associated with the current collapse are also accounted for with the drain lag effect.

4.2.3. Trapping Effects of Surface

In GaN HEMTs, the effects of surface trapping are commonly used to describe electron trapping on the AlGaN barrier surface. R. Vetury et al. [158] utilized the "virtual gate" idea to describe the current dispersion because of surface trapping on the ungated surface between the drain and gate contacts under negative V_{GS} and a positive drain bias. Thus, Figure 33a illustrates that, at first, due to the presence of a momentous initial drain current, the 2DEG channel must remain in the slot region; from the bilateral metal gates to the AlGaN surface in the slot area, the electrons are injected and captured by the surface states when the off-state gate voltage is applied. Afterwards, from both the bilateral gate edges to the middle of the slot region, the negative trapped surface charges steadily deplete the 2DEG conduction channel, which, acting as a "virtual gate" on both sides, converges at the middle of the slot region, and the entire 2DEG channel in the slot region becomes wholly depleted. Eventually, it has gradually become a complete "virtual gate", as presented in Figure 33b [159].

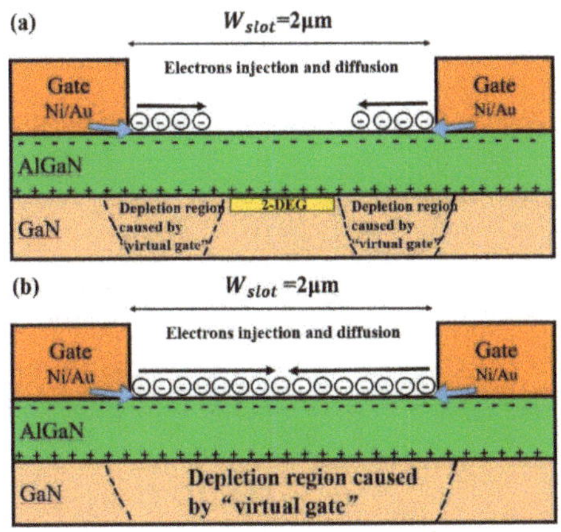

Figure 33. The device layout illustrates (**a**) electron injection in the slot region and the virtual gate in both areas. (**b**) the devices become fully virtual gates. Reproduced with permission [159]. Copyright 2019, the author (s), Published by AIP Publishing.

The impact of the virtual gate on the sheet carrier density is observed in Figure 34 [160], where the increased drain voltage at the gate edge of the drain side causes a linear reduction in the sheet carrier density. For this reason, electrons accumulate at the drain side's gate edge. Consequently, this accumulation of electrons not only influences the channel but also extends towards the lateral axis of the device, for instance, extending the depletion from 1.1 μm to 1.2 μm for the drain voltage from 0 V to 25 V.

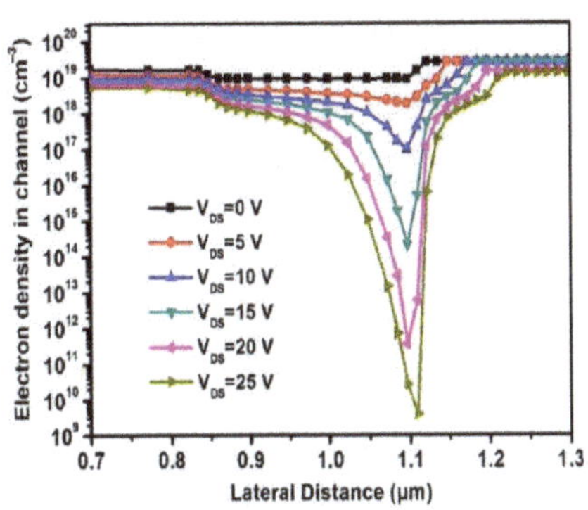

Figure 34. AlGaN/GaN HEMT sheet carrier density along the channel for different drain voltage. Reproduced with permission [160]. Copyright 2021, Elsevier.

4.2.4. Trapping Effect of Bulk

Bulk traps are traps of electrons which are found in the GaN buffer, AlGaN barrier, or other parts of the device heterostructure, as shown in Figure 27. J. Joh et al. [161] also reported that the injection of an electron into the bulk traps from the gate induced the bulk trapping effect to occur. Thus, X. Zheng et al. [162] illustrated that varying the V_{GS} has bulk trapping effects on the AlGaN barrier traps (TP2) and the GaN channel layer traps (TP1), presented in Figure 35. However, from the gate current (I_G), TP2 can capture more electrons if the V_{GS} increases the negative bias voltage. TP2s electron trapping reduces the channel carrier concentration, which lowers the drain current. Additionally, the trapping process of TP2 is speeded up by a higher negative bias voltage of V_{GS}, which decreases the action time of TP2. Meanwhile, TP1 is insensitive to the change in V_{GS}, causing little effect on TP1. Eventually, it is observed that the V_{GS} effect on TP1 is less than on TP2.

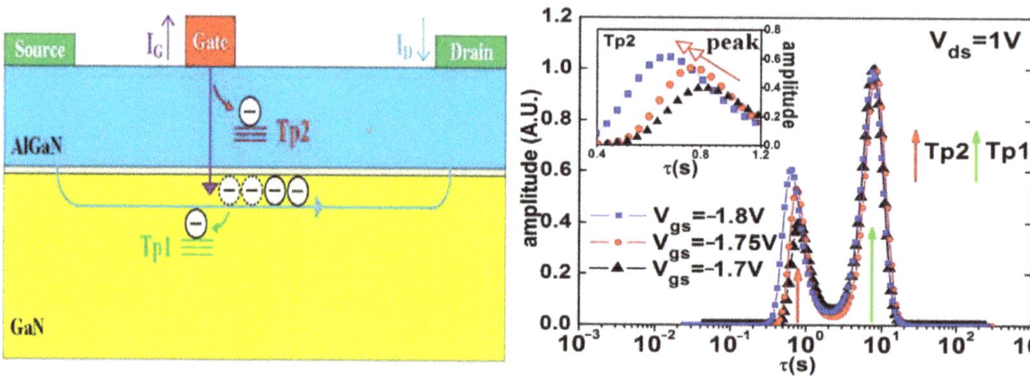

Figure 35. Trapping process of TP1 and TP2 with various values of V_{GS}. Reproduced with permission [162]. Copyright 2016, Elsevier.

P.V. Raja et al. [163] analyzed the buffer trapping effects on the AlGaN/GaN HEMT device. For the gate lag transient, the V_{GS} pulsed from −3 V to 0 V (from the off- to on-state) with a fixed V_{DS} = 10 V. The gate lag transient is simulated by changing the density of the buffer trap, N_{TB} (from 5×10^{16} to 5×10^{17} cm^{-3}), with a fixed electron trap at $E_{TB} = E_C - 0.47$ eV and an electron capture cross-section $\sigma_{nB} = 7 \times 10^{-17}$ cm^2. As a result, by increasing the N_{TB}, the signal magnitude of the gate transient is reduced, but variations of the transient are unchanged, as presented in Figure 36a. On the contrary, the V_{GS} remains at 0 V, and the V_{DS} is switched from 2 V to 10 V in the drain lag transient. The parameters of N_{TB} and σ_{nB} remain the same on the buffer trapping. Consequently, reducing the magnsitude of I_{DS} does not affect the current dispersion behavior, as displayed in Figure 36b.

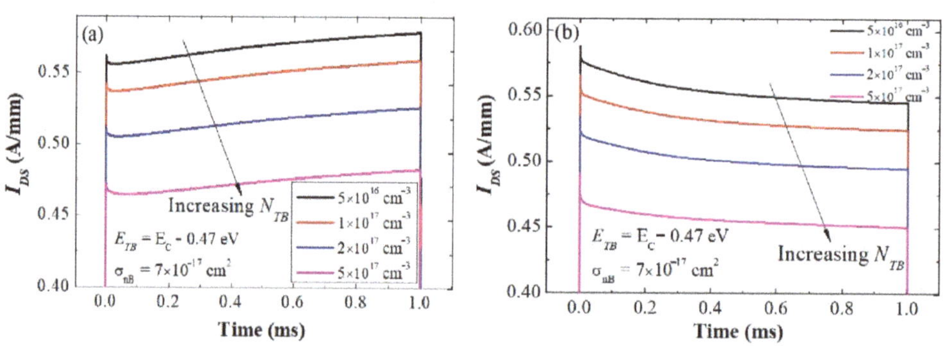

Figure 36. By varying buffer trap density, transient response of (**a**) gate lag and (**b**) drain lag are observed. Reproduced with permission [163]. Copyright 2021, Elsevier.

4.2.5. Kink Effect

The kink effect refers to the output I–V characteristics of an undesirable change phenomenon caused by changes in the conductivity of the drain [164]. This is determined by the working point of the device, as well as the temperature, voltage, and current. As is evident in Figure 37, lengthy integration durations while comparing the output characteristics were achieved with a downward and upward V_{DS} pumping. When the V_{DS}

is pushed down from a high value, the drain current drops at the V_{DS_kink} position, but it reappears when the V_{DS} is pumped up. As a result, the lower g_d is due to trapping the states and is connected among the source and drain electric field. This is also supported by the fact that the magnitude of the kink is unaffected by the running V_{GS} value, demonstrating that a large drain current is not required for a trap filling. The kink phenomenon has threshold characteristics related to the V_{DS_kink}. There is no kink in the I_{DS}. if the $V_{DS.max}$ does not exceed the V_{DS_kink}. As the gate voltage (V_{GS}) is increased from 0 V to $V_{GS.max}$, the nonmonotonic behavior of the V_{DS_kink} is noticed. In the region of a relatively high V_{GS}, the V_{DS_kink} rose with V_{GS}. Nevertheless, the V_{DS_kink} rose when V_{GS} reduced in the near pinch-off zone. The inset figure demonstrates the nonmonotonic behavior of the kink voltage (V_{DS_kink}) when the V_{GS} is stepping from 10 V to 0 with ΔV_{GS} =−0.2 V. Each sweep of V_{DS}, is followed by a 5 min rest [165].

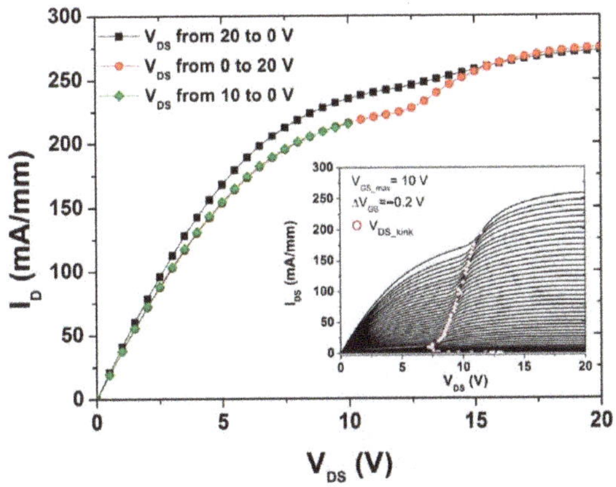

Figure 37. The kink effect in the I_D is clearly observed with a strong dependence on the V_{DS} sweep direction. Reproduced with permission [165]. Copyright 2016, the author(s), published by the American Vacuum Society.

4.2.6. Runaway Effect

When the drain voltage is raised in the saturation regime, the runaway effect occurs, defined by a simultaneous increase in the drain and gate currents [166,167]. Usually, the gate-to-drain diode component dominates the gate leakage, which rises as the V_{DS} approaches the threshold. The gate leakage is projected to decrease when the V_{GS} is minimized. As shown in Figure 38, the absolute gate current in the runaway mode is larger for the lower V_{GS} levels. After the runaway-inducing aging testing, the reliability characterizations might be severely detrimental since a continuous rise in the I_G and I_D could result in catastrophe failures, turning the runaway into an end-of-life mechanism. For the runaways, the significant drain and gate voltage will be noticed. Under open channel situations, the runaway shifts to a lower V_{DS} and greater negative V_G values. Lower activation voltages result from higher temperatures and longer aging durations. According to conduction mechanism research, the runaway seems to be linked to the Fowler–Nordheim tunnelling process.

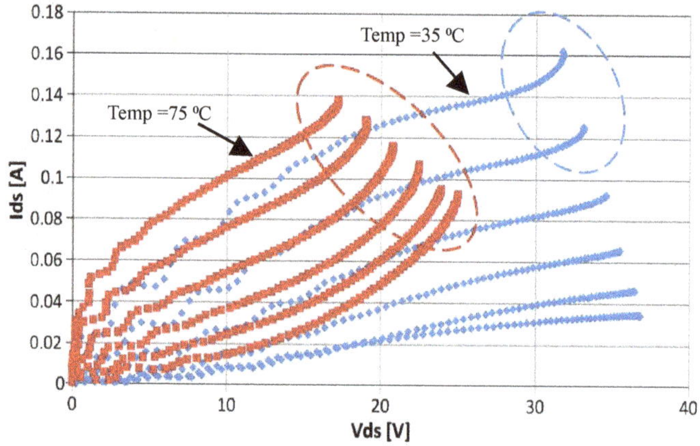

Figure 38. Output characteristics of the device reveal a runaway mechanism on the ellipse area with different temperatures. Reproduce with permission [167]. Copyright 2012, Elsevier.

4.2.7. Belly Shape Effect

According to Brunel et al. [168], after the HTOL or HTRB reliability testing, the electrical parasitic phenomena known as the belly shape effect (BS) in which on the Schottky forward characteristic, an excess of the gate leakage current resembling a "belly shape" character is noticed [168,169]. Figure 39 shows a standard belly shape. After just a few hours of aging, a belly shape forms, with varying magnitudes during the aging process, indicating degradations due to variations in the epitaxial layers, metal, or surface states. Nonetheless, the mechanics of its development are still unknown, and additional research is needed to address the dependability manifestations. Without compromising the overall dependability, it may result in high leakage currents for fascinating research on the situation and internal physics.

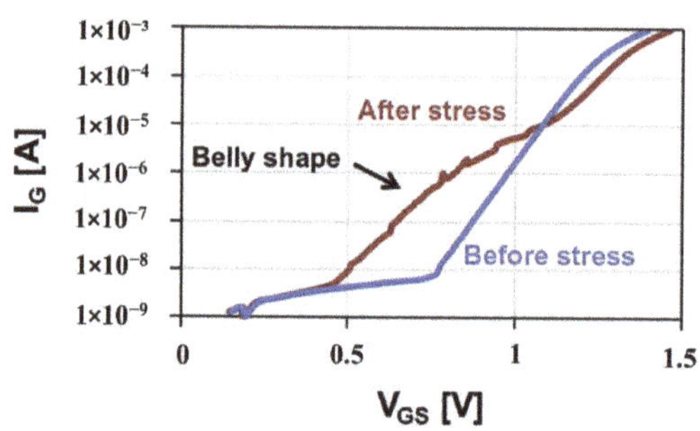

Figure 39. Before (**blue**) and after (**red**) 4000 h of HTRB stress, the forward characteristics of an HEMT with BS are shown. Reproduced with permission [168]. Copyright 2013, Elsevier.

4.3. Breakdown Voltage in GaN HEMTs Device

Gallium nitride transistors are expected to play an essential role in the next-generation power converters, thanks to the high expected breakdown voltage, which is a direct consequence of the high breakdown field of GaN (>300 V/μm). Thus, in switching the operation, the transistors alternate between the on-state, where the gate opens the channel and allows the carriers to flow through the device, and the off-state, where the gate closes the channel and blocks the carriers' current. At the off-state, the gate potential, V_{GS}, is lower than the device's threshold voltage, V_{TH}, and is considered a subthreshold condition. For an efficient switching, the off-state operation point conditions should be at a high positive drain voltage and negligible drain current; therefore, a strong gate-blocking capability is required. At very high positive drain voltage conditions, the blocking capability of the device degrades and gives rise to the subthreshold leakage (STL) current. The subthreshold leakage current will increase and become significant, thereby reducing the efficiency of the switching. A significant leakage current is three orders of magnitude lower than the device's maximal output current. At high voltages, currents higher than this value may initiate destructive processes in the device; therefore, it is considered the starting point of the device breakdown. In most cases, the breakdown voltage is either the sub-threshold drain leakage, or the gate leakage increases above 1 mA/mm.

The main physical reasons for sub-threshold current leakages (STL) increased at a higher drain bias voltage are sketched in Figure 40. The STL has to be taken into account for a high voltage GaN device engineering and it is explained as follows:

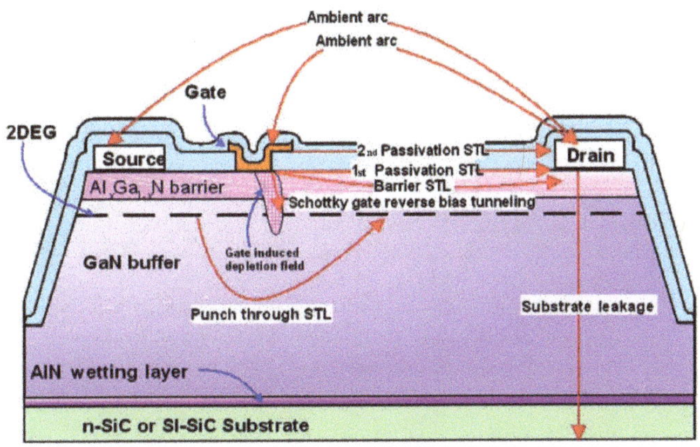

Figure 40. Illustration of most important sub-threshold current leakages (STL) and breakdown paths might appear during high voltage in GaN-based HEMTs. Reproduced with permission [170]. Copyright 2011, Elsevier.

First, the electrons punch through in the buffer underneath the gated channel region [171]: this effect depends on the magnitude of the vertical electric field in the gate region and the ability of the buffer structure to confine electrons to the channel, as shown in Figure 41a. Second, the Schottky gate reverses the bias tunnelling: this originates from the drain side edge of the gate and is triggered by the high electric fields present in this particular device region, as displayed in Figure 41b. Third, Figure 41c presents the vertical device breakdown or the substrate leakage across the epitaxial layers to the conductive substrates such as n-SiC or Si: this is mainly an issue of the buffering technology and can be prevented by suitable epitaxial concepts [171]. High-voltage GaN devices place very stringent demands on high-voltage buffer structures since, in most cases, these devices are fabricated on conductive substrates (Si, n-SiC) which are usually connected to either the drain or the source terminal of the power devices. Fourth, surface-related breakdown: this is mainly associated with the quality of device passivation, whether the interface is between the passivation layers or the semiconductor surface itself, as indicated in Figure 41d. Finally, an ambient arcing between the closely spaced device electrodes may also occur if the devices are operated in the air. Thus, this has to be taken into account by the device layout design, the technology, and design of the passivation, and also the device packaging.

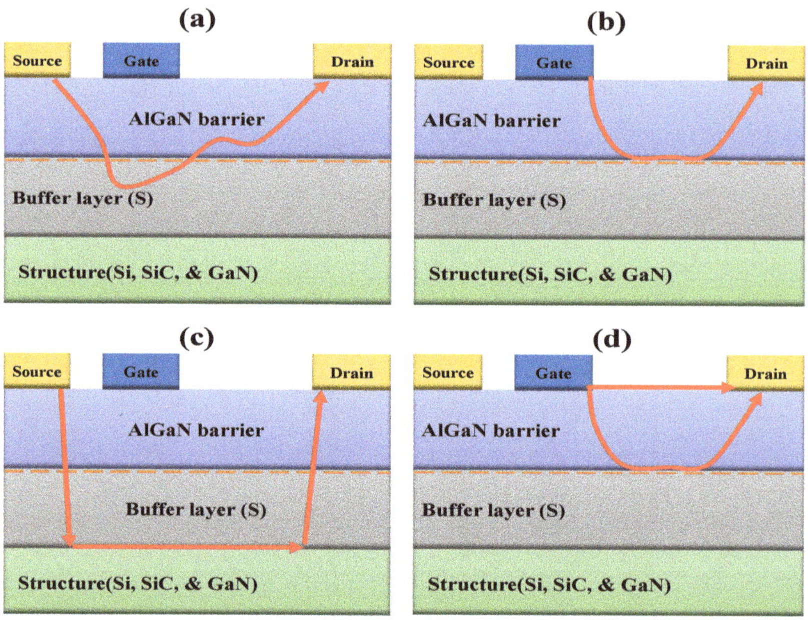

Figure 41. Schematic represents the main source of limiting the high voltage capability of AlGaN/GaN HEMT on substrates (Si, SiC, or GaN). (**a**) Punch through effect, (**b**) Breakdown along channel region, (**c**) Breakdown through buffer layer, (**d**) Leakage currents originating from the gate structure.

5. Challenges for GaN HEMTs Device

Next-generation power-efficient converters might be created using GaN-based power devices if the fundamental problems pertaining to the material's quality, device's fabrication, and its performance can be resolved. However, challenges arise, such as it is impossible to crystallize GaN from melted material because the bulk GaN crystal development process is more challenging compared to Si and SiC. Bulk GaN substrates require high temperatures, >2200 °C, and a nitrogen pressure >6GPa, plus they have tough growth methods, such as a hydride vapor-phase epitaxy (HVPE), sodium flux, and an acidic/basic ammonothermal method is required. On top of that, these GaN substrates have a defect density of between 10^4 and $10^6/\text{cm}^2$, which necessitates a careful examination before they can be used for commercial purposes [172]. Ion implantation is used to dope the conventional Si and SiC power devices selectively. However, for GaN, it is still a highly complex process requiring specialized instruments capable of high-temperature (>1200 °C) and high-pressure conditions (>1 GPa) [173]. Bulk GaNs commercialization is hampered not only by the high cost and small size of these substrates, but also by the enormous initial investment required to set up GaN-specific fabs, which could drive up the average selling price (ASP) of discrete devices and further impede the adoption of these devices. Therefore, achieving significant advancements in both the cost and substrate size of these bulk GaN substrates is crucial.

Moreover, some other challenges, such as the GaN HEMT PAs, operate at high-frequency and high-power conditions, where the drain and the gate metals meltdown because of the Eddy current triggered by the magnetic field from the nearby coil. On the contrary, the high operating current may also facilitate the issue. The meltdown occurs at the Au layer, and several hints of Au can be traced in either subsequent layer [174]. The heavy ion irradiation makes the challenge for the GaN HEMT because this ion irradiation induces a device damage by creating lattice defects and accelerating the degradations of the GaN HEMT under tests (CGHV1J006D, manufactured by Wolfspeed, Durham, NC, USA). On the other hand, a gate injection triggers an impact ionization in the channel, thus inducing failures under the off-state. Its breakdown drain bias becomes lower than the pristine one, owing to the defect generations between the channel layer and the buffer layer during the ion irradiation [175].

Furthermore, after wet etches are done to remove the ohmic contacts of electrically stressed devices; the formation of nano-cracks underneath the ohmic metal contacts of GaN HEMTs becomes challenging for the researchers [176]. This crack might grow from the metal inclusions under the influence of a vertically oriented electric field, which introduces biaxial stress in the channel layer. The hoop stresses introduced by the biaxial compressive stresses and the residual tensile stress introduced in the epitaxial process impel the cracks to extend from the alloyed S–D contacts into the channel.

6. Conclusions

GaN HEMTs appear to be a competitive semiconductor technology for the foreseeable future. The existing and projected demand for GaN-based products in power industries is outlined. The benefits and physics of III-N materials are explained to comprehend GaN-based device transistors better. The contest between Si, GaN, SiC, and β-Ga_2O_3 is anticipated to heat up in the following years since these devices are forecast to represent a crucial function in the next generation of power converters. Research at both the academic and industry levels will lead to a significant advancement in device technology. On the other hand, reliability is still a big issue. Thus, the main GaN HEMT reliability concerns caused by the trap effects like a drain, a gate lag, a current collapse, and a breakdown voltage are explained briefly. Finally, we discussed the lateral type AlGaN/GaN HEMT devices, but in the future, we will include the vertical type AlGaN/GaN HEMT devices. Overall, the future growth of the GaN research field will be aided by the continued improvement of all these features, thanks to the more sophisticated machinery and higher-quality materials.

Author Contributions: Conceptualization, M.F.P.M., M.S. and H.K.; methodology, N.I.; software, N.I.; validation, H.K. and M.S.; formal analysis, M.F.P.M. and M.F.A.J.K.; investigation, N.I. and M.F.P.M.; resources, N.I. and M.F.P.M.; data curation, N.I., H.K. and M.F.P.M.; writing—original draft preparation, N.I.; writing—review and editing, N.I., S.F., M.S. and M.F.P.M.; visualization, N.I., S.F., M.S., M.F.A.J.K. and M.F.P.M.; supervision, S.F., M.S., M.F.A.J.K. and M.F.P.M.; project administration, M.F.P.M.; funding acquisition, S.F. and M.F.P.M. All authors have read and agreed to the published version of the manuscript.

Funding: This research was funded by Universiti Sains Malaysia's Research University Incentive (RUI) grant "1001/PELECT/8014134".

Data Availability Statement: Not applicable.

Acknowledgments: The author would like to express his gratitude to USMs School of Electrical and Electronic Engineering, CEDEC, and INOR for providing the research facilities.

Conflicts of Interest: The authors declare no conflict of interest.

References

1. Fewson, D. *Introduction to Power Electronics*; Butterworth-Heinemann: Oxford, UK, 1998.
2. Baliga, B.J. 1—Introduction. In *Wide Bandgap Semiconductor Power Devices*; Baliga, B.J., Ed.; Woodhead Publishing: Sawston, UK, 2019; pp. 1–19. [CrossRef]
3. Mishra, U.K.; Parikh, P.; Wu, Y.-F. AlGaN/GaN HEMTs-an overview of device operation and applications. *Proc. IEEE* **2002**, *90*, 1022–1031. [CrossRef]
4. Power GaN 2021: Epitaxy, Devices, Application and Technology Trends Report. 2021. Available online: http://www.yole.fr/GaN_Power_Epitaxy_Devices_Applications_Technology_Trends_2021.aspx (accessed on 1 May 2021).
5. Compound Semiconductor Quarterly Market Monitor: Q3-2020. 2020. Available online: https://www.i-micronews.com/improved-reliability-in-power-gan-will-further-drive-its-market-growth-an-interview-with-transphorm/?cn-reloaded=1 (accessed on 5 November 2020).

6. RF GaN 2021: Applications, Players, Technology and Substrates 2021 Report. 2021. Available online: https://s3.i-micronews.com/uploads/2021/06/YINTR21191-GaN_RF_Market_June_2021_Yole_Flyer.pdf (accessed on 1 June 2021).
7. Millan, J.; Godignon, P.; Perpiñà, X.; Pérez-Tomás, A.; Rebollo, J. A survey of wide bandgap power semiconductor devices. *IEEE Trans. Power Electron.* **2013**, *29*, 2155–2163. [CrossRef]
8. Hudgins, J.L.; Simin, G.S.; Santi, E.; Khan, M.A. An assessment of wide bandgap semiconductors for power devices. *IEEE Trans. Power Electron.* **2003**, *18*, 907–914. [CrossRef]
9. Milligan, J.; Sheppard, S.; Pribble, W.; Wu, Y.-F.; Muller, G.; Palmour, J. SiC and GaN wide bandgap device technology overview. In Proceedings of the 2007 IEEE Radar Conference, Waltham, MA, USA, 17–20 April 2007; pp. 960–964.
10. Devices, P.G. *Materials, Applications and Reliability*; Springer International Publishing: Cham, Switzerland, 2017.
11. Schweber, B. *Use SiC-Based MOSFETs to Improve Power Conversion Efficiency*; Digi-Key Electronics: Thief River Falls, MN, USA, 2019.
12. Ballestín-Fuertes, J.; Muñoz-Cruzado-Alba, J.; Sanz-Osorio, J.F.; Laporta-Puyal, E. Role of Wide Bandgap Materials in Power Electronics for Smart Grids Applications. *Electronics* **2021**, *10*, 677. [CrossRef]
13. Marino, F.; Faralli, N.; Ferry, D.; Goodnick, S.; Saraniti, M. Figures of merit in high-frequency and high-power GaN HEMTs. *J. Phys. Conf. Ser.* **2009**, *193*, 012040. [CrossRef]
14. Baliga, B.J. Power semiconductor device figure of merit for high-frequency applications. *IEEE Electron Device Lett.* **1989**, *10*, 455–457. [CrossRef]
15. Shur, M.; Davis, R.F. *GaN-Based Materials and Devices: Growth, Fabrication, Characterization and Performance*; World Scientific: Toh Tuck, Singapore, 2004; Volume 33.
16. Nikolaev, V.I.; Stepanov, S.I.; Romanov, A.E.; Bougrov, V.E. 14–Gallium oxide. In *Single Crystals of Electronic Materials*; Fornari, R., Ed.; Woodhead Publishing: Sawston, UK, 2019; pp. 487–521. [CrossRef]
17. Bernardini, F.; Fiorentini, V.; Vanderbilt, D. Accurate calculation of polarization-related quantities in semiconductors. *Phys. Rev. B* **2001**, *63*, 193201. [CrossRef]
18. Rais-Zadeh, M.; Gokhale, V.J.; Ansari, A.; Faucher, M.; Théron, D.; Cordier, Y.; Buchaillot, L.J. Gallium nitride as an electromechanical material. *J. Microelectromech. Syst.* **2014**, *23*, 1252–1271. [CrossRef]
19. Bernardini, F.; Fiorentini, V.; Vanderbilt, D. Spontaneous polarization and piezoelectric constants of III-V nitrides. *Phys. Rev. B* **1997**, *56*, R10024. [CrossRef]
20. Asbeck, P.M. Electronic properties of III-nitride materials and basics of III-nitride FETs. In *Semiconductors and Semimetals*; Elsevier: Amsterdam, The Netherlands, 2019; Volume 102, pp. 1–40.
21. Ambacher, O.; Majewski, J.; Miskys, C.; Link, A.; Hermann, M.; Eickhoff, M.; Stutzmann, M.; Bernardini, F.; Fiorentini, V.; Tilak, V.J. Pyroelectric properties of Al (In) GaN/GaN hetero-and quantum well structures. *J. Phys. Condens. Matter* **2002**, *14*, 3399. [CrossRef]
22. Pearton, S.; Zolper, J.; Shul, R.; Ren, F.J. GaN: Processing, defects, and devices. *J. Appl. Phys.* **1999**, *86*, 1–78. [CrossRef]
23. Ambacher, O.; Foutz, B.; Smart, J.; Shealy, J.; Weimann, N.; Chu, K.; Murphy, M.; Sierakowski, A.; Schaff, W.; Eastman, L.J. Two dimensional electron gases induced by spontaneous and piezoelectric polarization in undoped and doped AlGaN/GaN heterostructures. *J. Appl. Phys.* **2000**, *87*, 334–344. [CrossRef]
24. Cros, A.; Garro, N.; Cantarero, A.; Coraux, J.; Renevier, H.; Daudin, B. Raman scattering as a tool for the evaluation of strain in Ga N/Al N quantum dots: The effect of capping. *Phys. Rev. B* **2007**, *76*, 165403. [CrossRef]
25. Pearton, S.; Ren, F.; Zhang, A.; Lee, K.; Reports, E.R. Fabrication and performance of GaN electronic devices. *Mater. Sci. Eng. R Rep.* **2000**, *30*, 55–212. [CrossRef]
26. Grabowski, S.P.; Schneider, M.; Nienhaus, H.; Mönch, W.; Dimitrov, R.; Ambacher, O.; Stutzmann, M. Electron affinity of AlxGa1−xN(0001) surfaces. *Appl. Phys. Lett.* **2001**, *78*, 2503–2505. [CrossRef]
27. Dabiran, A.M.; Wowchak, A.M.; Osinsky, A.; Xie, J.; Hertog, B.; Cui, B.; Look, D.C.; Chow, P.P. Very high channel conductivity in low-defect AlN/GaN high electron mobility transistor structures. *Appl. Phys. Lett.* **2008**, *93*, 082111. [CrossRef]
28. Kuzmik, J.; Pozzovivo, G.; Ostermaier, C.; Strasser, G.; Pogany, D.; Gornik, E.; Carlin, J.-F.; Gonschorek, M.; Feltin, E.; Grandjean, N. Analysis of degradation mechanisms in lattice-matched InAlN/GaN high-electron-mobility transistors. *J. Appl. Phys.* **2009**, *106*, 124503. [CrossRef]
29. Lee, S.R.; Wright, A.F.; Crawford, M.H.; Petersen, G.A.; Han, J.; Biefeld, R.M. The band-gap bowing of AlxGa1−xN alloys. *Appl. Phys. Lett.* **1999**, *74*, 3344–3346. [CrossRef]
30. Palacios, T.; Mishra, U.K. AlGaN/GaN High Electron Mobility Transistors. In *Nitride Semiconductor Devices: Principles and Simulation*; John Wiley & Sons: Hoboken, NJ, USA, 2007; pp. 211–233. [CrossRef]
31. He, X.-G.; Zhao, D.-G.; Jiang, D.-S. Formation of two-dimensional electron gas at AlGaN/GaN heterostructure and the derivation of its sheet density expression. *Chin. Phys. B* **2015**, *24*, 067301. [CrossRef]
32. Jones, E.A.; Wang, F.F.; Costinett, D. Review of Commercial GaN Power Devices and GaN-Based Converter Design Challenges. *IEEE J. Emerg. Sel. Top. Power Electron.* **2016**, *4*, 707–719. [CrossRef]
33. Khan, M.A.; Bhattarai, A.; Kuznia, J.N.; Olson, D.T. High electron mobility transistor based on a GaN-AlxGa1−xN heterojunction. *Appl. Phys. Lett.* **1993**, *63*, 1214–1215. [CrossRef]

34. Ueda, H.; Sugimoto, M.; Uesugi, T.; Fujishima, O.; Kachi, T. High current operation of GaN power HEMT. In Proceedings of the ISPSD'05 17th International Symposium on Power Semiconductor Devices and Ics, Santa Barbara, CA, USA, 23–26 May 2005; pp. 311–314.
35. Schwierz, F.; Ambacher, O. Recent advances in GaN HEMT development. In Proceedings of the 11th IEEE International Symposium on Electron Devices for Microwave and Optoelectronic Applications EDMO 2003, Orlando, FL, USA, 17–18 November 2003; pp. 204–209.
36. Amano, H.; Baines, Y.; Beam, E.; Borga, M.; Bouchet, T.; Chalker, P.R.; Charles, M.; Chen, K.J.; Chowdhury, N.; Chu, R.J. The 2018 GaN power electronics roadmap. *J. Phys. D Appl. Phys.* **2018**, *51*, 163001. [CrossRef]
37. Zhang, J.; Hao, Y.; Zhang, J.; Ni, J. The mobility of two-dimensional electron gas in AlGaN/GaN heterostructures with varied Al content. *Sci. China Ser. F Inf. Sci.* **2008**, *51*, 780–789. [CrossRef]
38. Miyoshi, M.; Egawa, T.; Ishikawa, H. Structural characterization of strained AlGaN layers in different Al content AlGaN/GaN heterostructures and its effect on two-dimensional electron transport properties. *J. Vac. Sci. Technol. B* **2005**, *23*, 1527–1531. [CrossRef]
39. Jones, E.A.; Wang, F.; Ozpineci, B. Application-based review of GaN HFETs. In Proceedings of the 2014 IEEE Workshop on Wide Bandgap Power Devices and Applications, Knoxville, TN, USA, 13–15 October 2014; pp. 24–29.
40. Derluyn, J.; Boeykens, S.; Cheng, K.; Vandersmissen, R.; Das, J.; Ruythooren, W.; Degroote, S.; Leys, M.; Germain, M.; Borghs, G.J. Improvement of AlGaN/GaN high electron mobility transistor structures by in situ deposition of a Si3N4 surface layer. *J. Appl. Phys.* **2005**, *98*, 054501. [CrossRef]
41. Hua, M.; Liu, C.; Yang, S.; Liu, S.; Fu, K.; Dong, Z.; Cai, Y.; Zhang, B.; Chen, K.J. GaN-based metal-insulator-semiconductor high-electron-mobility transistors using low-pressure chemical vapor deposition SiN x as gate dielectric. *IEEE Electron Device Lett.* **2015**, *36*, 448–450. [CrossRef]
42. Hashizume, T.; Nishiguchi, K.; Kaneki, S.; Kuzmik, J.; Yatabe, Z. State of the art on gate insulation and surface passivation for GaN-based power HEMTs. *Mater. Sci. Semicond. Process.* **2018**, *78*, 85–95. [CrossRef]
43. Sze, S.M.; Li, Y.; Ng, K.K. *Physics of Semiconductor Devices*; John Wiley & Sons: Hoboken, NJ, USA, 2021.
44. Asgari, A.; Kalafi, M. The control of two-dimensional-electron-gas density and mobility in AlGaN/GaN heterostructures with Schottky gate. *Mater. Sci. Eng. C* **2006**, *26*, 898–901. [CrossRef]
45. Ren, J.; Tang, C.W.; Feng, H.; Jiang, H.; Yang, W.; Zhou, X.; Lau, K.M.; Sin, J.K. A novel 700 V monolithically integrated Si-GaN 'ascaded field effect transistor. *IEEE Electron Device Lett.* **2018**, *39*, 394–396. [CrossRef]
46. Xue, P.; Maresca, L.; Riccio, M.; Breglio, G.; Irace, A. Experimental study on the short-circuit instability of cascode GaN HEMTs. *IEEE Trans. Electron Devices* **2020**, *67*, 1686–1692. [CrossRef]
47. Wu, C.-C.; Jeng, S.-L. Comparison of Parasitic Capacitances of Packaged Cascode Gallium Nitride Field-effect Transistors. *Sens. Mater.* **2018**, *30*, 453–461.
48. Elangovan, S.; Cheng, S.; Chang, E.Y. Reliability characterization of gallium nitride MIS-HEMT based cascode devices for power electronic applications. *Energies* **2020**, *13*, 2628. [CrossRef]
49. Lidow, A.; Strydom, J.; de Rooij, M.; Reusch, D. *GaN Transistors for Efficient Power Conversion*; Wiley: Hoboken, NJ, USA, 2014.
50. Cai, Y.; Zhou, Y.; Chen, K.J.; Lau, K.M. High-performance enhancement-mode AlGaN/GaN HEMTs using fluoride-based plasma treatment. *IEEE Electron Device Lett.* **2005**, *26*, 435–437.
51. Khan, M.A.; Chen, Q.; Sun, C.; Yang, J.; Blasingame, M.; Shur, M.; Park, H. Enhancement and depletion mode GaN/AlGaN heterostructure field effect transistors. *Appl. Phys. Lett.* **1996**, *68*, 514–516. [CrossRef]
52. Huang, S.; Wang, X.; Liu, X.; Wang, Y.; Fan, J.; Yang, S.; Yin, H.; Wei, K.; Wang, W.; Gao, H. Monolithic integration of E/D-mode GaN MIS-HEMTs on ultrathin-barrier AlGaN/GaN heterostructure on Si substrates. *Appl. Phys. Express* **2019**, *12*, 024001. [CrossRef]
53. Cai, Y.; Zhou, Y.; Lau, K.M.; Chen, K.J. Control of threshold voltage of AlGaN/GaN HEMTs by fluoride-based plasma treatment: From depletion mode to enhancement mode. *IEEE Trans. Electron Devices* **2006**, *53*, 2207–2215. [CrossRef]
54. Tang, X.; Li, B.K.; Moghadam, H.A.; Tanner, P.; Han, J.S.; Dimitrijev, S. Mechanism of Threshold Voltage Shift in p-GaN Gate AlGaN/GaN Transistors. *IEEE Electron Device Lett.* **2018**, *39*, 1145–1148. [CrossRef]
55. Uemoto, Y.; Hikita, M.; Ueno, H.; Matsuo, H.; Ishida, H.; Yanagihara, M.; Ueda, T.; Tanaka, T.; Ueda, D. Gate injection transistor (GIT)—A normally-off AlGaN/GaN power transistor using conductivity modulation. *IEEE Trans. Electron Devices* **2007**, *54*, 3393–3399. [CrossRef]
56. Jubadi, W.; Packeer, F.; Missous, M. Optimization of Empirical Modelling of Advanced Highly Strained In 0.7 Ga 0.3 As/In 0.52 Al 0.48 As pHEMTs for Low Noise Amplifier. *Int. J. Electr. Comput. Eng.* **2017**, *7*, 3393–3399.
57. Greco, G.; Iucolano, F.; Roccaforte, F. Review of technology for normally-off HEMTs with p-GaN gate. *Mater. Sci. Semicond. Process.* **2018**, *78*, 96–106. [CrossRef]
58. Roccaforte, F.; Greco, G.; Fiorenza, P.; Iucolano, F. An overview of normally-off GaN-based high electron mobility transistors. *Materials* **2019**, *12*, 1599. [CrossRef] [PubMed]
59. Hilt, O.; Knauer, A.; Brunner, F.; Bahat-Treidel, E.; Würfl, J. Normally-off AlGaN/GaN HFET with p-type Ga Gate and AlGaN buffer. In Proceedings of the 2010 22nd International Symposium on Power Semiconductor Devices & IC's (ISPSD), Hiroshima, Japan, 6–10 June 2010; pp. 347–350.

60. Meneghini, M.; Hilt, O.; Wuerfl, J.; Meneghesso, G. Technology and reliability of normally-off GaN HEMTs with p-type gate. *Energies* **2017**, *10*, 153. [CrossRef]
61. Efthymiou, L.; Longobardi, G.; Camuso, G.; Chien, T.; Chen, M.; Udrea, F. On the physical operation and optimization of the p-GaN gate in normally-off GaN HEMT devices. *Appl. Phys. Lett.* **2017**, *110*, 123502. [CrossRef]
62. Posthuma, N.; You, S.; Liang, H.; Ronchi, N.; Kang, X.; Wellekens, D.; Saripalli, Y.; Decoutere, S. Impact of Mg out-diffusion and activation on the p-GaN gate HEMT device performance. In Proceedings of the 2016 28th International Symposium on Power Semiconductor Devices and ICs (ISPSD), Žofín Palace, Prague, Czech Republic, 12–16 June 2016; pp. 95–98.
63. Islam, N.; Yusof, N.S.; Mohamed, M.F.P.; Syamsul, M.; Khan, M.F.A.J.; Ghazali, N.A.; Hairi, M.H. Optimization of 1-μm gate length InGaAs-InAlAs pHEMT. *Microelectron. Int.* **2022**. [CrossRef]
64. Kinoshita, T.; Obata, T.; Yanagi, H.; Inoue, S.-I. High p-type conduction in high-Al content Mg-doped AlGaN. *Appl. Phys. Lett.* **2013**, *102*, 012105. [CrossRef]
65. Roccaforte, F.; Frazzetto, A.; Greco, G.; Giannazzo, F.; Fiorenza, P.; Nigro, R.L.; Saggio, M.; Leszczyński, M.; Pristawko, P.; Raineri, V. Critical issues for interfaces to p-type SiC and GaN in power devices. *Appl. Surf. Sci.* **2012**, *258*, 8324–8333. [CrossRef]
66. Callsen, G.; Wagner, M.; Kure, T.; Reparaz, J.; Bügler, M.; Brunnmeier, J.; Nenstiel, C.; Hoffmann, A.; Hoffmann, M.; Tweedie, J. Optical signature of Mg-doped GaN: Transfer processes. *Phys. Rev. B* **2012**, *86*, 075207. [CrossRef]
67. Packeer, F.; Mohamad Isa, M.; Mat Jubadi, W.; Ian, K.W.; Missous, M. Fabrication and characterization of tensile In0.3Al0.7As barrier and compressive In0.7Ga0.3 As channel pHEMTs having extremely low gate leakage for low-noise applications. *J. Phys. D Appl. Phys.* **2013**, *46*, 264002. [CrossRef]
68. Kaneko, S.; Kuroda, M.; Yanagihara, M.; Ikoshi, A.; Okita, H.; Morita, T.; Tanaka, K.; Hikita, M.; Uemoto, Y.; Takahashi, S. Current-collapse-free operations up to 850 V by GaN-GIT utilizing hole injection from drain. In Proceedings of the 2015 IEEE 27th International Symposium on Power Semiconductor Devices & IC's (ISPSD), Hong Kong, China, 10–14 May 2015; pp. 41–44.
69. Sayadi, L.; Iannaccone, G.; Sicre, S.; Häberlen, O.; Curatola, G. Threshold voltage instability in p-GaN gate AlGaN/GaN HFETs. *IEEE Trans. Electron Devices* **2018**, *65*, 2454–2460. [CrossRef]
70. Packeer Mohamed, M.F.; Mohamed Omar, M.F.; Akbar Jalaludin Khan, M.F.; Ghazali, N.A.; Hairi, M.H.; Falina, S.; Samsol Baharin, M.S.N. New Submicron Low Gate Leakage In0.52Al0.48As-In0.7Ga0.3As pHEMT for Low-Noise Applications. *Micromachines* **2021**, *12*, 1497. [CrossRef]
71. Zhou, G.; Wan, Z.; Yang, G.; Jiang, Y.; Sokolovskij, R.; Yu, H.; Xia, G. Gate leakage suppression and breakdown voltage enhancement in p-GaN HEMTs using metal/graphene gates. *IEEE Trans. Electron Devices* **2020**, *67*, 875–880. [CrossRef]
72. Chiu, H.-C.; Liu, C.-H.; Huang, C.-R.; Chiu, C.-C.; Wang, H.-C.; Kao, H.-L.; Lin, S.-Y.; Chien, F.-T. Normally-Off p-GaN Gated AlGaN/GaN MIS-HEMTs with ALD-Grown Al2O3/AlN Composite Gate Insulator. *Membranes* **2021**, *11*, 727. [CrossRef]
73. Ngo, T.H.; Comyn, R.; Chenot, S.; Brault, J.; Damilano, B.; Vezian, S.; Frayssinet, E.; Cozette, F.; Defrance, N.; Lecourt, F. Combination of selective area sublimation of p-GaN and regrowth of AlGaN for the co-integration of enhancement mode and depletion mode high electron mobility transistors. *Solid-State Electron.* **2022**, *188*, 108210. [CrossRef]
74. Xu, N.; Hao, R.; Chen, F.; Zhang, X.; Zhang, H.; Zhang, P.; Ding, X.; Song, L.; Yu, G.; Cheng, K. Gate leakage mechanisms in normally off p-GaN/AlGaN/GaN high electron mobility transistors. *Appl. Phys. Lett.* **2018**, *113*, 152104. [CrossRef]
75. Zhang, W.; Liu, X.; Fu, L.; Huang, R.; Zhao, S.; Zhang, J.; Zhang, J.; Hao, Y. Investigation of normally-off GaN-based p-channel and n-channel heterojunction field-effect transistors for monolithic integration. *Results Phys.* **2021**, *24*, 104209. [CrossRef]
76. Chiu, H.-C.; Chang, Y.-S.; Li, B.-H.; Wang, H.-C.; Kao, H.-L.; Hu, C.-W.; Xuan, R. High-performance normally off p-GaN gate HEMT with composite AlN/Al 0.17 Ga 0.83 N/Al 0.3 Ga 0.7 N barrier layers design. *IEEE J. Electron Devices Soc.* **2018**, *6*, 201–206. [CrossRef]
77. Su, S.; Zhong, Y.; Zhou, Y.; Gao, H.; Zhan, X.; Chen, X.; Guo, X.; Sun, Q.; Zhang, Z.; Bi, W. A p-GaN-Gated Hybrid Anode Lateral Diode with a Thicker AlGaN Barrier Layer. *Phys. Status Solidi* **2020**, *217*, 1900781. [CrossRef]
78. Wang, H.-C.; Liu, C.-H.; Huang, C.-R.; Chiu, H.-C.; Kao, H.-L.; Liu, X. Hole Injection Effect and Dynamic Characteristic Analysis of Normally Off p-GaN HEMT with AlGaN Cap Layer on Low-Resistivity SiC Substrate. *Micromachines* **2022**, *13*, 807. [CrossRef]
79. Sun, C.; Hao, R.; Xu, N.; He, T.; Shi, F.; Yu, G.; Song, L.; Huang, Z.; Huang, R.; Zhao, Y. Normally-off p-GaN/AlGaN/GaN high-electron-mobility transistors using oxygen plasma treatment. *Appl. Phys. Express* **2019**, *12*, 051001. [CrossRef]
80. Chang, Y.-C.; Ho, Y.-L.; Huang, T.-Y.; Huang, D.-W.; Wu, C.-H. Investigation of Normally-Off p-GaN/AlGaN/GaN HEMTs Using a Self-Terminating Etching Technique with Multi-Finger Architecture Modulation for High Power Application. *Micromachines* **2021**, *12*, 432. [CrossRef]
81. Zhou, G.; Zeng, F.; Gao, R.; Wang, Q.; Cheng, K.; Li, L.; Xiang, P.; Du, F.; Xia, G.; Yu, H. p-GaN Gate HEMTs With 10.6 V Maximum Gate Drive Voltages by Mg Doping Engineering. *IEEE Trans. Electron Devices* **2022**, *69*, 2282–2286. [CrossRef]
82. Taube, A.; Kamiński, M.; Ekielski, M.; Kruszka, R.; Jankowska-Śliwińska, J.; Michałowski, P.P.; Zdunek, J.; Szerling, A. Selective etching of p-GaN over Al0.25Ga0.75N in Cl2/Ar/O2 ICP plasma for fabrication of normally-off GaN HEMTs. *Mater. Sci. Semicond. Process.* **2021**, *122*, 105450. [CrossRef]
83. Wang, H.; Mao, W.; Zhao, S.; He, Y.; Chen, J.; Du, M.; Zheng, X.; Wang, C.; Zhang, C.; Zhang, J. Unidirectional p-GaN gate HEMT with composite source-drain field plates. *Sci. China Inf. Sci.* **2022**, *65*, 129405. [CrossRef]
84. Zhong, Y.; Su, S.; Chen, X.; Zhou, Y.; He, J.; Gao, H.; Zhan, X.; Guo, X.; Liu, J.; Sun, Q.; et al. Normally-off HEMTs With Regrown p-GaN Gate and Low-Pressure Chemical Vapor Deposition SiNx Passivation by Using an AlN Pre-Layer. *IEEE Electron Device Lett.* **2019**, *40*, 1495–1498. [CrossRef]

85. Lükens, G.; Hahn, H.; Kalisch, H.; Vescan, A. Self-Aligned Process for Selectively Etched p-GaN-Gated AlGaN/GaN-on-Si HFETs. *IEEE Trans. Electron Devices* **2018**, *65*, 3732–3738. [CrossRef]
86. Wang, H.; Mao, W.; Zhao, S.; Chen, J.; Du, M.; Zheng, X.; Wang, C.; Zhang, C.; Zhang, J.; Hao, Y. Reverse blocking p-GaN gate AlGaN/GaN HEMTs with hybrid p-GaN ohmic drain. *Superlattices Microstruct.* **2021**, *156*, 106931. [CrossRef]
87. Huang, Y.-C.; Chiu, H.-C.; Kao, H.-L.; Wang, H.-C.; Liu, C.-H.; Huang, C.-R.; Chen, S.-W. High Thermal Dissipation of Normally off p-GaN Gate AlGaN/GaN HEMTs on 6-Inch N-Doped Low-Resistivity SiC Substrate. *Micromachines* **2021**, *12*, 509. [CrossRef]
88. Liu, K.; Wang, R.; Wang, C.; Zheng, X.; Ma, X.-h.; Bai, J.; Cheng, B.; Liu, R.; Li, A.; Zhao, Y.; et al. The influence of lightly-doped p-GaN cap layer on p-GaN/AlGaN/GaN HEMT. *Semicond. Sci. Technol.* **2022**, *37*, 075005. [CrossRef]
89. Rolland, G.; Rodriguez, C.; Gommé, G.; Boucherif, A.; Chakroun, A.; Bouchilaoun, M.; Pepin, M.C.; El Hamidi, F.; Maher, S.; Arès, R.; et al. High Power Normally-OFF GaN/AlGaN HEMT with Regrown p Type GaN. *Energies* **2021**, *14*, 6098. [CrossRef]
90. Kumar, V.; Kuliev, A.; Tanaka, T.; Otoki, Y.; Adesida, I. High transconductance enhancement-mode AlGaN/GaN HEMTs on SiC substrate. In *Electronics Letters*; Institution of Engineering and Technology: Stevenage, UK, 2003; Volume 39, pp. 1758–1760.
91. Pu, T.F. Study on Normally-off AlGaN/GaN Heterostructure Field-Effect Transistors with P-GaN Cap Layer. Ph.D. Thesis, Tokushima University, Tokushima, Japan, 2019.
92. Hu, Q.; Li, S.; Li, T.; Wang, X.; Li, X.; Wu, Y. Channel Engineering of Normally-OFF AlGaN/GaN MOS-HEMTs by Atomic Layer Etching and HighHigh- κ Dielectric. *IEEE Electron Device Lett.* **2018**, *39*, 1377–1380. [CrossRef]
93. He, Y.; Gao, H.; Wang, C.; Zhao, Y.; Lu, X.; Zhang, C.; Zheng, X.; Guo, L.; Ma, X.; Hao, Y. Comparative Study Between Partially and Fully Recessed-Gate Enhancement-Mode AlGaN/GaN MIS HEMT on the Breakdown Mechanism. *Phys. Status Solidi* **2019**, *216*, 1900115. [CrossRef]
94. Zhang, J.; He, L.; Li, L.; Ni, Y.; Que, T.; Liu, Z.; Wang, W.; Zheng, J.; Huang, Y.; Chen, J.; et al. High-Mobility Normally OFF Al2O3/AlGaN/GaN MISFET With Damage-Free Recessed-Gate Structure. *IEEE Electron Device Lett.* **2018**, *39*, 1720–1723. Available online: https://digital-library.theiet.org/content/journals/10.1049/el_20031124 (accessed on 30 September 2018). [CrossRef]
95. Fiorenza, P.; Greco, G.; Iucolano, F.; Patti, A.; Roccaforte, F. Channel Mobility in GaN Hybrid MOS-HEMT Using SiO_2 as Gate Insulator. *IEEE Trans. Electron Devices* **2017**, *64*, 2893–2899. [CrossRef]
96. He, J.; Cheng, W.C.; Wang, Q.; Cheng, K.; Yu, H.; Chai, Y. Recent Advances in GaN-Based Power HEMT Devices. *Adv. Electron. Mater.* **2021**, *7*, 2001045. [CrossRef]
97. Hua, M.; Cai, X.; Yang, S.; Zhang, Z.; Zheng, Z.; Wang, N.; Chen, K.J. Enhanced Gate Reliability in GaN MIS-FETs by Converting the GaN Channel into Crystalline Gallium Oxynitride. *ACS Appl. Electron. Mater.* **2019**, *1*, 642–648. [CrossRef]
98. Acurio, E.; Crupi, F.; Magnone, P.; Trojman, L.; Iucolano, F. Impact of AlN layer sandwiched between the GaN and the Al2O3 layers on the performance and reliability of recessed AlGaN/GaN MOS-HEMTs. *Microelectron. Eng.* **2017**, *178*, 42–47. [CrossRef]
99. Wang, B.; Huang, W.; Chi, L.; Al-Hashimi, M.; Marks, T.J.; Facchetti, A. High-k Gate Dielectrics for Emerging Flexible and Stretchable Electronics. *Chem. Rev.* **2018**, *118*, 5690–5754. [CrossRef]
100. Robertson, J.; Falabretti, B.J. Band offsets of high K gate oxides on III-V semiconductors. *J. Appl. Phys.* **2006**, *100*, 014111. [CrossRef]
101. Reddy, B.P.K.; Teja, K.B.R.; Kandpal, K. Investigation on High-κ Dielectric for Low Leakage AlGaN/GaN MIS-HEMT Device, Using Material Selection Methodologies. *Semiconductors* **2018**, *52*, 420–430. [CrossRef]
102. Khan, M.A.; Hu, X.; Sumin, G.; Lunev, A.; Yang, J.; Gaska, R.; Shur, M.S. AlGaN/GaN metal oxide semiconductor heterostructure field effect transistor. *IEEE Electron Device Lett.* **2000**, *21*, 63–65. [CrossRef]
103. Eller, B.S.; Yang, J.; Nemanich, R.J. Electronic surface and dielectric interface states on GaN and AlGaN. *J. Vac. Sci. Technol. A Vac. Surf. Film.* **2013**, *31*, 050807. [CrossRef]
104. Ochiai, M.; Akita, M.; Ohno, Y.; Kishimoto, S.; Maezawa, K.; Mizutani, T. AlGaN/GaN Heterostructure Metal-Insulator-Semiconductor High-Electron-Mobility Transistors with Si3N4 Gate Insulator. *Jpn. J. Appl. Phys.* **2003**, *42*, 2278–2280. [CrossRef]
105. Lee, C.-T.; Chen, H.-W.; Lee, H.-Y. Metal–oxide–semiconductor devices using Ga2O3 dielectrics on n-type GaN. *Appl. Phys. Lett.* **2003**, *82*, 4304–4306. [CrossRef]
106. Liu, S.; Yang, S.; Tang, Z.; Jiang, Q.; Liu, C.; Wang, M.; Chen, K.J. Al2O3/AlN/GaN MOS-Channel-HEMTs With an AlN Interfacial Layer. *IEEE Electron Device Lett.* **2014**, *35*, 723–725. [CrossRef]
107. Huang, S.; Jiang, Q.; Yang, S.; Zhou, C.; Chen, K.J. Effective Passivation of AlGaN/GaN HEMTs by ALD-Grown AlN Thin Film. *IEEE Electron Device Lett.* **2012**, *33*, 516–518. [CrossRef]
108. Hsu, C.-H.; Shih, W.-C.; Lin, Y.-C.; Hsu, H.-T.; Hsu, H.-H.; Huang, Y.-X.; Lin, T.-W.; Wu, C.-H.; Wu, W.-H.; Maa, J.-S.; et al. Improved linearity and reliability in GaN metal–oxide–semiconductor high-electron-mobility transistors using nanolaminate La2O3/SiO2 gate dielectric. *Jpn. J. Appl. Phys.* **2016**, *55*, 04EG04. [CrossRef]
109. Deen, D.A.; Storm, D.F.; Bass, R.; Meyer, D.J.; Katzer, D.S.; Binari, S.C.; Lacis, J.W.; Gougousi, T. Atomic layer deposited Ta_2O_5 gate insulation for enhancing breakdown voltage of AlN/GaN high electron mobility transistors. *Appl. Phys. Lett.* **2011**, *98*, 023506. [CrossRef]
110. Rawat, A.; Meer, M.; Surana, V.k.; Bhardwaj, N.; Pendem, V.; Garigapati, N.S.; Yadav, Y.; Ganguly, S.; Saha, D. Thermally Grown TiO2 and Al2O3 for GaN-Based MOS-HEMTs. *IEEE Trans. Electron Devices* **2018**, *65*, 3725–3731. [CrossRef]
111. Jiang, H.; Liu, C.; Ng, K.W.; Tang, C.W.; Lau, K.M. High-Performance AlGaN/GaN/Si Power MOSHEMTs With ZrO_2 Gate Dielectric. *IEEE Trans. Electron Devices* **2018**, *65*, 5337–5342. [CrossRef]

112. Li, M.; Wang, J.; Zhang, B.; Tao, Q.; Wang, H.; Cao, Q.; Huang, C.; Liu, J.; Mo, J.; Wu, W.; et al. Improved fabrication of fully-recessed normally-off SiN/SiO2/GaN MISFET based on the self-terminated gate recess etching technique. *Solid-State Electron.* **2021**, *177*, 107927. [CrossRef]
113. Jiang, H.; Tang, C.W.; Lau, K.M. Enhancement-Mode GaN MOS-HEMTs With Recess-Free Barrier Engineering and High-*k* ZrO_2 Gate Dielectric. *IEEE Electron Device Lett.* **2018**, *39*, 405–408. [CrossRef]
114. Asubar, J.T.; Kawabata, S.; Tokuda, H.; Yamamoto, A.; Kuzuhara, M. Enhancement-Mode AlGaN/GaN MIS-HEMTs With High V_{TH} and High I_{Dmax} Using Recessed-Structure With Regrown AlGaN Barrier. *IEEE Electron Device Lett.* **2020**, *41*, 693–696. [CrossRef]
115. He, J.; Wang, Q.; Zhou, G.; Li, W.; Jiang, Y.; Qiao, Z.; Tang, C.; Li, G.; Yu, H. Normally-OFF AlGaN/GaN MIS-HEMTs With Low R ON and V th Hysteresis by Functioning In-situ SiN x in Regrowth Process. *IEEE Electron Device Lett.* **2022**, *43*, 529–532. [CrossRef]
116. Hua, M.; Zhang, Z.; Wei, J.; Lei, J.; Tang, G.; Fu, K.; Cai, Y.; Zhang, B.; Chen, K.J. Integration of LPCVD-SiNx gate dielectric with recessed-gate E-mode GaN MIS-FETs: Toward high performance, high stability and long TDDB lifetime. In Proceedings of the 2016 IEEE International Electron Devices Meeting (IEDM), San Francisco, CA, USA, 3–7 December 2016; pp. 11–14.
117. Zhang, J.; He, L.; Li, L.; Ni, Y.; Que, T.; Liu, Z.; Wang, W.; Zheng, J.; Huang, Y.; Chen, J.; et al. A balancing method for low Ron and high Vth normally-off GaN MISFET by preserving a damage-free thin AlGaN barrier layer. In Proceedings of the 2018 IEEE 30th International Symposium on Power Semiconductor Devices and ICs (ISPSD), Chicago, IL, USA, 13–17 May 2018; pp. 224–227.
118. Cai, X.; Hua, M.; Zhang, Z.; Yang, S.; Zheng, Z.; Cai, Y.; Chen, K.J.; Wang, N. Atomic-scale identification of crystalline GaON nanophase for enhanced GaN MIS-FET channel. *Appl. Phys. Lett.* **2019**, *114*, 053109. [CrossRef]
119. Zhang, B.; Wang, J.; Wang, X.; Wang, C.; Huang, C.; He, J.; Wang, M.; Mo, J.; Hu, Y.; Wu, W. Improved Performance of Fully-recessed High-threshold-voltage GaN MIS-HEMT with In Situ H2/N2 Plasma Pretreatment. *IEEE Electron Device Lett.* **2022**, *43*, 1021–1024. [CrossRef]
120. Azam, F.; Tanneeru, A.; Lee, B.; Misra, V. Engineering a Unified Dielectric Solution for AlGaN/GaN MOS-HFET Gate and Access Regions. *IEEE Trans. Electron Devices* **2020**, *67*, 881–887. [CrossRef]
121. He, L.; Li, L.; Yang, F.; Zheng, Y.; Zhang, J.; Que, T.; Liu, Z.; Zhang, J.; Wu, Q.; Liu, Y. Correlating device behaviors with semiconductor lattice damage at MOS interface by comparing plasma-etching and regrown recessed-gate Al_2O_3/GaN MOS-FETs. *Appl. Surf. Sci.* **2021**, *546*, 148710. [CrossRef]
122. Greco, G.; Fiorenza, P.; Iucolano, F.; Severino, A.; Giannazzo, F.; Roccaforte, F. Conduction Mechanisms at Interface of AlN/SiN Dielectric Stacks with AlGaN/GaN Heterostructures for Normally-off High Electron Mobility Transistors: Correlating Device Behavior with Nanoscale Interfaces Properties. *ACS Appl. Mater. Interfaces* **2017**, *9*, 35383–35390. [CrossRef]
123. Cai, Y.; Zhang, Y.; Liang, Y.; Mitrovic, I.Z.; Wen, H.; Liu, W.; Zhao, C. Low ON-State Resistance Normally-OFF AlGaN/GaN MIS-HEMTs With Partially Recessed Gate and ZrO_x Charge Trapping Layer. *IEEE Trans. Electron Devices* **2021**, *68*, 4310–4316. [CrossRef]
124. Zhang, X.; Wei, X.; Zhang, P.; Zhang, H.; Zhang, L.; Deng, X.; Fan, Y.; Yu, G.; Dong, Z.; Fu, H. Low Threshold Voltage Shift in AlGaN/GaN MIS-HEMTs on Si Substrate Using SiNx/SiON as Composite Gate Dielectric. *Electronics* **2022**, *11*, 895. [CrossRef]
125. Zhou, Q.; Liu, L.; Zhang, A.; Chen, B.; Jin, Y.; Shi, Y.; Wang, Z.; Chen, W.; Zhang, B. 7.6 V Threshold Voltage High-Performance Normally-Off Al2O3 GaN MOSFET Achieved by Interface Charge Engineering. *IEEE Electron Device Lett.* **2016**, *37*, 165–168. [CrossRef]
126. Biswas, D.; Tsuboi, T.; Egawa, T. GaN/InGaN double quantum well (DQW) gate structure for GaN-on-Si based normally-off AlGaN/GaN high electron mobility transistors (HEMTs). *Mater. Sci. Semicond. Process.* **2021**, *135*, 106109. [CrossRef]
127. Wang, H.; Wang, J.; Liu, J.; Li, M.; He, Y.; Wang, M.; Yu, M.; Wu, W.; Zhou, Y.; Dai, G. Normally-off fully recess-gated GaN metal–insulator–semiconductor field-effect transistor using Al2O3/Si3N4 bilayer as gate dielectrics. *Appl. Phys. Express* **2017**, *10*, 106502. [CrossRef]
128. Ma, Y.; Xiao, M.; Du, Z.; Yan, X.; Cheng, K.; Clavel, M.; Hudait, M.K.; Kravchenko, I.; Wang, H.; Zhang, Y. Tri-gate GaN junction HEMT. *Appl. Phys. Lett.* **2020**, *117*, 143506. [CrossRef]
129. Nela, L.; Zhu, M.; Ma, J.; Matioli, E. High-Performance Nanowire-Based E-Mode Power GaN MOSHEMTs With Large Work-Function Gate Metal. *IEEE Electron Device Lett.* **2019**, *40*, 439–442. [CrossRef]
130. Xue, H.; Hussain, K.; Talesara, V.; Razzak, T.; Gaevski, M.; Mollah, S.; Rajan, S.; Khan, A.; Lu, W. High-Current-Density Enhancement-Mode Ultrawide-Bandgap AlGaN Channel Metal–Insulator–Semiconductor Heterojunction Field-Effect Transistors with a Threshold Voltage of 5 V. *Phys. Status Solidi–Rapid Res. Lett.* **2021**, *15*, 2000576. [CrossRef]
131. An, H.D.; Min, S.R.; Lee, S.H.; Park, J.; Kim, G.U.; Yoon, Y.J.; Seo, J.H.; Cho, M.S.; Jang, J.; Bae, J.-H.J.; et al. Fabrication and Performances of Recessed Gate AlGaN/GaN MOSFETs with Si_3N_4/TiO_2 Stacked Dual Gate Dielectric. *J. Semicond. Technol. Sci.* **2022**, *22*, 1598–1657. [CrossRef]
132. Medjdoub, F.; Carlin, J.; Gaquiere, C.; Grandjean, N.; Kohn, E. Status of the emerging InAlN/GaN power HEMT technology. *Open Electr. Electron. Eng. J.* **2008**, *2*, 1–7. [CrossRef]
133. Saito, W.; Takada, Y.; Kuraguchi, M.; Tsuda, K.; Omura, I. Recessed-gate structure approach toward normally off high-Voltage AlGaN/GaN HEMT for power electronics applications. *IEEE Trans. Electron Devices* **2006**, *53*, 356–362. [CrossRef]
134. Saito, W.; Nitta, T.; Kakiuchi, Y.; Saito, Y.; Tsuda, K.; Omura, I.; Yamaguchi, M. Suppression of Dynamic On-Resistance Increase and Gate Charge Measurements in High-Voltage GaN-HEMTs With Optimized Field-Plate Structure. *IEEE Trans. Electron Devices* **2007**, *54*, 1825–1830. [CrossRef]

135. Karmalkar, S.; Mishra, U.K. Enhancement of breakdown voltage in AlGaN/GaN high electron mobility transistors using a field plate. *IEEE Trans. Electron Devices* **2001**, *48*, 1515–1521. [CrossRef]
136. Okamoto, Y.; Ando, Y.; Nakayama, T.; Hataya, K.; Miyamoto, H.; Inoue, T.; Senda, M.; Hirata, K.; Kosaki, M.; Shibata, N.; et al. High-power recessed-gate AlGaN-GaN HFET with a field-modulating plate. *IEEE Trans. Electron Devices* **2004**, *51*, 2217–2222. [CrossRef]
137. Zanoni, E.; Meneghini, M.; Chini, A.; Marcon, D.; Meneghesso, G. AlGaN/GaN-Based HEMTs Failure Physics and Reliability: Mechanisms Affecting Gate Edge and Schottky Junction. *IEEE Trans. Electron Devices* **2013**, *60*, 3119–3131. [CrossRef]
138. Meneghesso, G.; Meneghini, M.; Tazzoli, A.; Stocco, A.; Chini, A.; Zanoni, E. Reliability issues of gallium nitride high electron mobility transistors. *Int. J. Microw. Wirel. Technol.* **2010**, *2*, 39–50. [CrossRef]
139. Hashizume, T.; Kotani, J.; Hasegawa, H. Leakage mechanism in GaN and AlGaN Schottky interfaces. *Appl. Phys. Lett.* **2004**, *84*, 4884–4886. [CrossRef]
140. Miller, E.J.; Dang, X.Z.; Yu, E.T. Gate leakage current mechanisms in AlGaN/GaN heterostructure field-effect transistors. *J. Appl. Phys.* **2000**, *88*, 5951–5958. [CrossRef]
141. Mizuno, S.; Ohno, Y.; Kishimoto, S.; Maezawa, K.; Mizutani, T. Large Gate Leakage Current in AlGaN/GaN High Electron Mobility Transistors. *Jpn. J. Appl. Phys.* **2002**, *41*, 5125–5126. [CrossRef]
142. Cao, X.A.; Stokes, E.B.; Sandvik, P.M.; LeBoeuf, S.F.; Kretchmer, J.; Walker, D. Diffusion and tunneling currents in GaN/InGaN multiple quantum well light-emitting diodes. *IEEE Electron Device Lett.* **2002**, *23*, 535–537. [CrossRef]
143. Karmalkar, S.; Sathaiya, D.M.; Shur, M.S. Mechanism of the reverse gate leakage in AlGaN/GaN high electron mobility transistors. *Appl. Phys. Lett.* **2003**, *82*, 3976–3978. [CrossRef]
144. Liu, Z.H.; Ng, G.I.; Arulkumaran, S.; Maung, Y.K.T.; Zhou, H. Temperature-dependent forward gate current transport in atomic-layer-deposited Al2O3/AlGaN/GaN metal-insulator-semiconductor high electron mobility transistor. *Appl. Phys. Lett.* **2011**, *98*, 163501. [CrossRef]
145. Tan, W.; Houston, P.; Parbrook, P.; Wood, D.; Hill, G.; Whitehouse, C. Gate leakage effects and breakdown voltage in metalorganic vapor phase epitaxy AlGaN/GaN heterostructure field-effect transistors. *Appl. Phys. Lett.* **2002**, *80*, 3207–3209. [CrossRef]
146. Xu, W.; Rao, H.; Bosman, G. Evidence of space charge limited flow in the gate current of AlGaN/GaN high electron mobility transistors. *Appl. Phys. Lett.* **2012**, *100*, 223504. [CrossRef]
147. Arslan, E.; Altındal, Ş.; Özçelik, S.; Ozbay, E. Dislocation-governed current-transport mechanism in (Ni/Au)–AlGaN/AlN/GaN heterostructures. *J. Appl. Phys.* **2009**, *105*, 023705. [CrossRef]
148. Binari, S.C.; Ikossi, K.; Roussos, J.A.; Kruppa, W.; Doewon, P.; Dietrich, H.B.; Koleske, D.D.; Wickenden, A.E.; Henry, R.L. Trapping effects and microwave power performance in AlGaN/GaN HEMTs. *IEEE Trans. Electron Devices* **2001**, *48*, 465–471. [CrossRef]
149. Ye, R.; Cai, X.; Du, C.; Liu, H.; Zhang, Y.; Duan, X.; Zhu, J. An Overview on Analyses and Suppression Methods of Trapping Effects in AlGaN/GaN HEMTs. *IEEE Access* **2021**, *10*, 21759–21773. [CrossRef]
150. Khan, M.A.; Shur, M.S.; Chen, Q.C.; Kuznia, J.N. Current/voltage characteristic collapse in AlGaN/GaN heterostructure insulated gate field effect transistors at high drain bias. In *Electronics Letters*; Institution of Engineering and Technology: Stevenage, UK, 1994; Volume 30, pp. 2175–2176. Available online: https://digital-library.theiet.org/content/journals/10.1049/el_19941461 (accessed on 1 September 2022).
151. Wen, C.P. Proposed GaN HFET current collapse mechanism. In Proceedings of the 2005 Asia-Pacific Microwave Conference Proceedings, Suzhou, China, 4–7 December 2005; p. 4.
152. Hasegawa, H.; Inagaki, T.; Ootomo, S.; Hashizume, T.J. Mechanisms of current collapse and gate leakage currents in AlGaN/GaN heterostructure field effect transistors. *J. Vac. Sci. Technol. B Microelectron. Nanometer Struct. Process. Meas. Phenom.* **2003**, *21*, 1844–1855. [CrossRef]
153. Zhang, S.; Wei, K.; Ma, X.-H.; Hou, B.; Liu, G.-G.; Zhang, Y.-C.; Wang, X.-H.; Zheng, Y.-K.; Huang, S.; Li, Y.-K. Reduced reverse gate leakage current for GaN HEMTs with 3 nm Al/40 nm SiN passivation layer. *Appl. Phys. Lett.* **2019**, *114*, 013503. [CrossRef]
154. Tirado, J.M.; Sanchez-Rojas, J.L.; Izpura, J.I. Simulation of surface state effects in the transient response of AlGaN/GaN HEMT and GaN MESFET devices. *Semicond. Sci. Technol.* **2006**, *21*, 1150–1159. [CrossRef]
155. Tirado, J.M.; Sanchez-Rojas, J.L.; Izpura, J.I. 2D simulation of static surface states in AlGaN/GaN HEMT and GaN MESFET devices. *Semicond. Sci. Technol.* **2005**, *20*, 864–869. [CrossRef]
156. Mitrofanov, O.; Manfra, M. Mechanisms of gate lag in GaN/AlGaN/GaN high electron mobility transistors. *Superlattices Microstruct.* **2003**, *34*, 33–53. [CrossRef]
157. Hu, W.D.; Chen, X.S.; Yin, F.; Zhang, J.B.; Lu, W. Two-dimensional transient simulations of drain lag and current collapse in GaN-based high-electron-mobility transistors. *J. Appl. Phys.* **2009**, *105*, 084502. [CrossRef]
158. Vetury, R.; Zhang, N.Q.; Keller, S.; Mishra, U.K. The impact of surface states on the DC and RF characteristics of AlGaN/GaN HFETs. *IEEE Trans. Electron Devices* **2001**, *48*, 560–566. [CrossRef]
159. Cao, Q.; Wang, J.; Li, M.; Wang, H.; Tao, Q.; Zhang, B.; Wu, W. Study on the charging current of surface traps in AlGaN/GaN high electron mobility transistors with a slot gate structure. *Appl. Phys. Lett.* **2019**, *115*, 152105. [CrossRef]
160. Godfrey, D.; Nirmal, D.; Arivazhagan, L.; Godwinraj, D.; Mohan Kumar, N.; Chen, Y.; Yeh, W. Current collapse degradation in GaN High Electron Mobility Transistor by virtual gate. *Microelectron. J.* **2021**, *118*, 105293. [CrossRef]
161. Joh, J.; Alamo, J.A.d. A Current-Transient Methodology for Trap Analysis for GaN High Electron Mobility Transistors. *IEEE Trans. Electron Devices* **2011**, *58*, 132–140. [CrossRef]

162. Zheng, X.; Feng, S.; Zhang, Y.; Yang, J. Identifying the spatial position and properties of traps in GaN HEMTs using current transient spectroscopy. *Microelectron. Reliab.* **2016**, *63*, 46–51. [CrossRef]
163. Vigneshwara Raja, P.; Nallatamby, J.-C.; DasGupta, N.; DasGupta, A. Trapping effects on AlGaN/GaN HEMT characteristics. *Solid-State Electron.* **2021**, *176*, 107929. [CrossRef]
164. Kaushik, J.K.; Balakrishnan, V.R.; Panwar, B.S.; Muralidharan, R. On the Origin of Kink Effect in Current–Voltage Characteristics of AlGaN/GaN High Electron Mobility Transistors. *IEEE Trans. Electron Devices* **2013**, *60*, 3351–3357. [CrossRef]
165. Kang, Y.; Sung, H.-k.; Kim, H. Investigation of kink effect in normally-off AlGaN/GaN recessed-gate MOS-heterostructure FETs. *J. Vac. Sci. Technol. B Nanotechnol. Microelectron. Mater. Process. Meas. Phenom.* **2016**, *34*, 052202. [CrossRef]
166. Brunel, L.; Lambert, B.; Carisetti, D.; Malbert, N.; Curutchet, A.; Labat, N. Electrical Runaway in AlGaN/GaN HEMTs: Physical Mechanisms and Impact on Reliability. *IEEE Trans. Electron Devices* **2017**, *64*, 1548–1553. [CrossRef]
167. Lambert, B.; Thorpe, J.; Behtash, R.; Schauwecker, B.; Bourgeois, F.; Jung, H.; Bataille, J.; Mezenge, P.; Gourdon, C.; Ollivier, C. Reliability data's of 0.5 µm AlGaN/GaN on SiC technology qualification. *Microelectron. Reliab.* **2012**, *52*, 2200–2204. [CrossRef]
168. Brunel, L.; Lambert, B.; Mezenge, P.; Bataille, J.; Floriot, D.; Grünenpütt, J.; Blanck, H.; Carisetti, D.; Gourdel, Y.; Malbert, N.; et al. Analysis of Schottky gate degradation evolution in AlGaN/GaN HEMTs during HTRB stress. *Microelectron. Reliab.* **2013**, *53*, 1450–1455. [CrossRef]
169. Rzin, M.; Curutchet, A.; Labat, N.; Malbert, N.; Brunel, L.; Lambert, B. Schottky gate of AlGaN/GaN HEMTs: Investigation with DC and low frequency noise measurements after 7000 hours HTOL test. In Proceedings of the 2015 International Conference on Noise and Fluctuations (ICNF), Xian, China, 2–6 June 2015; pp. 1–4.
170. Wuerfl, J.; Bahat-Treidel, E.; Brunner, F.; Cho, E.; Hilt, O.; Ivo, P.; Knauer, A.; Kurpas, P.; Lossy, R.; Schulz, M.; et al. Reliability issues of GaN based high voltage power devices. *Microelectron. Reliab.* **2011**, *51*, 1710–1716. [CrossRef]
171. Bahat-Treidel, E.; Hilt, O.; Brunner, F.; Wurfl, J.; Trankle, G. Punchthrough-Voltage Enhancement of AlGaN/GaN HEMTs Using AlGaN Double-Heterojunction Confinement. *IEEE Trans. Electron Devices* **2008**, *55*, 3354–3359. [CrossRef]
172. Kucharski, R.; Sochacki, T.; Lucznik, B.; Bockowski, M. Growth of bulk GaN crystals. *J. Appl. Phys.* **2020**, *128*, 050902. [CrossRef]
173. Sakurai, H.; Omori, M.; Yamada, S.; Furukawa, Y.; Suzuki, H.; Narita, T.; Kataoka, K.; Horita, M.; Bockowski, M.; Suda, J.; et al. Highly effective activation of Mg-implanted p-type GaN by ultra-high-pressure annealing. *Appl. Phys. Lett.* **2019**, *115*, 142104. [CrossRef]
174. Sangwan, V.; Tan, C.M.; Kapoor, D.; Chiu, H.C. Electromagnetic Induced Failure in GaN-HEMT High-Frequency Power Amplifier. *IEEE Trans. Ind. Electron.* **2020**, *67*, 5708–5716. [CrossRef]
175. Islam, Z.; Paoletta, A.L.; Monterrosa, A.M.; Schuler, J.D.; Rupert, T.J.; Hattar, K.; Glavin, N.; Haque, A. Heavy ion irradiation effects on GaN/AlGaN high electron mobility transistor failure at off-state. *Microelectron. Reliab.* **2019**, *102*, 113493. [CrossRef]
176. Whiting, P.G.; Rudawski, N.G.; Holzworth, M.R.; Pearton, S.J.; Jones, K.S.; Liu, L.; Kang, T.S.; Ren, F. Nanocrack formation in AlGaN/GaN high electron mobility transistors utilizing Ti/Al/Ni/Au ohmic contacts. *Microelectron. Reliab.* **2017**, *70*, 41–48. [CrossRef]

crystals

Review

Surface Dispersion Suppression in High-Frequency GaN Devices

Pengfei Zhu, Xianfeng Ni, Qian Fan and Xing Gu *

Institute of Next Generation Semiconductor Materials, Southeast University, Suzhou 215123, China
* Correspondence: xinggu@seu.edu.cn

Abstract: GaN-based high electron mobility transistors (HEMTs) are shown to have excellent properties, showing themselves to perform well among the throng of solid-state power amplifiers. They are particularly promising candidates for next-generation mobile communication applications due to their high power density, frequency, and efficiency. However, the radio-frequency (RF) dispersion aroused by a high surface-state density inherent in nitrides causes the degradation of GaN devices' performance and reliability. Although various dispersion suppression strategies have been proposed successively—including surface treatment, passivation, field plate, cap layer, and Si surface doping—outcomes were not satisfactory for devices with higher frequencies until the emergence of a novel N-polar deep recess (NPDR) structure broke this deadlock. This paper summarizes the generation of dispersion, several widespread dispersion containment approaches, and their bottlenecks under high frequencies. Subsequently, we highlight the NPDR structure as a potential substitute, evaluate its technical benefits, and review the continuous exertions in recent years.

Keywords: GaN-based HEMTs; RF dispersion; suppression; NPDR structure; high-frequency

Citation: Zhu, P.; Ni, X.; Fan, Q.; Gu, X. Surface Dispersion Suppression in High-Frequency GaN Devices. *Crystals* **2022**, *12*, 1461. https://doi.org/10.3390/cryst12101461

Academic Editor: Dmitri Donetski

Received: 27 September 2022
Accepted: 14 October 2022
Published: 16 October 2022

1. Introduction

Millimeter-wave technology is undoubtedly an ongoing research focus for fifth-generation mobile communication technology (5G) and beyond. GaN-based high electron mobility transistors (HEMTs) are taking the lead in the radio-frequency (RF) industry, especially for those applications that require abundant power density, thanks to the unparallel combined material properties of a wide bandgap (E_g ~ 3.4 eV), large critical breakdown electric field, high electron saturation velocity, high electron mobility and sheet density achieved by 2-Dimensional electron gas (2DEG), and excellent thermal properties when prepared on silicon carbide (SiC) substrates [1–4]. However, the existence of direct current (DC)-RF dispersion (also known as the "current collapse") leads to a significant decrease in output power density and causes reliability-related issues in GaN-based HEMTs, especially at higher frequencies, which limits the development and application of such devices. Research has shown that the widespread presence of trapping states (covering the semiconductor surface, barrier layer, interface, channel, and buffer layer) in III/V group compound semiconductors should be mainly responsible for the dispersion of GaN HEMTs [5]. In high-frequency devices, the closer proximity of 2DEG to the surface highlights the role of surface traps (surface states) on dispersion [6,7]. These surface states trapping electrons to form a "virtual gate" (similar to a gate function) further deplete the 2DEG channel, which strongly impacts the device output characteristics and reliability. Engineers have tried to solve this problem by using multiple processes, including surface treatment [8], passivation [9], field plate [10], cap layer [11], and Si surface doping [12] in recent years. With the increasing demand for higher operating frequencies, the reduced gate length and the aspect ratio in need of being maintained drive the 2DEG much closer to the surface compared with the lower frequency HEMT epitaxial structure, thus limiting the effectiveness of those techniques in improving the dispersion. In this vein, researchers working on high-frequency GaN devices have sought novel technologies to better solve this aporia.

In this paper, we explain the origin of surface states and the mechanism causing dispersion in GaN devices. The current dominant methods for suppression dispersion and their bottlenecks in the high-frequency domain are summarized. Finally, we focus on the N-polar deep recess (NPDR) structure, addressing its uniqueness in controlling dispersion and reviewing the research advances in NPDR metal-oxide-semiconductor (MIS)-HEMTs.

2. Surface States and Virtual Gate

In GaN-based HEMTs, surface states are critical sources of 2DEG in the channel [13], allowing the device to achieve a very high electron concentration without the need for intentional doping (unlike GaAs-based 2DEG), which dramatically reduces ionized impurity scattering and enhances carrier mobility. On the other hand, GaN-based HEMTs are also plagued by the dispersion effect generated by surface states. This section mainly discusses the creation of surface states and the principle of dispersion caused by them.

2.1. Sources of Surface States

Surface states, a type of electron-bound state, can be categorized into two sorts based on their formation: intrinsic and extrinsic.

Intrinsic surface states arise due to the termination of the ideal crystal structure at the surface. It is common knowledge that GaN, like any crystal, cannot be extended infinitely in space. Therefore, the periodic potential field will be interrupted at the surface and cause additional energy levels in the forbidden band known as surface energy levels, whose corresponding electronic states are named surface states. Chemical bonding theory explains that the chemical bonds of GaN crystal break at the surface, and the unbonded electron pairs form dangling bonds, the corresponding electronic states being surface states [14]. Besides, those created by threading dislocations accessible at the surface, native oxide, N-vacancies, ions adsorbed from the surrounding environment (such as oxygen, carbon, and hydrogen, each with its unique energy level), and surface damage in the processing are referred to as extrinsic surface states [14–16]. These two different forms of generation pathways are shown in Figure 1.

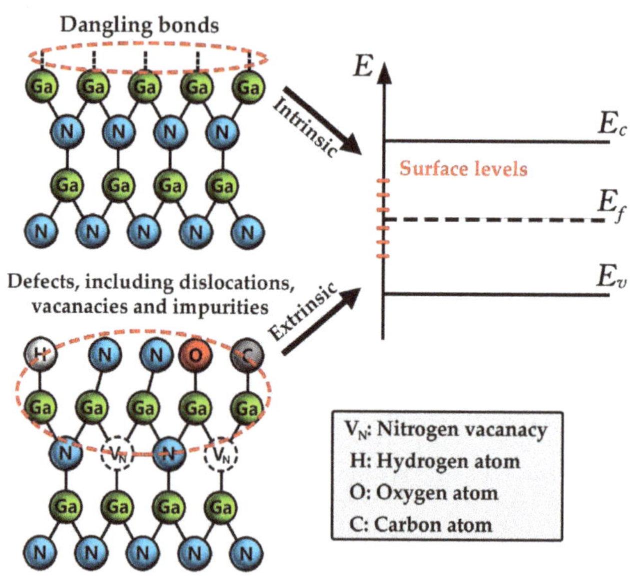

Figure 1. Mechanisms for the generation of intrinsic and extrinsic surface states.

2.2. Virtual Gate Model

Depending on their behaviors, surface states can be classified as either acceptors or donors. Table 1 shows the charge situations of these two categories of surface states in the empty and filled (after trapping the electrons) conditions. Under the strong influence of the polarized electric field, the donor surface state filled with electrons below the Fermi level will be partly ionized, allowing the electrons to escape into the channel to produce a high concentration of 2DEG and leave large positive charges on the surface [17]. As such, the surface of the GaN device should not show a negative surface potential. Vetury et al. [18] observed that a significant number of negative charges reside on the device's surface, conflicting with the previous theory when they employed the floating gate approach to investigate the surface state impact of AlGaN/GaN HEMTs in 2001. In response, they explained that these positive charge centers (empty surface states) located between the gate and drain act like traps to capture gate leakage electrons or those channel hot electrons escaping via the barrier to the surface [15,19,20]. During the process from off-state to on-state, electron emission from surface states with a slower period differs from RF signals, so the trapped electrons are not released when the device is turned on, which is equivalent to another gate existing between the gate and the drain electrodes, i.e., virtual gate [21,22]. As shown in Figure 2, the trapped electrons move and jump on the surface to form a negative charge region, causing the gate depletion region to expand laterally, depleting the 2DEG in the channel and resulting in the current collapse. This RF dispersion is generally determined by pulsed I_{DS}–V_{DS} measurement, depending on different stressed conditions, and can be defined as:

$$\text{Dispersion}\ (\%) = \frac{I_{DS,DC} - I_{DS,Pulsed}}{I_{DS,DC}} \times 100\% \quad (1)$$

Table 1. Electrical properties of donor and acceptor surface states.

Type	Empty	Filled
Donor surface states	Positive	Neutral
Acceptor surface states	Neutral	Negative

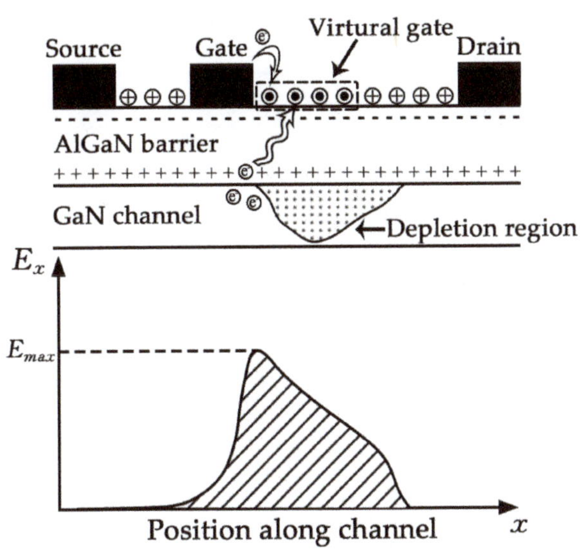

Figure 2. The device model shows the virtual gate's location, schematic representation, and the electric field distribution in the channel.

3. Methods for Suppression Surface Dispersion and Their Bottlenecks

Based on the principle of extra depletion of the 2DEG channel by the virtual gate model, it can be concluded that one of the most critical tasks in dispersion control is to minimize trapping electron action attributable to surface states. Common solutions in AlGaN/GaN HEMTs are summarized in Figure 3. Five categories of methods have been developed and demonstrated to suppress the RF dispersion with mixed success: (1) surface treatment that intends to diminish surface state density to reduce the odds of surface states trapping electrons; (2) passivation, which intends to bury surface traps by depositing dielectrics on the device surface to prevent gate leakage electrons from being trapped in them; (3) the uses of field plate structure intend to alleviate the gate edge electric field crowding and enhance the emission rate of trapped electrons; (4) adding a GaN cap intends to control the gate-lag caused by the leakage current; (5) si surface doping intends to compensate for the trapping energy level to screen the 2DEG from the traps. This section summarizes the research advances of the above technologies and the bottlenecks they encounter at high frequencies.

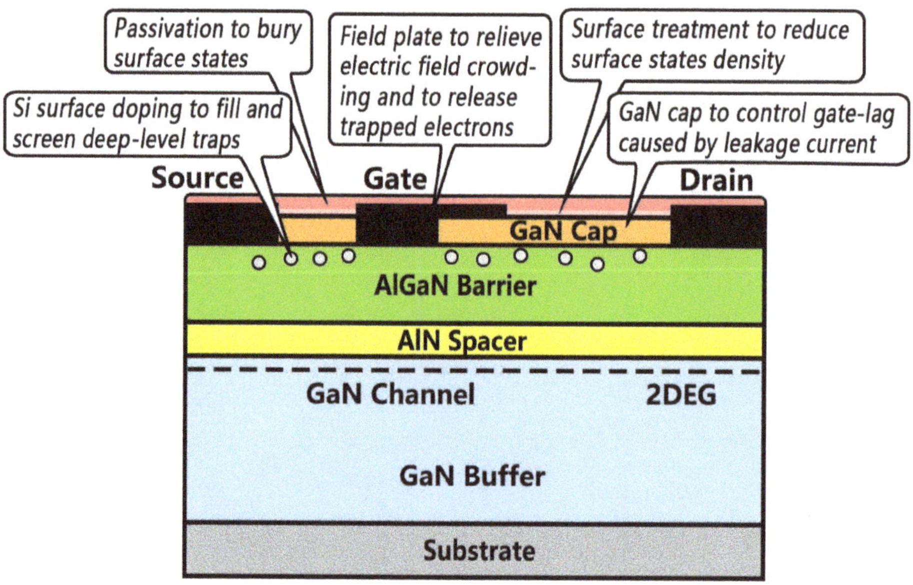

Figure 3. Common methods for suppressing RF dispersion in AlGaN/GaN HEMTs.

3.1. *Common Methods for Suppression Surface Dispersion*

3.1.1. Surface Treatment

Natural oxides and organic matter adsorbed on the (Al)GaN surface create many interface states between the dielectric used for passivation and the barrier, which reduces the effectiveness of passivation. Consequently, surface treatment prior to dielectric deposition is of extraordinary importance to obtain a high-quality interface.

Organic matter and oxides are mainly composed of carbon and oxygen. UV/ozone surface treatment can proficiently remove carbon [23], while oxygen can be treated by some wet chemical cleaning methods. King et al. [24] found that HF and HCl were able to minimize oxygen coverage on AlN and GaN surfaces, and residual halogen ions (cl^- and F^-) tie up the dangling bonds preventing the surface from being oxidized again. The same result was found in the experiments of Diale et al. [25], who respectively compared the methods of cleaning GaN surfaces with HCl, KOH, and $(NH_4)_2S$. They concluded that KOH was effective for carbon removal but contributed to a rough surface. The sample cleaned in $(NH_4)_2S$, a chemical avoiding surface re-oxidation, acquired the optimum cleaning result with the lowest values of both C and O, root mean square (RMS) roughness, and Ga/N ratio. Other wet chemical cleaning efforts mainly focus on HNO_3, H_2SO_4, NH_4OH, and NaOH.

Surface treatment using plasma like NH_3, N_2, and N_2O is also valuable for suppressing dispersion to enhance performance [26–29]. By filling nitrogen vacancies and removing impurity oxygen atoms, nitrogen-related plasma treatment considerably enhances the quality of the SiN_x/GaN interface [28]. Romero et al. [27] treated the AlGaN/GaN HEMT (which had been wet cleaned by NH_4OH) with in situ N_2 plasma (N_2PP) at 200 °C for 1 min prior to SiN_x passivation of the device. The treated GaN HEMT obtained an $I_{DS,}$ which was 10% higher than the untreated one, and achieved a more advanced knee voltage. A 65% drop in the density of interface states was discovered after applying N_2PP, which was responsible for reducing gate leakage current by two to three times (from 7.7 to 2.3 × 10^{-2} A/cm^2 at V_{GD} = −20 V), utilizing capacitance-voltage (C–V) and conductance-frequency (G-F) measurements. Furthermore, NH_3 plasma-treated performs better than N_2 in GaN HEMTs [26], with RF dispersion respectively decreasing from 63% (without pretreatment) to 1% and 9% (shown in Figure 4). This is probably due to the fact that NH_3 plasma eliminates oxide and carbon residues while reducing the density of surface state defects.

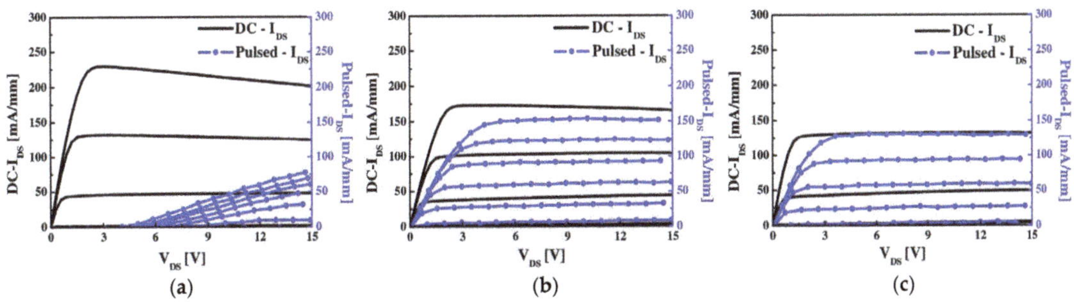

Figure 4. DC and pulsed I_{DS}–V_{DS} characteristics for the GaN device. (**a**) Without pretreatment; (**b**) with NH_4OH/N_2 pretreatment; (**c**) with NH_4OH/NH_3 pretreatment. DC and pulsed I_{DS}–V_{DS} were measured at V_{DS} = 0 to 15 V and V_{GS} = −3 to 1 V. The pulse width and period were 500 ns and 1 ms. Adapted with permission from Ref. [26]. Copyright 2010, The Japan Society of Applied Physics.

In addition, plasma can also be combined with heat treatments such as high-temperature annealing in vacuum [30] or N_2 [31] and NH_3 [32] atmospheres to enhance the desorption of contaminants further and assist in the elimination of remaining surface defects, thereby improving the surface quality of the device. It has to be emphasized, however, that the surface treatment also increases the risk of unreliability and is, therefore, mostly involved in laboratory research efforts.

3.1.2. Passivation

Since 2000, when Green et al. [15] deposited a layer of Si_3N_4 for passivation on the surface of AlGaN/GaN HEMT to obtain the highest power density (4 W/mm at 4 GHz) during that period, passivation has become one of the most commonly used and effective techniques to solve RF dispersion after more than 20 years of development.

Previous studies have reported that SiN_x, AlN, SiO_2, SiNO, Al_2O_3, and HfO_2 (high-κ dielectric) can all be applied to GaN-based HEMTs [28,31,33–35]. However, some researchers believe that nitrides (compared to oxides) probably are more suitable for GaN devices because oxygen impurities can easily diffuse into the GaN crystal and generate a high density of deep (or slow) interface states, producing additional current collapse [36,37]. This conclusion is consistent with the report of Geng et al. [31], which evaluated AlGaN/GaN MIS-HEMTs utilizing PECVD deposition of 20 nm SiN_x, SiO_2, and SiNO as passivation layers, respectively. It is found that the interface-trapped charge density (or the deep energy level defects density with long emission time of electrons) for $\Delta E > 0.657$ eV in SiO_2 is 4.13 × 10^{12} cm^{-2} (three times larger than that of SiN_x, ~1.38×10^{12} cm^{-2} in SiN_x). Compared to only 11.06% dispersion in SiN_x, the current collapse in SiO_2 reaches 84.14% at $V_{gs} = -18$ V, proving the success of SiN_x passivation and the fact that SiO_2 would not be appropriate for GaN devices. Based on taking SiN_x as a dielectric, Huang et al. [38] proposed an improved structure using bilayer LPCVD SiN_x; a Si-rich SiN_x layer (Si/N = 59/41) with a thickness of ~10 nm was first deposited, followed by a ~67 nm Si-poor SiN_x layer (Si/N = 50/50, high-resistive). Figure 5 exhibits a TEM cross-sectional image of this bilayer SiN_x passivation, and the black dotted line refers to the boundary between the first and second layers. The Si-rich SiN_x layer upper to surface is argued to remove the deep-level traps at the AlGaN/SiN_x interface, which makes microwave GaN HEMTs show a much smaller degradation of about 9.7% (~30% dispersion in 70 nm Si-poor SiN_x sample) while maintaining a low gate leakage current owing to the second high-resistive layer.

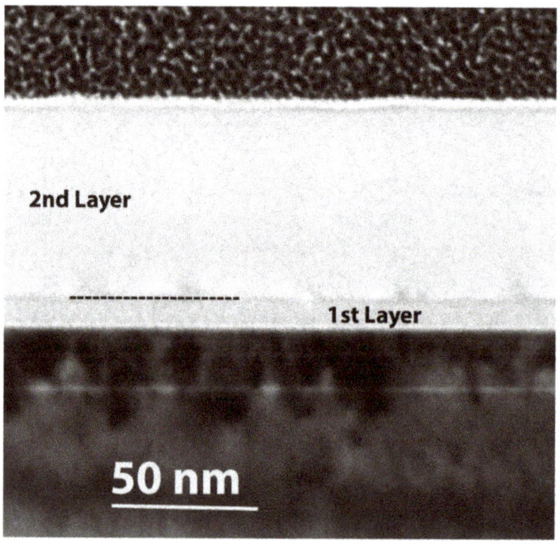

Figure 5. TEM cross-sectional image shows a distinguishable bilayer structure of the passivation layer. Reprinted with permission from Ref. [38]. Copyright 2018 by IEEE.

3.1.3. Field Plate

The field plate is applied in GaN-based HEMTs to reduce the electric field crowding at the gate corner of the drain side, preventing electron injection from the gate into surface states, and forming a virtual gate. Moreover, the horizontal electric field near the end of the field plate will strengthen the process of electrons escaping from the traps and speed up the electrons hopping along the AlGaN surface, further lowering the trapped electron density and, therefore, substantially suppressing current collapse [22]. The effectiveness of the field plate in suppressing current collapse is more robust with increasing field plate length. The trapped charges on the device surface decreased from 1.58×10^{12} cm^{-2} (no field plate) to 0.8×10^{12} cm^{-2} after using a 1.5 μm field plate structure [39], the pulsed I_{DS}–V_{DS} characteristics and trapped charge density are displayed in Figure 6. The field plate this method has now been widely adopted and greatly influenced the high-voltage devices. Despite this, it is worth noting that it is thrown into an awkward situation in high-frequency applications. Because this design not only reduces the speed of the channel electrons owing to the weakened electric field but also introduces additional external capacitance that limits the efficiency and gain of the device. Accordingly, in order to ensure anticipant RF performance, the field plate has found very limited usage in high-frequency industries.

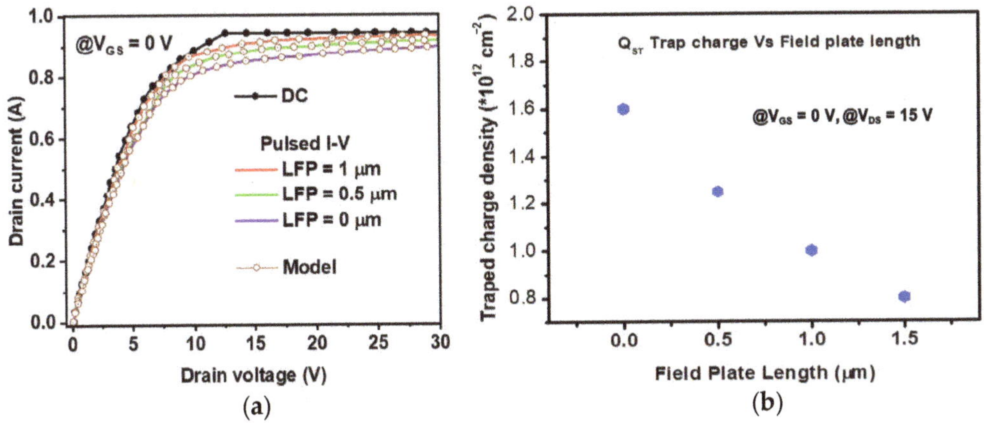

Figure 6. AlGaN/GaN HEMT characteristics with different field plate lengths. (**a**) Pulsed I_{DS}-V_{DS} characteristics with Quiescent Voltage ($V_{GS,Q}$ = −8 V and $V_{DS,Q}$ = 50 V); (**b**) trapped charge (Q_{ST}). Adapted with permission from Ref. [39]. Copyright 2021 by Elsevier.

3.1.4. GaN Cap

GaN cap, a thin layer of GaN (about a few nanometers) on top of the barrier that forms a sandwich structure (e.g., GaN/AlGaN/GaN), is another well-established technique in industry for managing the current collapse. This cap can be doped GaN (typically n-type doping by SiH_4 or p-type doping by Cp_2Mg) or i-GaN. No matter what type of GaN cap is used, it degrades gate leakage levels [40], both vertically and horizontally. Vertically, raising the effective Schottky barrier by the GaN cap provides a stronger limitation on the emission and tunneling of hot electrons [11,41]. Horizontally, as presented by Sarkar et al. [11], the GaN cap passivates the access region to suppress lateral electrons hopping via surface traps, significantly reducing the total leakage current (shown in Figure 7a). In other words, the GaN cap can effectively screen 2DEG from dispersion-related traps. As shown in Figure 7b,c, 2 nm of GaN cap layer effectively controls the current collapse in GaN-based HEMTs, while enhancing the device's frequency performance (62% improving in f_T) at the expense of a small amount of gain.

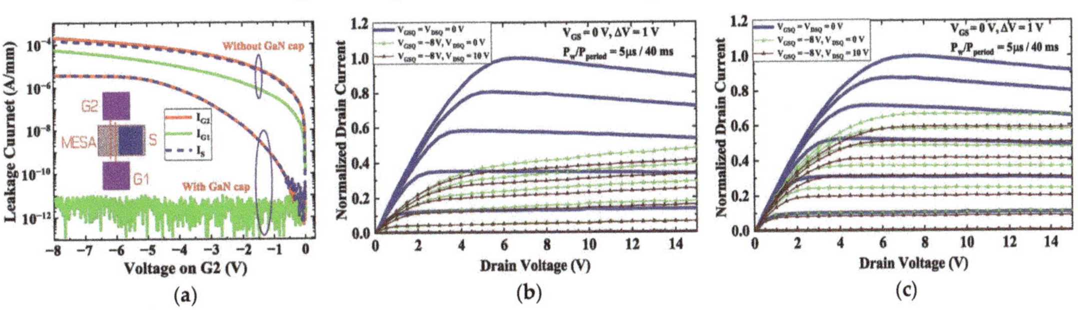

Figure 7. (**a**) Characteristics of the double gate. I_S is the vertical leakage current through 2DEG, and I_{G1} is the horizontal leakage current by electron hopping between surface states. I_{G2} is the total leakage current ($I_{G2} = I_S + I_{G1}$). Pulsed output characteristics of HEMTs, (**b**) without GaN cap, and (**c**) with GaN cap. Adapted with permission from Ref. [11]. Copyright 2022 by Elsevier.

3.1.5. Si Surface Doping

Surface doping of GaN HEMTs with Si also contributes to the suppression of dispersion [42]. Si (as an electron donor) surface doping can prevent the slow process of re-emitting trapped channel electrons from surface states by filling and screening the deep-level traps [12]. However, the method also causes an escalation in leakage levels (electrons can tunnel through the thinned potential barriers more easily), so attention to a concentration should be paid when doping the surface of the device to avoid the impact of poor leakage current and breakdown resistance on device reliability.

3.2. Bottlenecks Encountered at High Frequencies

At lower frequencies (below Ka-band), the previously mentioned techniques can be combined to control RF dispersion in GaN HEMTs. As the gate length is further decreased to increase the operating frequency, the distance from the polarized surface to the 2DEG channel also has to be reduced to maintain the desired aspect ratio and contain the short channel effect, whereas the corresponding cost is the 2DEG channel suffers from the surface RF dispersion more seriously, rendering the effect of all surface engineering techniques less effective. When the operating frequency is pushed over 30 GHz (which means just entering the millimeter-wave range), the capacitance penalty imposed by the field plate causes this structure to be disabled because of the severely degraded high-frequency performance. Therefore, researchers urgently need to find new techniques to alleviate the problem of surface state trapping electrons to ensure the device's reliability at high-frequency operation.

4. NPDR MIS-HEMTs for High-Frequency Device

Due to the comparatively good material quality and electrical features, Ga-polar GaN has dominated power devices for a long time. In recent years, the material growth technique has constantly been progressing on N-polar GaN; thus, related device research gradually gained significant interest [43,44]. On account of the opposing polarity, N-polar GaN HEMTs offer a series of natural advantages for higher frequency epitaxy and device design over the Ga-polar [45], such as (a) devices with a narrower top layer bandgap achieve lower contact resistance; (b) a natural back-barrier improves 2DEG confinement and minimize short channel effect while giving rise to higher transconductance. Additionally, the presence of the back-barrier allows the device to keep leakage current (from channel to substrate) at a low range even when there is no doping (e.g., Fe, C) in the buffer, avoiding the dispersion introduced by slow deep energy level discharge [46]; (c) flexible device aspect ratios accommodate shorter gate lengths resulting from increased frequency. These features are tailored to the needs of high-frequency devices so that the N-polar GaN transistor stood distinctly above all others in 2012, with a f_T = 270 GHz and f_{max} = 370 GHz [47]. However, the most outstanding value of N-polar GaN materials is the ability to make the deep recess structure dramatically mitigate RF dispersion to guarantee reliability at higher frequencies [48]. Here, the basic structure of NPDR is displayed, the reason why deep recess can only be applied on the N-polar GaN (rather than Ga-polar) is discussed, and the research advances on NPDR are summarized.

4.1. What and Why Is NPDR?

The most basic structure of NPDR is to directly thicken the GaN channel above the back-barrier, as shown in Figure 8a, and the 2DEG at the bottom of the channel is isolated from the surface to control dispersion. The gate is buried deep in the GaN channel to considerably boost vertical scaling and strengthen the gate's modulation of 2DEG. To avoid gate leakage currents and regulate the dispersion created by the sidewall, a thin layer of dielectric (normally MOCVD SiN_x) is commonly placed for initial passivation before depositing gate metal. In the improved design of the NPDR structure (shown in Figure 8b), the insertion of an AlGaN cap layer (as etch stop layer) is considered necessary because the difference in components allows for more precise selective etching, while the cap provides better Schottky contact for the device. Apart from these two causes, the polarization electric field generated by N-polar AlGaN also mitigates the injection of gate leakage electrons during reverse bias [49]. Since the thick GaN cap keeps the surface away from the channel (differs from passivation, thick dielectric does not change the surface position due to the absence of polarization), even if a virtual gate is formed on the surface of the access region, there is almost no depletion of charge in the channel. The scattering associated with the surface is also effectively controlled, thus substantially increasing the 2DEG mobility. As mentioned earlier, the traditional passivation of HEMTs using ex situ SiN_x dielectrics still has massive trapping states, whether in bulk or at the interface. However, no significant interfacial states are presented in the NPDR structure because the GaN cap has no difference in lattice constants from the channel. There should also be no more bulk traps within the UID GaN cap than in the channel, which ensures an ultra-low trap density near the 2DEG. To sum up, the deep recess structure is impressive for eliminating the RF dispersion in GaN HEMTs.

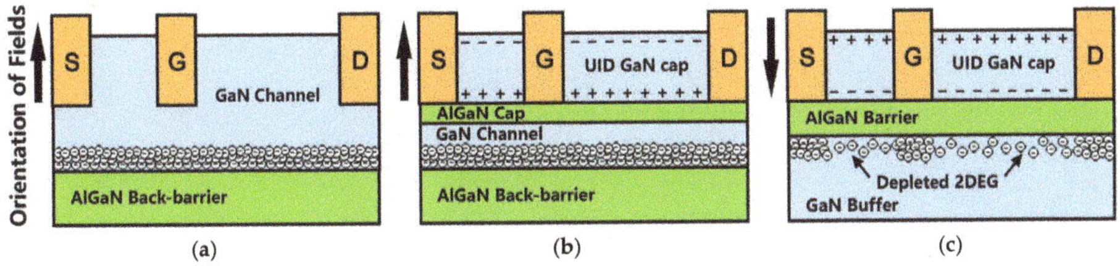

Figure 8. Schematic diagrams of on-state device operation for deep recess GaN HEMT with different structures. (**a**) N-polar; (**b**) N-polar with AlGaN cap; (**c**) Ga-polar.

In fact, deep recess structures were first proposed for Ga-polar devices [50–52], but the results did not meet the expectations. As the polarization of the GaN cap caused the conduction band to be pulled up, the 2DEG would be depleted below the access region (shown in Figure 8c). Shen et al. speculated that suitable Si doping could compensate for the negative polarization charge and prevent the accumulation of holes. They proposed direct Si delta-doping at the GaN/AlGaN interface [52], but unfortunately, the introduction of Si would cause higher gate leakage and, thus, a reduction in breakdown voltage. Another method that insets a Si-doped graded AlGaN between the UID GaN cap and the AlGaN barrier is introduced for similar motivations [50]. Although the lower Al component of AlGaN can partially alleviate the 2DEG depletion by polarization, it is difficult to overcome the process because of the high requirements for precise etching. The additional graded AlGaN is also detrimental to the device aspect ratio. Hence, it can be confirmed that using a deep recess structure on the Ga-polar must be compromised for device performance.

In contrast to the Ga polarity, the electric field created in the access region of the N-polar GaN HEMTs partially counteracts the electric field provided by the gate. It pushes the conduction band to the Fermi energy level (Figure 9), thus augmenting the 2DEG concentration [53]. This provides an intrinsic advantage wherein the Ga-polar devices cannot be duplicated, and demonstrates that only the N-polar GaN HEMTs are appropriate for the deep recess.

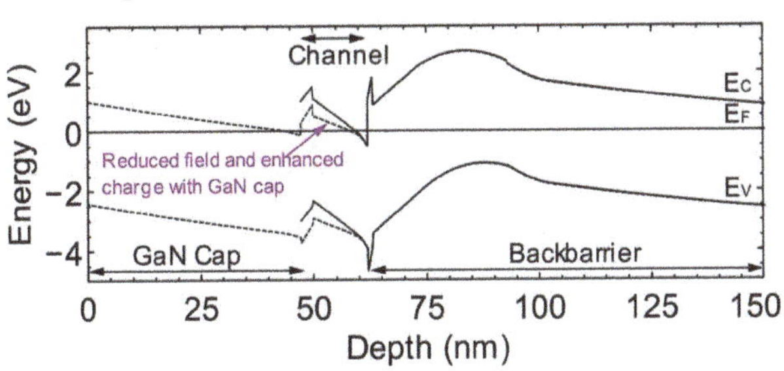

Figure 9. Energy band diagram comparing the N-polar GaN HEMT with and without a GaN cap. Reprinted with permission from Ref. [53]. Copyright 2018 by IEEE.

4.2. Research Advances

In 2011, Kolluri et al. [54] first designed and fabricated an N-polar GaN HEMT with a thick unintentionally doped (UID) GaN cap, i.e., NPDR MIS-HEMT, and found that a 120 nm UID GaN cap was sufficient to remove any RF dispersion in the device. Since the last decade, extensive efforts have been devoted to NPDR structures to eliminate RF dispersion in solid-state power devices.

Wienecke et al. [55] from UCSB broadened the operating frequency to W-band (80-100 GHz) by fabricating NPDR MIS-HEMT on a sapphire substrate in 2016, with a 75 nm gate length and T-shaped gate. After implementing a T-gate design, weakened parasitic capacitance enhances device RF performance (both f_T and f_{max}) to some extent [56].

The controlling dispersion ability is assessed using DC and 200 ns pulsed I–V tests. Figure 10a supports that the device's on-resistance (R_{on}) and I_{max} are 0.58 Ω and 1.5 A/mm, respectively, under the DC condition. Only a minimal current collapse occurs even in the pulsed measurement, demonstrating the deep recess structure's efficient dispersion administration in high-frequency devices. Thanks to this, the device achieves a 2.9 W/mm P_{out} with an associated PAE of 15.5% at a load of 94 GHz. The following year, this team significantly improved the NPDR structure design, i.e., gate-sidewall overlapping [49]; the 3D model is presented in Figure 11. A 30 nm overlap on the sidewall was introduced when depositing the gate metal (the gate was aligned to the bottom of the notch structure in the previous design). This design is believed to assist the drain-side gate in relieving the edge electric field while allowing the device to handle more significant voltage fluctuations (over 12 V) and raise this device's power density to double the previous density. As can be seen from the pulse diagram (Figure 10b), the wrapping of the metal gate around the sidewall does bring about amazing dispersion management (details of which will be discussed later), resulting in the absence of any current collapse and knee-out phenomenon. Furthermore, in the small-signal RF measurement (shown in Figure 10c), the f_T and f_{max} of this HEMT are extrapolated to be 113 and 323 GHz, respectively, when the V_{DS} and V_{GS} bias are set to 13 V and −1.75 V. N-polar GaN HEMTs passivated by MOCVD and PECVD SiN_x during the same period, on the contrary, tend to exhibit an undesirable dispersion behavior [57,58].

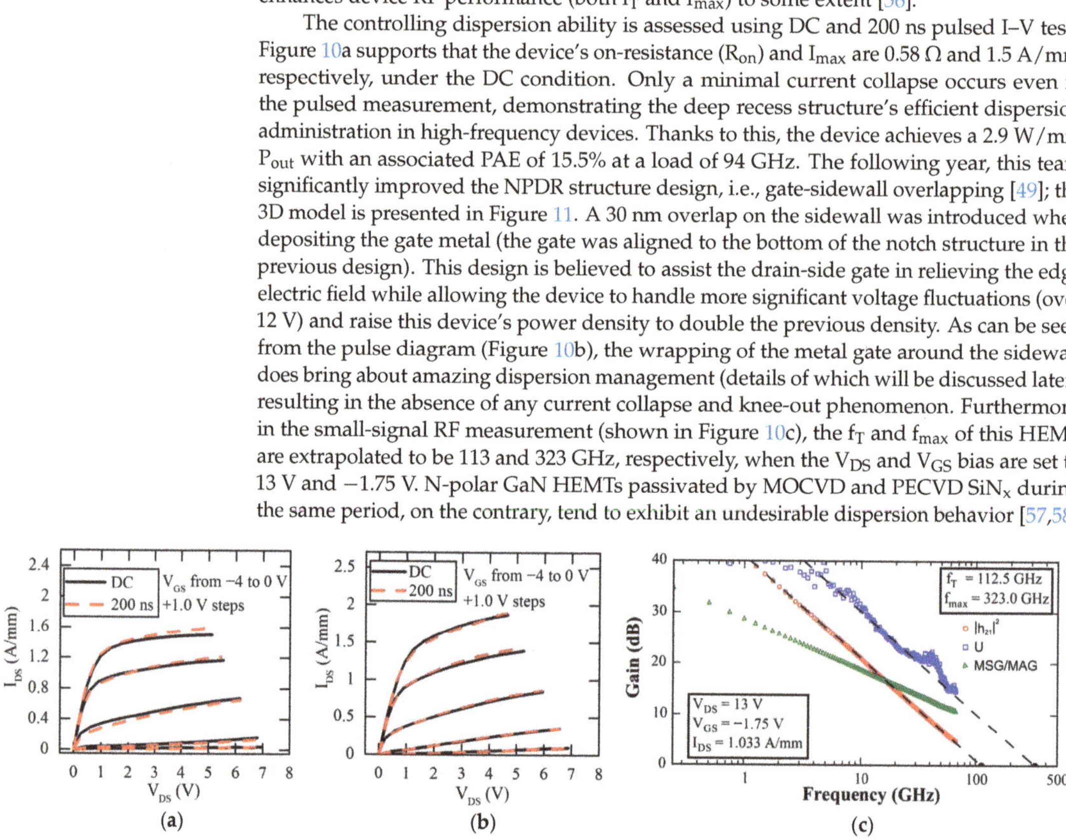

Figure 10. (a) DC and pulsed I_{DS}–V_{DS} measurements of the NPDR MIS-HEMT, fabricated by Wienecke et al. in 2016. Adapted with permission from Ref. [55]. Copyright 2016 by IEEE. (b) DC and pulsed I_{DS}–V_{DS} measurements and (c) small-signal RF performance of the NPDR MIS-HEMT with gate-sidewall overlapping, fabricated by Steven et al. in 2017. Adapted with permission from Ref. [49]. Copyright 2017 by IEEE.

The relationship between gate metal coverage of recessed regions and dispersion was further discussed by Wienecke [59]. In the case of a fully exposed sidewall, the dispersion aroused by the gap (refers to the space between the bottom of the gate metal and the groove, schematic diagram as shown in Figure 12) was first determined. Figure 13 exhibits the dispersion vs. the drain-side and source-side gap and the effect of different gate coverage on the output power and gate leakage current, respectively. When the gate metal covers the entire gate depression area (no gap and the sidewall is completely covered), the dispersion induced by those surface states on the sidewall is wholly removed. If the gap is minute (for example, 5 nm), the produced dispersion is negligible and can be acceptable. Once the gap approaches 50 nm or more, the RF dispersion will increase dramatically; hence, wide gaps over 50 nm are forbidden in NPDR structures. The gap discussed above relates to the drain side, whereas the source-side sidewall is completely covered, and a 50 nm field plate is used so that the experimental dispersion is mainly from the drain side. The same test procedure is applied to the source side, and a similar dispersion behavior is obtained. Results of gate coverage contribution to device performance suggest that the full metal coverage sample achieved the highest RF P_{out} of any transistor under the drain bias. However, it will cause a slight degradation in f_T/f_{max} due to the larger C_{gs} and C_{gd} [60]. The I_g drops rapidly with the increase of sidewall coverage, but the device with a 15 nm gap is even lower than that of the one with half of the sidewall covered; thus, the mechanism of gate metal deposition coverage on the gate leakage current should be further investigated. Weighing the device's frequency characteristics, breakdown and output power performance, Full Metal Coverage is considered the best trade-off solution. Most reports on NPDR structures currently use a self-alignment process to cover the entire gate depression region.

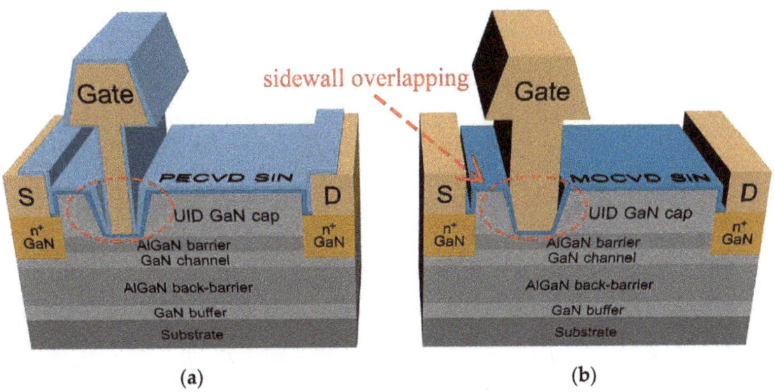

Figure 11. 3D model of NPDR MIS-HEMTs. (**a**) Without gate-sidewall overlapping; (**b**) with gate-sidewall overlapping.

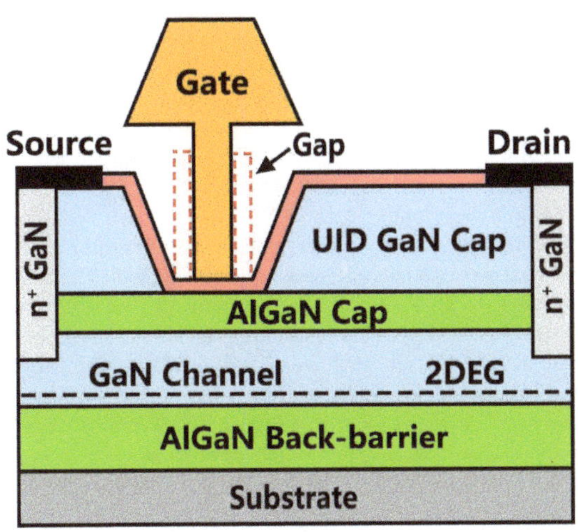

Figure 12. Schematic diagram of the gap between the gate metal and the GaN cap sidewalls.

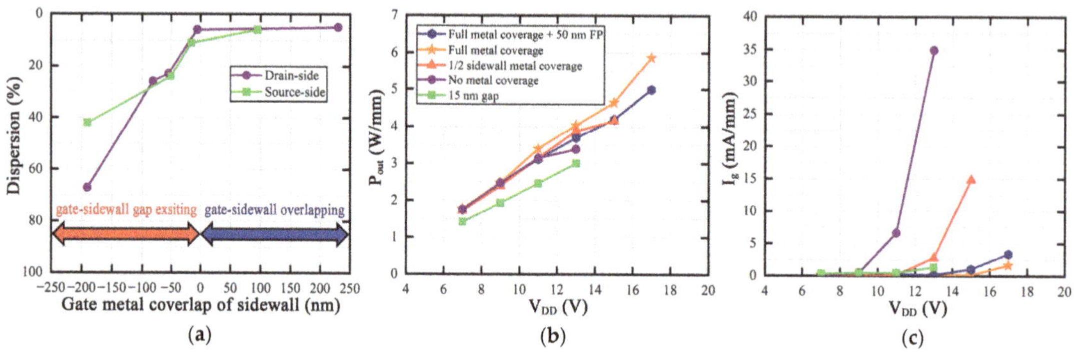

Figure 13. (**a**) Plots of the dispersion vs. source-side and drain-side gap. (**b**,**c**) The performance of output power and gate leakage current at different gate coverage. Reworked from Ref. [59].

The UID GaN cap thickness is also one of the crucial parameters in the NPDR structure. It is clear that the thicker the cap is, the further the channel is from the device surface, which minimizes the surface charge impact on the 2DEG and achieves superior dispersion control. 47.5 nm GaN cap thickness was found to be the most suited in earlier research until lately, this classic thickness has been replaced by 20 nm, primarily to reduce the fringing capacitance while still maintaining the high access region channel conductivity [61]. This action helps improve the device's gain and f_T/f_{max} combination but perhaps leads to incomplete dispersion control. While additional PECVD passivation can be used to compensate for the reduced ability to regulate dispersion and obtain the desired results.

UID GaN cap with PECVD SiN_x passivation to commonly suppress dispersion has been demonstrated and fabricated many times in N-polar GaN devices [55,61–63]. Utilization of the double-layer structure makes it possible to achieve excellent performance without the original thick cap, preventing parasitic channels caused by the strong polarization effect (PECVD SiN_x does not produce polarization). At the same time, gate misalignment inside the trench is inevitable during the gate deposition, and PECVD SiN_x is also effective in eliminating the remaining dispersion left by gate misalignment [55]. Regardless of the rationale, the additional SiN_x passivation does have beneficial effects on the breakdown and power performance of the device. Romanczyk found that depositing 40 nm PECVD SiN_x on the surface of a typical 47.5 nm UID GaN cap yielded higher output power at 10 GHz and allowed the HEMT to operate at higher $V_{DS,Q}$ (the SiN_x passivation increased $V_{T,access}$ so that the channel can withstand higher voltages without being depleted) [62]. Under pulsed I_{DS}–V_{DS} measurement (650 ns pulse with and 0.065% duty cycle), the I_D of the device drops only 8.5%, even with the stress as high as 23 V. Figure 14 compares the performance of this NPDR GaN HEMT with the superior Ga-polar monolithic microwave integrated circuit (MMIC) at W-band. When the load-pull measure is boosted to 94 GHz, a peak output power of 8.84 W/mm with 27% associated PAE (V_D = 23 V) made it the best-performing device among the N-polar GaN device family. A slight reduction in voltage (to 20 V) would show a peak PAE of 30.7%, also unprecedented in N-polar GaN devices to date. This result far exceeds the leading level in the W-band reported for Ga-polar devices (an MMIC reported by Niida et al. [64] achieving 3.6 W/mm P_{out} with a maximum PAE of 12.3%, and an AlN/GaN HEMT reported by Harrouche et al. [65] achieving 4 W/mm P_{out} with a PAE of 14.3% in 2019), proving the absolute dominance of NPDR structures in the millimeter-wave RF field.

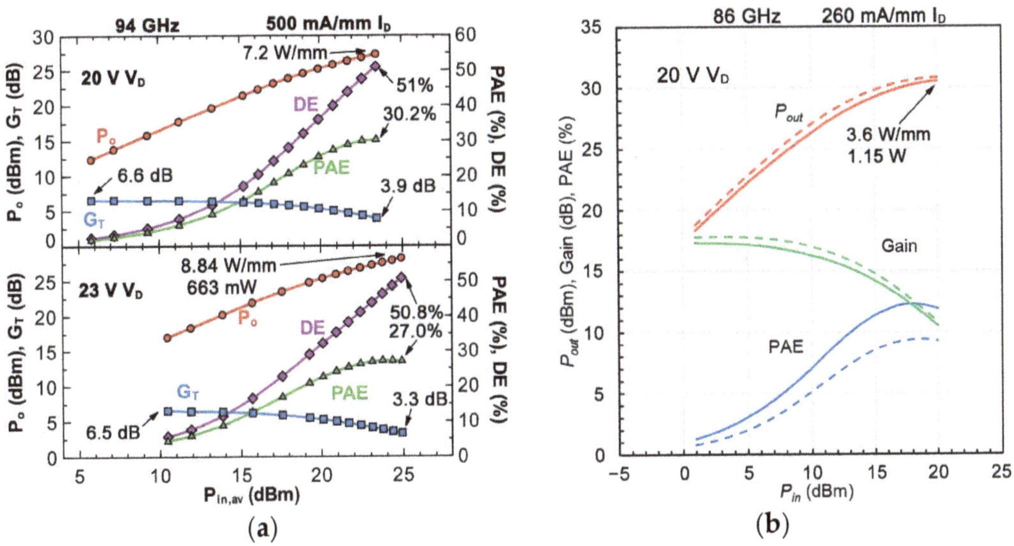

Figure 14. Power characteristics of devices in the W-band. (**a**) NPDR GaN HEMT reported by Romanczyk. Adapted with permission from Ref. [62]. Copyright 2020 by IEEE. (**b**) Advanced Ga-polar MMIC, the solid and dashed lines are the results in the actual measurement and simulation scenarios, respectively. Adapted with permission from Ref. [64]. Copyright 2016 by IEEE.

A thin double-cap layer structure, 2.6 nm of AlGaN deposited on top of 10 nm UID GaN as the cap layer, facilitates gain enhancement and suppresses parasitic channels [66]. At a 1 μs gate pulse width and an 800 ns drain pulse width, the test findings demonstrate that only around 10% dispersion ($V_{GS,Q}$ = −6 V, $V_{DS,Q}$ = 0 V) is observed. The gain enhancement due to the thin double-cap design facilitates the modulation effect of the gate on the 2DEG. It reduces the signal distortion of the device, enabling it to be more suitable for receivers that operate at low drain bias voltages.

More recently, Transphorm Inc. [67] has grown NPDR epitaxial stack on a 4-inch off-axis sapphire and a 4-inch off-axis C-face SiC substrates, respectively, and obtained excellent material quality characteristics for both, which demonstrates the feasibility of commercializing GaN NPDR MIS-HEMTS in the future. The STEM and typical AFM topography of the epitaxial stack are indicated in Figure 15. It can be seen in the STEM that the GaN buffer gradually filters out the linear dislocations starting from the nucleation layer to ensure the material quality in the upper HEMT region. In addition, there is a good hierarchical structure between the GaN and Al(Ga)N, with a low center-to-edge non-uniformity of about 0.3%. The low roughness average (R_a) of only 1.1 nm (on 100 mm sapphire) and 1.5 nm (on 100 mm SiC) reveals that the N-polar epitaxial layer has a smooth surface, implying success in the large-size N-polar GaN wafers' growth. Researchers at Transphorm Inc. also found a low average R_{sh} of 234.3 Ω/sq for the 2DEG channel (with a 2% non-uniformity) owing to the 2DEG with an electron density of about 1.44×10^{13} cm^{-2} and high electron mobility (is approximately 1850 cm^2/V·s). It is worth mentioning that these values are the average mobility measured in both the parallel and perpendicular directions to the miscut steps. Thus, higher mobility and lower R_{sh} may be obtained if the current moves precisely in the parallel direction (along the terrace direction) of the actual deep recess devices.

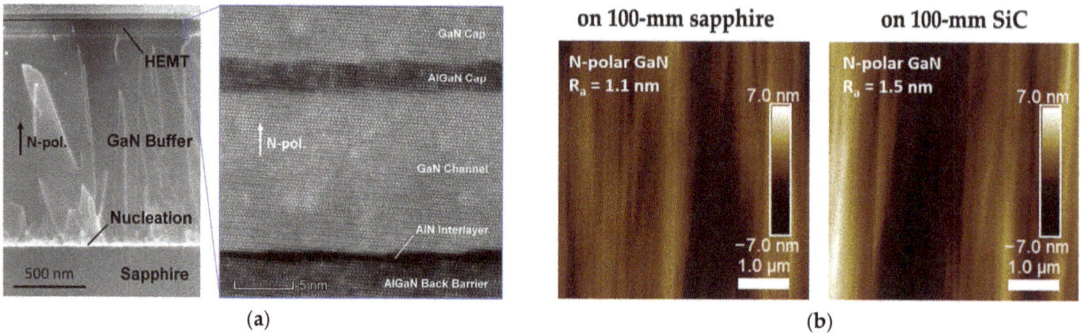

Figure 15. (**a**) STEM cross-section of the N-polar GaN epitaxy and NPDR MIS-HEMT layers; (**b**) AFM topography image of N-polar GaN epitaxy on 100 mm sapphire and 100 mm miscut C-face SiC substrates. Adapted with permission from Ref. [67]. Copyright 2021 by IEEE.

The parameters and the performance of NPDR MIS-HEMTs operating in a range of frequencies are summarized in Table 2 for reference. The superior results imply that the NPDR structure is an efficient solution for suppressing surface dispersion of high-frequency devices, which may be commercially available and employed in the future RF industry, especially in the growing demand of the 5G market. However, it is necessary first to consider the poor heat dissipation caused by the current recovery. GaN-on-diamond may be a prospective option to address the thermal aggregation of NPDR GaN HEMTs in the future [68–70].

Table 2. Characterization of NPDR MIS-HEMTs applied to various wave bands, including epitaxial structures, output performances, and dispersion suppression.

Ref.	Band	Freq. (GHz)	Peak P_o			Peak PAE			Disp. (%)	Cap Thk. (nm)	SiN_x Thk. (nm)	Substrate
			V_D(V)	P_o (W/mm)	PAE (%)	V_D(V)	P_o (W/mm)	PAE (%)				
[55]	W	94	10	2.9	15.5	8	1.7	20		110	18	Sapphire
[60]			14	4.2		8	1.6	16.5		22	NO	Sapphire
[61]			16	5.5	20.6	12	3.7	25.9	<5	20	20	Sapphire
[71]			11	3.7		9	3.0	27.8	<5	47.5	NO	SiC
[72]			18	6.2	33.8	16	5.6	34.8		47.5	24	SiC
[49]			16	6.7	14.4	15	4.8	16.9	NO	47.5	NO	SiC
[73]			20	7.1	25.1	14	4.6	27.5		47.5	NO	SiC
[53]			20	7.9	26.9	16	5.3	28.8	<10	47.5	NO	SiC
[62]			23	8.8	27	20	7.2	30.2	8.5	47.5	40	SiC
[74]		87				8	2.5	34.2	NO	47.5	NO	SiC
[66]	Ka	30							10	12.6	20	SiC
[53]			20	8.1	52.5	14	5.6	55.9	<10	47.5	NO	SiC
[63]			26	10.3	47.4					47.5	40	SiC
[53]	X	10	20	7.8	54.9	14	5.1	58.1	<10	47.5	NO	SiC
[54]	C	4	24	5.5	74				NO	120	NO	Sapphire

5. Conclusions

GaN-based HEMTs have shown great performance in the microwave RF field. Nevertheless, the inherent high-density surface states of GaN and the inevitable defects during fabricating are potential sources of surface traps. In the switching transition of the device, the behaviors of trapping and delayed releasing electrons by the surface states are the leading causes of RF surface dispersion and pose a significant challenge to HEMTs to maintain high output power and stability. Accordingly, how to suppress RF dispersion in GaN devices is an interesting research problem.

Surface treatment, passivation, field plate, GaN cap layer, and Si surface doping are demonstrated to effectively suppress RF dispersion and are widely used in the industry.

While these methods provide varying degrees of dispersion suppression, as higher frequencies are explored, the exacerbated surface trapping effect or excessive parasitic capacitance makes these techniques progressively less effective.

N-polar GaN has many special advantages in high-frequency applications due to the opposite polarity, especially allowing the growth of a thick GaN channel (or thick UID GaN cap on the access region), providing an alternative way to minimize the dispersion caused by surface trapping at high frequencies. The utilization of a thick GaN layer keeps the surface far from the 2DEG channel to control the surface states and minimize the impact of surface charging on device operation. The isolation between the channel and the surface enables the NPDR structure to be considered the most promising approach for solving RF dispersion in GaN-based high-frequency power devices. In the past decade, the structural design of NPDR has been continuously improved, such as self-alignment for full coverage of the depression region with gate metal, simultaneous use of deep recess structure and SiN_x passivation, and a double-layer structure for the receiver side. From the results, NPDR is expected to significantly contribute to the elimination of RF dispersion in high-frequency GaN HEMTs, restore their stability and inherent high power density, and is likely to be used in commercial 5G millimeter-wave RF devices in the future. However, the high price of SiC substrates, the high-quality requirements for N-polar GaN epitaxy, and heat dissipation are the primary impediments to large-scale industrial fabrication. Therefore, if more substantial progress can be achieved in addressing these obstacles, the application prospects of high-frequency GaN HEMTs will extend further.

Author Contributions: Conceptualization, P.Z. and X.G.; investigation, X.N. and Q.F.; resources, X.N. and Q.F; data curation, P.Z. and X.G.; visualization, P.Z.; writing—original draft preparation, P.Z.; writing—review and editing, X.G.; funding acquisition, X.G. All authors have read and agreed to the published version of the manuscript.

Funding: This research was funded by the National Nature Science Foundation of China, grant number 62074033.

Data Availability Statement: Not applicable.

Acknowledgments: The authors would like to thank Maxwell QH from Dulwich College Suzhou for his contribution to the English language and style in this work.

Conflicts of Interest: The authors declare no conflict of interest.

References

1. Fujita, S. Wide-bandgap semiconductor materials: For their full bloom. *Jpn. J. Appl. Phys.* **2015**, *54*, 030101. [CrossRef]
2. Roccaforte, F.; Fiorenza, P.; Greco, G.; Lo Nigro, R.; Giannazzo, F.; Iucolano, F.; Saggio, M. Emerging trends in wide band gap semiconductors (SiC and GaN) technology for power devices. *Microelectron. Eng.* **2018**, *187–188*, 66–77. [CrossRef]
3. Dubey, S.K.; Mishra, M.; Islam, A. Characterization of ALGAN / GAN based HEMT for low noise and high frequency application. *Int. J. Numer. Model.* **2022**, *35*, e2932. [CrossRef]
4. Chen, S.; Witkowski, L.; Gu, X. (Invited) Important Recent Advancement of GaN Technology and MMICs for W-Band Applications. *ECS Trans.* **2017**, *80*, 945–952. [CrossRef]
5. Ye, R.; Cai, X.; Du, C.; Liu, H.; Zhang, Y.; Duan, X.; Zhu, J. An Overview on Analyses and Suppression Methods of Trapping Effects in AlGaN/GaN HEMTs. *IEEE Access* **2022**, *10*, 21759–21773. [CrossRef]
6. Rongming, C.; Likun, S.; Fichtenbaum, N.; Zhen, C.; Keller, S.; DenBaars, S.P.; Mishra, U.K. Correlation Between DC–RF Dispersion and Gate Leakage in Deeply Recessed GaN/AlGaN/GaN HEMTs. *IEEE Electron Device Lett.* **2008**, *29*, 303–305. [CrossRef]
7. Rawal, D.S.; Amit; Sharma, S.; Kapoor, S.; Laishram, R.; Vinayak, S. Current collapse scaling in GaN/AlGaN/SiC high electron mobility transistors. *Solid State Electron. Lett.* **2019**, *1*, 30–37. [CrossRef]
8. Shuxun, L.; Maoju, W.; Bing, X.; Wen, C.P.; Min, Y.; Jinyan, W.; Yilong, H.; Wengang, W.; Sen, H.; Chen, K.J. Bo Shen Reduction of Current Collapse in GaN High-Electron Mobility Transistors Using a Repeated Ozone Oxidation and Wet Surface Treatment. *IEEE Electron Device Lett.* **2015**, *36*, 757–759. [CrossRef]
9. Madan, J.; Pandey, R.; Arora, H.; Chaujar, R. Analysis of Varied Dielectrics as Surface Passivation on AlGaN/GaN HEMT for Analog Applications. In Proceedings of the 2018 6th Edition of International Conference on Wireless Networks & Embedded Systems (WECON), Rajpura, India, 16–17 November 2018; pp. 15–18.

10. Hasan, M.T.; Asano, T.; Tokuda, H.; Kuzuhara, M. Current Collapse Suppression by Gate Field-Plate in AlGaN/GaN HEMTs. *IEEE Electron Device Lett.* **2013**, *34*, 1379–1381. [CrossRef]
11. Sarkar, S.; Khade, R.P.; DasGupta, A.; DasGupta, N. Effect of GaN cap layer on the performance of AlInN/GaN-based HEMTs. *Microelectron. Eng.* **2022**, *258*, 111756. [CrossRef]
12. Song, L.; Fu, K.; Zhang, Z.; Sun, S.; Li, W.; Yu, G.; Hao, R.; Fan, Y.; Shi, W.; Cai, Y.; et al. Interface Si donor control to improve dynamic performance of AlGaN/GaN MIS-HEMTs. *AIP Adv.* **2017**, *7*, 125023. [CrossRef]
13. Longobardi, G.; Udrea, F.; Sque, S.; Hurkx, G.A.M.; Croon, J.; Napoli, E.; Sonsky, J. Impact of Donor Traps on the 2DEG and Electrical Behavior of AlGaN/GaN MISFETs. *IEEE Electron Device Lett.* **2014**, *35*, 27–29. [CrossRef]
14. Miczek, M.; Mizue, C.; Hashizume, T.; Adamowicz, B. Effects of interface states and temperature on the C-V behavior of metal/insulator/AlGaN/GaN heterostructure capacitors. *J. Appl. Phys.* **2008**, *103*, 104510. [CrossRef]
15. Green, B.M.; Chu, K.K.; Chumbes, E.M.; Smart, J.A.; Shealy, J.R.; Eastman, L.F. The effect of surface passivation on the microwave characteristics of undoped AlGaN/GaN HEMTs. *IEEE Electron Device Lett.* **2000**, *21*, 268–270. [CrossRef]
16. Jogai, B. Influence of surface states on the two-dimensional electron gas in AlGaN/GaN heterojunction field-effect transistors. *J. Appl. Phys.* **2003**, *93*, 1631–1635. [CrossRef]
17. Ibbetson, J.P.; Fini, P.T.; Ness, K.D.; DenBaars, S.P.; Speck, J.S.; Mishra, U.K. Polarization effects, surface states, and the source of electrons in AlGaN/GaN heterostructure field effect transistors. *Appl. Phys. Lett.* **2000**, *77*, 250–252. [CrossRef]
18. Vetury, R.; Zhang, N.Q.; Keller, S.; Mishra, U.K. The impact of surface states on the DC and RF characteristics of AlGaN/GaN HFETs. *IEEE Trans. Electron Devices* **2001**, *48*, 560–566. [CrossRef]
19. Hwang, I.; Kim, J.; Chong, S.; Choi, H.-S.; Hwang, S.-K.; Oh, J.; Shin, J.K.; Chung, U.-I. Impact of Channel Hot Electrons on Current Collapse in AlGaN/GaN HEMTs. *IEEE Electron Device Lett.* **2013**, *34*, 1494–1496. [CrossRef]
20. Meneghesso, G.; Verzellesi, G.; Pierobon, R.; Rampazzo, F.; Chini, A.; Mishra, U.K.; Canali, C.; Zanoni, E. Surface-Related Drain Current Dispersion Effects in AlGaN–GaN HEMTs. *IEEE Trans. Electron Devices* **2004**, *51*, 1554–1561. [CrossRef]
21. Patanvariya, D.G.; Kumar, S.; Giri, L.I. Influence of Current Collapse in AlGaN/GaN High Electron Mobility Transistors. In Proceedings of the 2019 10th International Conference on Computing, Communication and Networking Technologies (ICCCNT), Kanpur, India, 6–8 July 2019; pp. 1–5.
22. Huang, H.; Liang, Y.C.; Samudra, G.S.; Chang, T.-F.; Huang, C.-F. Effects of Gate Field Plates on the Surface State Related Current Collapse in AlGaN/GaN HEMTs. *IEEE Trans. Power Electron.* **2014**, *29*, 2164–2173. [CrossRef]
23. Tingting, Y.; Xinyu, L.; Yingkui, Z.; Chengzhan, L.; Ke, W.; Guoguo, L. Impact of UV/ozone surface treatment on AlGaN/GaN HEMTs. *J. Semicond.* **2009**, *30*, 124001. [CrossRef]
24. King, S.W.; Barnak, J.P.; Bremser, M.D.; Tracy, K.M.; Ronning, C.; Davis, R.F.; Nemanich, R.J. Cleaning of AlN and GaN surfaces. *J. Appl. Phys.* **1998**, *84*, 5248–5260. [CrossRef]
25. Diale, M.; Auret, F.D.; van der Berg, N.G.; Odendaal, R.Q.; Roos, W.D. Analysis of GaN cleaning procedures. *Appl. Surf. Sci.* **2005**, *246*, 279–289. [CrossRef]
26. Kim, J.H.; Choi, H.G.; Ha, M.-W.; Song, H.J.; Roh, C.H.; Lee, J.H.; Park, J.H.; Hahn, C.-K. Effects of Nitride-Based Plasma Pretreatment Prior to SiN_x Passivation in AlGaN/GaN High-Electron-Mobility Transistors on Silicon Substrates. *Jpn. J. Appl. Phys.* **2010**, *49*, 04DF05.
27. Romero, M.F.; Jimenez, A.; Miguel-Sanchez, J.; Brana, A.F.; Gonzalez-Posada, F.; Cuerdo, R.; Calle, F.; Munoz, E. Effects of N-2 plasma pretreatment on the SiN passivation of AlGaN/GaN HEMT. *IEEE Electron Device Lett.* **2008**, *29*, 209–211. [CrossRef]
28. Shih-Chien, L.; Bo-Yuan, C.; Yueh-Chin, L.; Ting-En, H.; Huan-Chung, W.; Chang, E.Y. GaN MIS-HEMTs With Nitrogen Passivation for Power Device Applications. *IEEE Electron Device Lett.* **2014**, *35*, 1001–1003. [CrossRef]
29. Mi, M.; Lu, Y.; Hao, Y.; Ma, X.; Yang, L.; Hou, B.; Zhang, M. Improved f_{max} and breakdown voltage in AlGaN/GaN HEMT with plasma treatment. In Proceedings of the 2018 1st Workshop on Wide Bandgap Power Devices and Applications in Asia (WiPDA Asia), Xi'an, China, 16–18 May 2018; pp. 208–211.
30. Machuca, F.; Liu, Z.; Sun, Y.; Pianetta, P.; Spicer, W.E.; Pease, R.F.W. Simple method for cleaning gallium nitride (0001). *J. Vac. Sci. Technol. A Vac. Surf. Film.* **2002**, *20*, 1784–1786. [CrossRef]
31. Geng, K.; Chen, D.; Zhou, Q.; Wang, H. AlGaN/GaN MIS-HEMT with PECVD SiNx, SiON, SiO2 as Gate Dielectric and Passivation Layer. *Electronics* **2018**, *7*, 416. [CrossRef]
32. Grabow, L.C.; Uhlrich, J.J.; Kuech, T.F.; Mavrikakis, M. Effectiveness of in situ NH3 annealing treatments for the removal of oxygen from GaN surfaces. *Surf. Sci.* **2009**, *603*, 387–399. [CrossRef]
33. Cheng, K.-Y.; Wu, S.-C.; Yu, C.-J.; Wang, T.-W.; Liao, J.-H.; Wu, M.-C. Comparative study on performance of AlGaN/GaN MS-HEMTs with SiNx, SiOx, and SiNO surface passivation. *Solid-State Electron.* **2020**, *170*, 107824. [CrossRef]
34. Huang, T.; Malmros, A.; Bergsten, J.; Gustafsson, S.; Axelsson, O.; Thorsell, M.; Rorsman, N. Suppression of Dispersive Effects in AlGaN/GaN High-Electron-Mobility Transistors Using Bilayer SiN_x Grown by Low Pressure Chemical Vapor Deposition. *IEEE Electron Device Lett.* **2015**, *36*, 537–539. [CrossRef]
35. Zhao, S.-X.; Liu, X.-Y.; Zhang, L.-Q.; Huang, H.-F.; Shi, J.-S.; Wang, P.-F. Impacts of Thermal Atomic Layer-Deposited AlN Passivation Layer on GaN-on-Si High Electron Mobility Transistors. *Nanoscale Res. Lett.* **2016**, *11*, 137. [CrossRef] [PubMed]
36. Huang, S.; Jiang, Q.; Yang, S.; Zhou, C.; Chen, K.J. Effective Passivation of AlGaN/GaN HEMTs by ALD-Grown AlN Thin Film. *IEEE Electron Device Lett.* **2012**, *33*, 516–518. [CrossRef]

37. Liu, S.C.; Wong, Y.-Y.; Lin, Y.-C.; Chang, E.Y. Low Current Collapse and Low Leakage GaN MIS-HEMT Using AlN/SiN as Gate Dielectric and Passivation Layer. *ECS Trans.* **2014**, *61*, 211–214. [CrossRef]
38. Huang, T.; Bergsten, J.; Thorsell, M.; Rorsman, N. Small- and Large-Signal Analyses of Different Low-Pressure-Chemical-Vapor-Deposition SiN_x Passivations for Microwave GaN HEMTs. *IEEE Trans. Electron Devices* **2018**, *65*, 908–914. [CrossRef]
39. Godfrey, D.; Nirmal, D.; Arivazhagan, L.; Godwinraj, D.; Mohan Kumar, N.; Chen, Y.; Yeh, W. Current collapse degradation in GaN High Electron Mobility Transistor by virtual gate. *Microelectron. J.* **2021**, *118*, 105293. [CrossRef]
40. Arulkumaran, S.; Egawa, T.; Ishikawa, H. Studies on the Influences of i -GaN, n -GaN, p -GaN and InGaN Cap Layers in AlGaN/GaN High-Electron-Mobility Transistors. *Jpn. J. Appl. Phys.* **2005**, *44*, 2953–2960. [CrossRef]
41. Kordoš, P.; Bernát, J.; Marso, M. Impact of layer structure on performance of unpassivated AlGaN/GaN HEMT. *Microelectron. J.* **2005**, *36*, 438–441. [CrossRef]
42. Mitrofanov, O.; Manfra, M.; Weimann, N. Impact of Si doping on radio frequency dispersion in unpassivated GaN/AlGaN/GaN high-electron-mobility transistors grown by plasma-assisted molecular-beam epitaxy. *Appl. Phys. Lett.* **2003**, *82*, 4361–4363. [CrossRef]
43. Gambin, V.; Gu, X.; Heying, B. AIN buffer N-polar GaN HEMT Profile. Patent US8710511B2, 4 April 2014.
44. Gambin, V.; GU, X. Ingan channel n-polar gan hemt Profile. Patent WO2014031229A1, 27 February 2014.
45. Wong, M.H.; Keller, S.; Nidhi, S.D.; Denninghoff, D.J.; Kolluri, S.; Brown, D.F.; Lu, J.; Fichtenbaum, N.A.; Ahmadi, E.; Singisetti, U.; et al. N-polar GaN epitaxy and high electron mobility transistors. *Semicond. Sci. Technol.* **2013**, *28*, 074009. [CrossRef]
46. Nakajima, A.; Fujii, K.; Horio, K. Numerical Analysis of Buffer-Trap Effects on Gate Lag in AlGaN/GaN High Electron Mobility Transistors. *Jpn. J. Appl. Phys.* **2011**, *50*, 104303. [CrossRef]
47. Smorchkova, I.; Grundbacher, R.; Gambin, V.; Gu, X.; Lewis, G.; Oshita, F.; Namba, C.; Liu, P.; Poust, S.; Nam, P.; et al. N-Face GaN HEMT Technology for Next-Generation Mixed-Signal Electronics. In Proceedings of the 37th Annual GOMACTech Conference, Las Vegas, NV, USA, 19–22 March 2012.
48. Gu, X.; Xie, J.; Lee, C. Process of fabricating high efficiency, high linearity N-polar gallium-nitride (GaN) Transistors. Patent US10749009B1, 18 August 2020.
49. Steven, W.; Romanczyk, B.; Guidry, M.; Li, H.; Ahmadi, E.; Hestroffer, K.; Zheng, X.; Keller, S.; Mishra, U.K. N-Polar GaN Cap MISHEMT With Record Power Density Exceeding 6.5 W/mm at 94 GHz. *IEEE Electron Device Lett.* **2017**, *38*, 359–362.
50. Shen, L.; Coffie, R.; Buttari, D.; Heikman, S.; Chakraborty, A.; Chini, A.; Keller, S.; DenBaars, S.P.; Mishra, U.K. High-Power Polarization-Engineered GaN/AlGaN/GaN HEMTs Without Surface Passivation. *IEEE Electron Device Lett.* **2004**, *25*, 7–9. [CrossRef]
51. Shen, L.; Palacios, T.; Poblenz, C.; Corrion, A.; Chakraborty, A.; Fichtenbaum, N.; Keller, S.; Denbaars, S.P.; Speck, J.S.; Mishra, U.K. Unpassivated high power deeply recessed GaN HEMTs with fluorine-plasma surface treatment. *IEEE Electron Device Lett.* **2006**, *27*, 214–216. [CrossRef]
52. Shen, L.; Pei, Y.; McCarthy, L.; Poblenz, C.; Corrion, A.; Fichtenbaum, N.; Keller, S.; Denbaars, S.P.; Speck, J.S.; Mishra, U.K. Deep-recessed GaN HEMTs using selective etch technology exhibiting high microwave performance without surface passivation. In Proceedings of the 2007 IEEE/MTT-S International Microwave Symposium, Honolulu, HI, USA, 3–8 June 2007; pp. 623–626.
53. Romanczyk, B.; Wienecke, S.; Guidry, M.; Li, H.; Ahmadi, E.; Zheng, X.; Keller, S.; Mishra, U.K. Demonstration of Constant 8 W/mm Power Density at 10, 30, and 94 GHz in State-of-the-Art Millimeter-Wave N-Polar GaN MISHEMTs. *IEEE Trans. Electron Devices* **2018**, *65*, 45–50. [CrossRef]
54. Kolluri, S.; Brown, D.F.; Wong, M.H.; Dasgupta, S.; Keller, S.; DenBaars, S.P.; Mishra, U.K. RF Performance of Deep-Recessed N-Polar GaN MIS-HEMTs Using a Selective Etch Technology Without Ex Situ Surface Passivation. *IEEE Electron Device Lett.* **2011**, *32*, 134–136. [CrossRef]
55. Wienecke, S.; Romanczyk, B.; Guidry, M.; Li, H.; Zheng, X.; Ahmadi, E.; Hestroffer, K.; Megalini, L.; Keller, S.; Mishra, U.K. N-Polar Deep Recess MISHEMTs with Record 2.9 W/mm at 94 GHz. *IEEE Electron Device Lett.* **2016**, *37*, 713–716. [CrossRef]
56. Mohapatra, M.; De, T.; Panda, A.K. Performance analysis of AlGaN/GaN based HEMT for different gate structure (Normal gate and T-shaped gate). In Proceedings of the 2017 Devices for Integrated Circuit (DevIC), Kalyani, India, 23–24 March 2017; pp. 144–148.
57. Zheng, X.; Guidry, M.; Li, H.; Romanczyk, B.; Ahmadi, E.; Hestroffer, K.; Wienecke, S.; Keller, S.; Mishra, U.K. N-polar GaN MIS-HEMTs on sapphire with a proposed figure of merit fmax·VDS, Q of 9.5 THz.V. In Proceedings of the 2017 75th Annual Device Research Conference (DRC), South Bend, IN, USA, 25–28 June 2017; pp. 1–2.
58. Zheng, X.; Li, H.; Ahmadi, E.; Hestroffer, K.; Guidry, M.; Romanczyk, B.; Wienecke, S.; Keller, S.; Mishra, U.K. High frequency N-polar GaN planar MIS-HEMTs on sapphire with high breakdown and low dispersion. In Proceedings of the 2016 Lester Eastman Conference (LEC), Bethlehem, PA, USA, 2–4 August 2016; pp. 42–45.
59. Wienecke, S.M. N-Polar Deep Recess MISHEMTs for mm-Wave Applications. Ph.D. Thesis, University of California-Santa Barbara, Los Angeles, CA, USA, 2018.
60. Romanczyk, B.; Wienecke, S.; Guidry, M.; Li, H.; Hestroffer, K.; Ahmadi, E.; Zheng, X.; Keller, S.; Mishra, U.K. mm-Wave N-polar GaN MISHEMT with a self-aligned recessed gate exhibiting record 4.2 W/mm at 94 GHz on Sapphire. In Proceedings of the 2016 74th Annual Device Research Conference (DRC), Newark, DE, USA, 19–22 June 2016; pp. 1–2.
61. Romanczyk, B.; Li, W.; Guidry, M.; Hatui, N.; Krishna, A.; Wurm, C.; Keller, S.; Mishra, U.K. N-Polar GaN-on-Sapphire Deep Recess HEMTs With High W-Band Power Density. *IEEE Electron Device Lett.* **2020**, *41*, 1633–1636. [CrossRef]

62. Romanczyk, B.; Zheng, X.; Guidry, M.; Li, H.; Hatui, N.; Wurm, C.; Krishna, A.; Ahmadi, E.; Keller, S.; Mishra, U.K. W-Band Power Performance of SiN-Passivated N-Polar GaN Deep Recess HEMTs. *IEEE Electron Device Lett.* **2020**, *41*, 349–352. [CrossRef]
63. Romanczyk, B.; Guidry, M.; Zheng, X.; Shrestha, P.; Li, H.; Ahmadi, E.; Keller, S.; Mishra, U.K. Evaluation of linearity at 30 GHz for N-polar GaN deep recess transistors with 10.3 W/mm of output power and 47.4% PAE. *Appl. Phys. Lett.* **2021**, *119*, 072105. [CrossRef]
64. Niida, Y.; Kamada, Y.; Ohki, T.; Ozaki, S.; Makiyama, K.; Minoura, Y.; Okamoto, N.; Sato, M.; Joshin, K.; Watanabe, K. 3.6 W/mm high power density W-band InAlGaN/GaN HEMT MMIC power amplifier. In Proceedings of the 2016 IEEE Topical Conference on Power Amplifiers for Wireless and Radio Applications (PAWR), Austin, TX, USA, 24–27 January 2016; pp. 24–26.
65. Harrouche, K.; Kabouche, R.; Okada, E.; Medjdoub, F. High Performance and Highly Robust AlN/GaN HEMTs for Millimeter-Wave Operation. *IEEE J. Electron Devices Soc.* **2019**, *7*, 1145–1150. [CrossRef]
66. Shrestha, P.; Guidry, M.; Romanczyk, B.; Hatui, N.; Wurm, C.; Krishna, A.; Pasayat, S.S.; Karnaty, R.R.; Keller, S.; Buckwalter, J.F.; et al. High Linearity and High Gain Performance of N-Polar GaN MIS-HEMT at 30 GHz. *IEEE Electron Device Lett.* **2020**, *41*, 681–684. [CrossRef]
67. Bisi, D.; Romanczyk, B.; Liu, X.; Gupta, G.; Brown-Heft, T.; Birkhahn, R.; Lal, R.; Neufeld, C.J.; Keller, S.; Parikh, P.; et al. Commercially Available N-polar GaN HEMT Epitaxy for RF Applications. In Proceedings of the 2021 IEEE 8th Workshop on Wide Bandgap Power Devices and Applications (WiPDA), Redondo Beach, CA, USA, 7–9 November 2021; pp. 250–254.
68. Gu, X.; Lee, C.; Xie, J.; Beam, E.; Becker, M.; Grotjohn, T.; Anaya, J.; Kuball, M. *GaN-on-Diamond with Ultra-Low Thermal Barrier Resistance*; Qorvo Inc.: Richardson, TX, USA, 2016.
69. Zhou, Y.; Anaya, J.; Pomeroy, J.; Sun, H.; Gu, X.; Xie, A.; Beam, E.; Becker, M.; Grotjohn, T.A.; Lee, C.; et al. Barrier-Layer Optimization for Enhanced GaN-on-Diamond Device Cooling. *ACS Appl. Mater. Interfaces* **2017**, *9*, 34416–34422. [CrossRef] [PubMed]
70. Yates, L.; Anderson, J.; Gu, X.; Lee, C.; Bai, T.; Mecklenburg, M.; Aoki, T.; Goorsky, M.S.; Kuball, M.; Piner, E.L.; et al. Low Thermal Boundary Resistance Interfaces for GaN-on-Diamond Devices. *ACS Appl. Mater. Interfaces* **2018**, *10*, 24302–24309. [CrossRef] [PubMed]
71. Romanczyk, B.; Guidry, M.; Wienecke, S.; Li, H.; Ahmadi, E.; Zheng, X.; Keller, S.; Mishra, U.K. W-band N-polar GaN MISHEMTs with high power and record 27.8% efficiency at 94 GHz. In Proceedings of the 2016 IEEE International Electron Devices Meeting (IEDM), San Francisco, CA, USA, 3–7 December 2016; pp. 3.5.1–3.5.4.
72. Liu, W.; Romanczyk, B.; Guidry, M.; Hatui, N.; Wurm, C.; Li, W.; Shrestha, P.; Zheng, X.; Keller, S.; Mishra, U.K. 6.2 W/Mm and Record 33.8% PAE at 94 GHz From N-Polar GaN Deep Recess MIS-HEMTs With ALD Ru Gates. *IEEE Microw. Wirel. Compon. Lett.* **2021**, *31*, 748–751. [CrossRef]
73. Romanczyk, B.; Guidry, M.; Zheng, X.; Li, H.; Ahmadi, E.; Keller, S.; Mishra, U.K. Bias-Dependent Electron Velocity Extracted From N-Polar GaN Deep Recess HEMTs. *IEEE Trans. Electron Devices* **2020**, *67*, 1542–1546. [CrossRef]
74. Romanczyk, B.; Guidry, M.; Wienecke, S.; Li, H.; Ahmadi, E.; Zheng, X.; Keller, S.; Mishra, U.K. Record 34.2% efficient mm-wave N-polar AlGaN/GaN MISHEMT at 87 GHz. *Electron. Lett.* **2016**, *52*, 1813–1814. [CrossRef]

crystals

MDPI

Review

Process of Au-Free Source/Drain Ohmic Contact to AlGaN/GaN HEMT

Lin-Qing Zhang [1], Xiao-Li Wu [1], Wan-Qing Miao [1], Zhi-Yan Wu [1,*], Qian Xing [2,*] and Peng-Fei Wang [3]

1 College of Electronic and Electrical Engineering, Henan Normal University, No. 46 East of Construction Road, Xinxiang 453007, China; 14110720069@fudan.edu.cn (L.-Q.Z.); kuankuan7693@163.com (X.-L.W.); miaowanqing1113@163.com (W.-Q.M.)
2 College of Intelligent Engineering, Henan Institute of Technology, No. 699 Pingyuan Road (East Section), Xinxiang 453003, China
3 State Key Laboratory of ASIC and System, School of Microelectronics, Fudan University, No. 220 Handan Road, Shanghai 200043, China; pfw@fudan.edu.cn
* Correspondence: 2018010@htu.edu.cn (Z.-Y.W.); xingqian8911@163.com (Q.X.)

Abstract: AlGaN/GaN high electron mobility transistors (HEMTs) are regarded as promising candidates for a 5G communication system, which demands higher frequency and power. Source/drain ohmic contact is one of the key fabrication processes crucial to the device performance. Firstly, Au-contained metal stacks combined with RTA high-temperature ohmic contact schemes were presented and analyzed, including process conditions and contact formation mechanisms. Considering the issues with the Au-contained technique, the overview of a sequence of Au-free schemes is given and comprehensively discussed. In addition, in order to solve various problems caused by high-temperature conditions, novel annealing techniques including microwave annealing (MWA) and laser annealing (LA) were proposed to form Au-free low-temperature ohmic contact to AlGaN/GaN HEMT. The effects of the annealing method on surface morphology, gate leakage, dynamic on-resistance (R_{ON}), and other device characteristics are investigated and presented in this paper. By using a low-temperature annealing atmosphere or selective annealing method, gate-first Si-CMOS compatible AlGaN/GaN HEMT technology can be realized for high frequency and power application.

Keywords: AlGaN/GaN; HEMT; ohmic contact; Au-free low-temperature

Citation: Zhang, L.-Q.; Wu, X.-L.; Miao, W.-Q.; Wu, Z.-Y.; Xing, Q.; Wang, P.-F. Process of Au-Free Source/Drain Ohmic Contact to AlGaN/GaN HEMT. *Crystals* **2022**, *12*, 826. https://doi.org/10.3390/cryst12060826

Academic Editors: Peng Chen, Zhizhong Chen and Dmitri Donetski

Received: 24 April 2022
Accepted: 6 June 2022
Published: 10 June 2022

1. Introduction

Traditional Si-based radio frequency (RF) and power devices, due to the physical limitation of Si material, cannot meet the requirements of higher-speed and higher-power in a 5G generation communication system. GaN-based HEMTs are regarded as promising candidates for high-frequency and high-power electronic devices due to their superior material properties [1–4]. The GaN-based materials are always heterogeneous and grown on other substrates, such as Si, SiC, and sapphire [5–8]. Among the heterojunction device structures, the AlGaN/GaN HEMT exhibits the best electronic characteristics [9–11]. Furthermore, the AlGaN/GaN-on-Si is the structure most commonly selected for power application for the following reasons [12–15]: the fabrication process is compatible with Si CMOS process flow, which helps to reduce the manufacturing cost. Realization of a large-size AlGaN/ GaN-on-Si wafer can also further reduce the cost. The regular epitaxial AlGaN/GaN layers are grown on a high-resistivity Si substrate by metal organic chemical vapor deposition (MOCVD), which is shown in Figure 1. From Si bottom to top, a GaN buffer layer with a thickness of about 3 μm is deposited for reducing stress caused by lattice mismatch. After that, the device layer is formed by depositing an unintentional doped (UID) GaN device layer (~100 nm), AlN interlayer (~1 nm), and $Al_xGa_{1-x}N$ barrier layer (~25 nm, x varies from 0.15–0.4). The two-dimensional electron gas (2DEG) is formed at the AlGaN/GaN interface due to the polarization effect [16,17]. Finally, the GaN cap layer (~3 nm) is deposited for protecting the device surface.

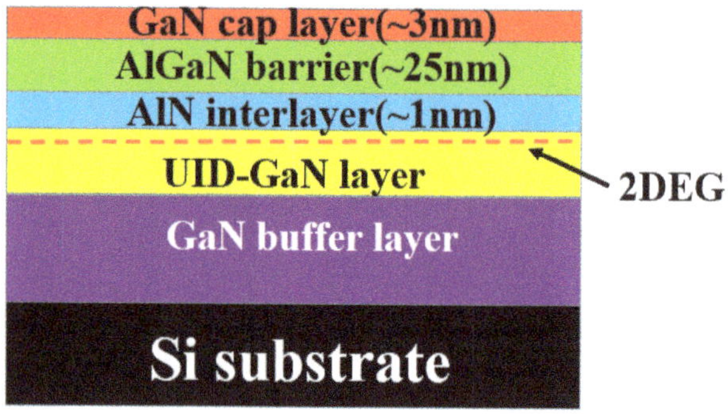

Figure 1. Schematic AlGaN/GaN-on Si device structure.

For RF application, the gate of the fabricated device is always formed in a T-shape for smaller gate resistance [18–20]. The process flow, which can be seen in Figure 2, consists of mesa isolation, source/drain ohmic contact formation, SiN passivation, T-shape gate formation, gate electrode deposition, via formation, and testing electrode deposition.

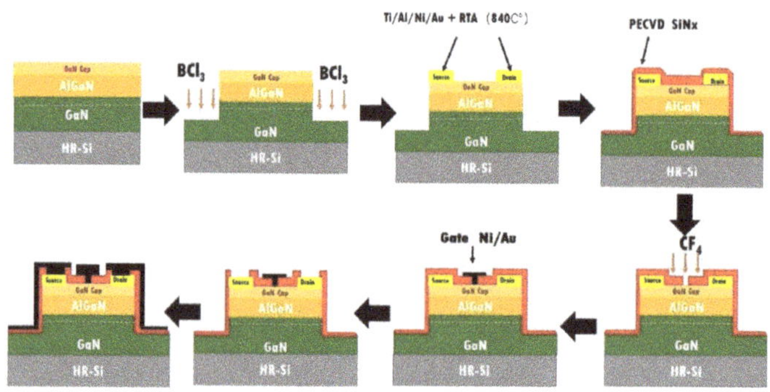

Figure 2. Schematic process flow of T-shaped AlGaN/GaN HEMT.

For power application, the gate is fabricated in a regular shape, as shown in Figure 3. The fabrication process consists of mesa isolation, source/drain ohmic contact formation, gate electrode deposition, SiN passivation, via formation, and testing electrode deposition.

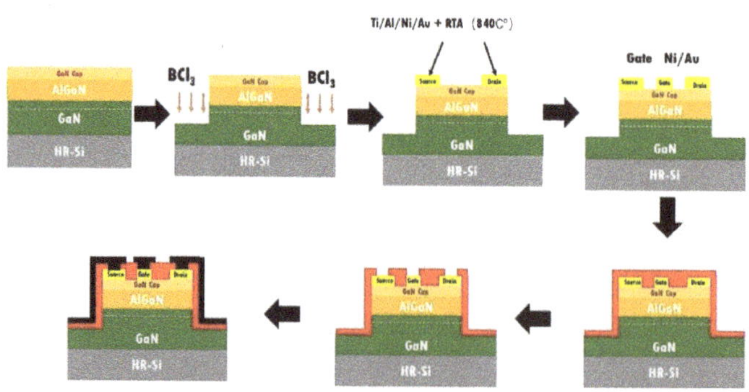

Figure 3. Schematic process flow of regular AlGaN/GaN HEMT.

The quality of source/drain ohmic contact has a great impact on the performance of AlGaN/GaN HEMT, such as on-resistance (R_{on}), output current (I_{DS}), extrinsic transconductance (G_m), and current gain cut-off frequency (f_T) [11,21–25]. Always, the source/drain regions form ohmic contact to reduce R_{ON} and improve output current and f_T characteristics. Thus, the source/drain ohmic contact technique should be investigated. Over the years, researchers have systematically investigated ohmic contact to n^+-GaN and AlGaN/GaN for achieving low contact resistance (R_C).

The most popular metallization schemes of ohmic contact in AlGaN/GaN HEMT is Ti/Al/X/Au, where X can be Ni, Mo, Pt, Ta, Ir, etc. [26–30]. Ti/Al/X/Au Au-contained schemes will be discussed in Section 2. However, the Au-contained metal scheme is not suitable for Si-based CMOS fabrication lines because Au has a high diffusion coefficient in silicon, which can lead to heavy metal contamination [31,32]. The use of Au will increase the fabrication cost, and long-term Au diffusion will lead to ohmic contact degradation [31,33]. Thus, an Au-free ohmic contact metal scheme should be proposed and investigated for achieving the purpose. In Section 3.1, the most advanced Au-free ohmic contact techniques for AlGaN/GaN HEMTs are displayed and discussed. Thermal budget is another issue that should be considered in AlGaN/GaN HEMT source/drain ohmic contact formation. Traditional rapid thermal annealing (RTA) usually needs a high-temperature atmosphere to form ohmic contact between a metal stack and an AlGaN/GaN heterojunction, which will lead to the oxidation of the AlGaN surface and create N defects [34], degrading the dynamic performance of AlGaN/GaN-based HEMTs. Also, TiN protrusions of random sizes take shape during the high-temperature annealing process [35]. Rough surface morphology and unideal ohmic metal edge sharpness can be observed to easily form short contact between the source/drain metal stack and gate metal electrode [36]. This phenomenon will be more severe for the Q band and Ka band small-sized AlGaN/GaN HEMT. In addition, the high-temperature process is incompatible with the gate-first fabrication technique [37]. Thus, instead of pursuing the optimized annealing temperature of the RTA technique, a novel annealing technique may be employed as an alternative to form low-temperature Au-free AlGaN/GaN HEMT ohmic contact with the existing Au-free metal schemes, including microwave annealing (MWA) and laser annealing (LA), etc. Details about MWA and LA techniques to form ohmic contact to AlGaN/GaN HEMTs will be discussed in Section 3.2, which is promising for the realization of gate-first technology and a Si-CMOS compatible GaN HEMT technique.

2. Ti/Al/X/Au Au-Contained Ohmic Contact Technique

Considering the wide bandgap of (Al)GaN, two methods are adopted to form good quality ohmic contact: (1) reduce the barrier height(Φ_B) by selecting the appropriate metal and (2) increase the possibility of tunneling by forming an n^+-(Al)GaN surface. The energy band diagram of metal/(Al)GaN by two methods are shown in Figure 4.

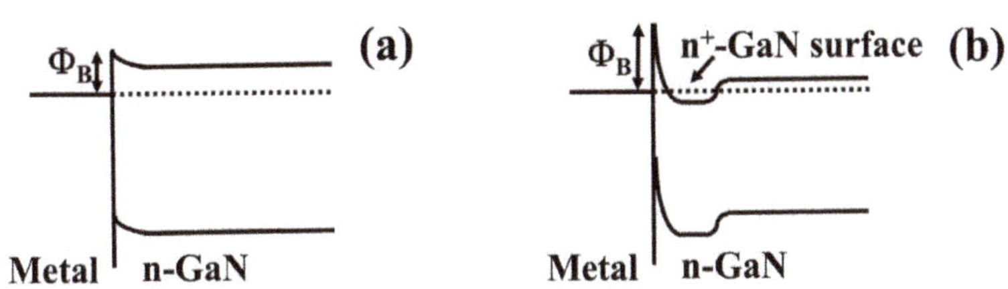

Figure 4. Energy band diagram of metal/(Al)GaN by reducing the barrier height(Φ_B) (**a**) and increasing the possibility of tunneling by forming n^+-(Al)GaN surface (**b**).

For the first method, researchers selected Al, Ti, Au, Ni, Mo, Pd, Pt, Cr, etc. as contact metals and investigated their contact characteristics with (Al)GaN [38–46]. The work function (W_m), specific contact resistivity (ρ_c), and Φ_B characteristics of different metal materials can be seen in Table 1.

Table 1. W_m, ρ_c, Φ_B characteristics of different metals to (Al)GaN.

Metal	W_m (eV)	ρ_c ($\times 10^{-6}$ Ω/cm^2)	Φ_B (eV)	Refs.
Al	4.28	8	0.8	[38,45]
Ti	4.33	3	0.58	[38–41,43]
Au	5.1	2.35	0.9	[38,43,44]
Ni	5.15	6.99	1.4	[38,41,44]
Mo	4.6	5.3	0.81	[42]
Pd	5.12	9.78	1.9	[38,46]
Pt	5.65	11	1.6	[38,41]
Cr	4.5	11.8	0.39	[41]

The results indicate that it is challenging to form good quality ohmic contact between single layer metal and (Al)GaN due to the higher W_m of the selected metal material. Also, the Ti, Al, etc. single layers are prone to oxidation during the high-temperature annealing process. Thus, a multilayer metal stack combined with a high-temperature annealing scheme became the common choice for GaN-based devices' source/drain ohmic contact formation [47]. The multilayer metal stack, as shown in Figure 5, always consists of four metal layers: the barrier layer, coating layer, diffusion barrier layer, and cap layer. The most popular metal scheme is Ti/Al/X/Au, where X can be Ni, Mo, Pt, Ta, Ir, etc. During the annealing process, the barrier layer Ti reacts with the AlGaN layer to form TiN [48]. TiN carries out lower work function [49], which lowers the Schottky barrier height and therefore helps in ohmic contact formation. Also, the created N vacancies formed in the reaction process make the AlGaN layer underneath the contact metal n^+-doped, which makes electron tunneling easy for ohmic contact formation [50]. Coating layer Al reacts with Ti under the high-temperature atmosphere to form Al_3Ti, which helps in ohmic contact formation [51]. The diffusion layer Ni, Mo, Pt, and Ta etc., which owns high melt point prevents Au indiffusion and Al outdiffusion [52]. Furthermore, the surface morphology of the contact metal can be affected by the metal layer. Au, which acts as a cap layer, prevents the Ti, Al layer oxidation within the high-temperature annealing atmosphere. Also, the Au layer improves the contact conductivity [53].

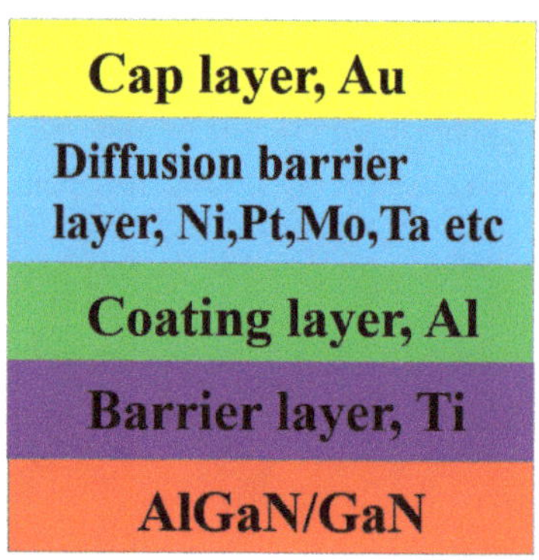

Figure 5. Schematic structure of metal stacks used for forming ohmic contact to AlGaN/GaN HEMT.

Ohmic contact resistance R_C, ρ_c, surface morphology, and thermal stability are the important indexes of ohmic contact quality. The R_C value can be affected by the selected metal stack, the metal thickness, ohmic recessing of the AlGaN layer, surface pre-treatment prior to ohmic metallization, annealing time and temperature, n-type doped in the semiconductor, etc. [52,54–63]. These affecting factors were investigated and optimized by universities and research institutions. Jacobs B et al. [54] have systematically studied the influence of the metal stack, Ti, Al, Ni thickness, Ti/Al ratio, and annealing condition on R_C. In their results, the achieved optimized R_C is 0.2 $\Omega{\cdot}$mm with Ti thickness of 30 nm, Ti/Al ratio of 6, Ni thickness of 40 nm, and annealing temperature of 900 °C for 30 s. Yan et al. [59] demonstrated that the R_C can be further reduced by a multi-step annealing process. With the improvement of AlGaN/GaN material growth, ohmic contact formation for metal and AlGaN/GaN becomes challenging. Buttari D et al. [60] developed a low-damage Cl_2 reactive ion etching (RIE) recess etch on an AlGaN layer (7nm), decreasing R_C from 0.45 Ω·mm to 0.27 Ω·mm. By Si ion implantation in the source/drain region, Nomoto K et al. [61] proved that the on-resistance can be decreased dramatically. By etching holes that were 0.8 μm × 0.8 μm in size and etching a depth of 10 nm, Wang et al. [62] developed a pattern of square holes technique in the source/drain AlGaN layer and achieved an R_C as low as 0.1 Ω·mm, as shown in Figure 6. The fabricated AlGaN/GaN HEMTs exhibits comparable output current, peak transconductance and threshold voltage characteristics with conventional AlGaN/GaN HEMTs. Before ohmic contact metal deposition, Fujishima T et al. [63] developed a $SiCl_4$ and BCl_3 plasma treatment on the source/drain surface and achieved low R_C values of 0.41 Ω·mm and 0.17 Ω·mm, respectively. By optimizing the affecting factors of Ti/Al/X/Au schemes, the achieved R_C can be as low as 0.3 Ω·mm, and even much lower. The obtained low R_C can meet the requirements for the RF and power applications of AlGaN/GaN HEMT by using Ti/Al/Ni/Au metal schemes.

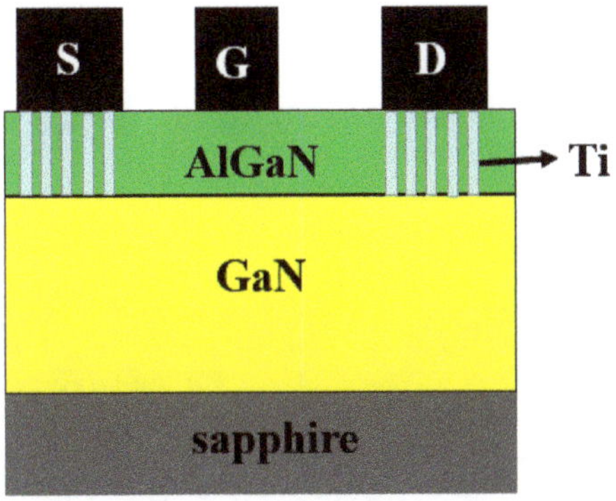

Figure 6. Schematic structure of AlGaN/GaN HEMT with holes etching in the ohmic region.

Universities and industries have conducted studies on the ohmic contact formation mechanism. Two methods are regarded as the mechanism to form ohmic contact to AlGaN/GaN HEMT: (1) a field emission (FE) tunneling mechanism [64–66] via the formation of a thinner high doping or a lower barrier; and (2) a spike mechanism [67,68] by TiN direct electron path formation along the dislocation. In the FE tunneling mechanism, Ti reacts with the GaN or AlGaN layer to form TiN. N should be extracted from GaN or AlGaN, generating a highly n-doped interface region, which is responsible for the occurrence of FE tunneling. The work function of TiN (3.74 eV) is lower than Ti (4.33 eV), which is believed to lower the Schottky barrier for low-resistance ohmic contact formation. In the spike mechanism, the TiN projections forming along the dislocation places penetrate through the AlGaN layer under a high temperature, which sets up a direct electron path connecting the 2DEG and the ohmic metal. This method is believed to be more efficient in electron transferring than the FE tunneling mode. With the improvement of AlGaN/GaN material growth, the spike phenomenon, which penetrates through the whole AlGaN layer, becomes less likely to happen. The cross-sectional diagrams of both physical models are shown in Figure 7, which reveals the ohmic contact formation mechanism.

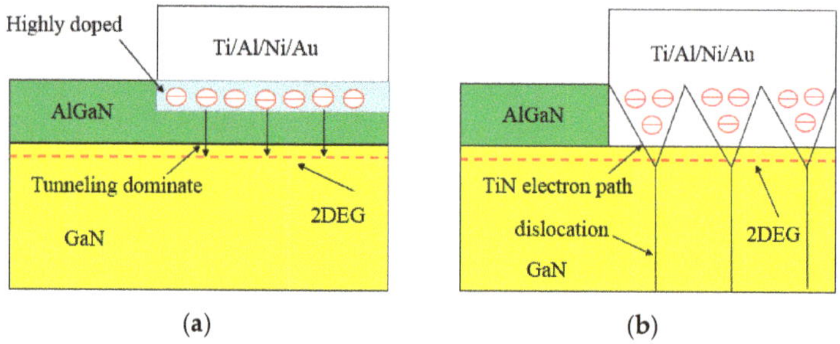

Figure 7. Physical models of Ti/Al/Ni(x)/Au ohmic contact to AlGaN/GaN HEMT: FE tunneling mechanism (**a**) and spike mechanism (**b**).

3. Au-Free Ohmic Contact Technique

3.1. Existing Advanced Au-Free Ohmic Contact Technology

The cap layer of Au is proposed to prevent the oxidation of ohmic metal under a high-temperature atmosphere. Au is also proposed to form Ga vacancies in AlGaN or GaN layers to decrease the contact resistance [31]. It is challenging to form high-quality ohmic contact with low contact resistance by using Au-free metal schemes. However, as mentioned in the introduction section, Au-contained metal schemes suffer from Si-based CMOS technology non-compatibility, rough surface morphology, and degradation of dynamic device performance. Also, the use of Au will increase the fabrication cost. Thus, Au-free ohmic contact metal schemes should be proposed and investigated for achieving the purpose. Over the years, groups from all over the world have studied and proposed a number of metal schemes, such as Ti/Al/W [33], Ta/Al/Ta [69], Ti/Al/Ti/TiN [70,71], Ta/Si/Ti/Al/Ni/Ta [72], Ti/Al/Ni/Pt [73], Ti/Al/TiN [74], Ti/TiN [75], Ti/Al/NiV [76], Ti/Al/Ti/W [77,78], Ti_xAl_y [79], and Ti/Al/Ni/Ti [80]. In these schemes, the cap layer of Au is replaced by W, TiN, and Ta, etc. For the Ti/Al-based method, Lee H S et al. [81] used a fully recessed AlGaN structure in the source/drain region, followed by a Ti/Al/W deposition and 870 °C annealing process, reducing the R_C to 0.49 Ω·mm. This method enables the direct contact between Ti/Al/W and 2 DEG. The Ti/Al/W metal stacks also exhibit much smoother morphology and a much lower voltage drop property for the same output current level, which is a very important characteristic for high-voltage power electronics. The source/drain AlGaN layer can be partially recessed to achieve ohmic contact with an R_C of 0.21 Ω·mm using Ti/Al/Ti/TiN at the low temperature of 550 °C [71]. By optimizing the Ti thickness (2.5 nm) and annealing temperature (550 °C), the low R_C ohmic contact can be realized. High-resolution transmission electron microscopy (HR-TEM) and electron dispersive X-ray spectroscopy (EDS) were adopted to investigate alloy distribution and the ohmic contact formation mechanism. From the TEM and EDS results as shown in Ref. [71], they deduced that Ti plays a catalytic role for the Al-N reaction and N extraction. The formation of a continuous AlN thin layer at the contact interface accounts for the low resistance of Au-free ohmic contacts on AlGaN/GaN at a low annealing temperature.

Yoshida T et al. [33] reduced the R_C to 0.358 Ω·mm at the low temperature of 500 °C by optimizing the thickness (a thin Ti layer of 2.7 nm) of Ti using Ti/Al/W, as shown in Figure 8. The fabricated devices have a maximum drain current density (I_{DS}), R_{ON}, and peak transconductance (G_m) of 0.422 A/mm, 9.7 Ω·mm and 73.2 mS/mm, respectively. These characteristics are comparable to Au-contained AlGaN/GaN HEMTs.

Furthermore, the cap layer can be replaced by Pt, Ni, and NiV. Liu et al. [73] used a Ti/Al/Ni/Pt Au-free metal stack and achieved an R_C of 0.6 Ω·mm at 975 °C annealing temperature. The maximum drain current and G_m achieved 700 mA/mm and 140 mS/mm, respectively. Tham W H et al. [76] used a Ti/Al/NiV source/drain metal stack scheme and achieved an R_C of 0.8 Ω·mm featuring the low temperature of 500 °C. The AlGaN/GaN MIS-HEMTs were fabricated and measured. The achieved specific on-state resistivity and maximum saturation drain current were 2.3 mΩ cm^2 and 0.21 A/mm. Gao et al. [80] demonstrated a two-step Ti/Al/Ni/Ti Au-free ohmic contact technique (Ti/Al deposited by EBE and Ni/Ti grown by magnetron sputtering) and achieved an ρ_c of 5.59 × 10^{-5} Ω/cm^2. The achieved maximum saturation drain current and R_{ON} reached around 0.4 A/mm and 15.6 Ω·mm.

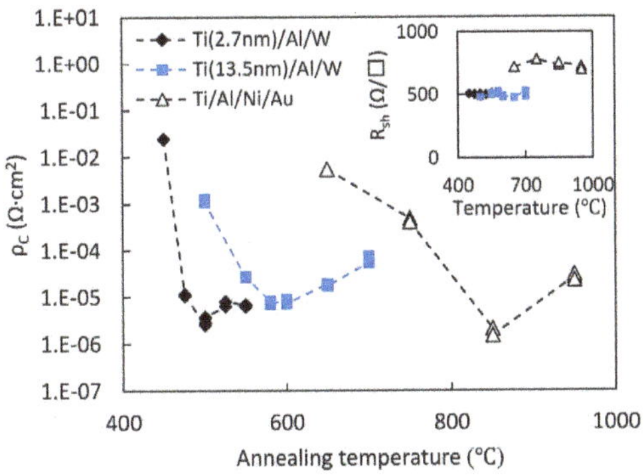

Figure 8. Specific contact resistivity ρ_C vs. annealing temperature. All samples were annealed for 1 min, expect for samples with Ti (2.7 nm)/Al/W, which were annealed for 10 min. The inset shows the corresponding channel sheet resistances Rsh. Reprinted with permission from Ref. [33]. Copyright 2018, Wiley.

Ta-based schemes were also proposed to form ohmic contact to AlGaN/GaN heterojunctions. Li et al. [82] deposited a Ta/Si/Ti/Al/Ni/Ta metal stack and achieved a low R_C of 0.22 Ω·mm by optimizing the annealing temperature. The ohmic contact formation was also investigated by using X-ray diffraction (XRD), TEM, and EDS analysis, as shown in Figure 9. It was deduced that the formation of Ti_xSi_y alloy with a low work function and intermixing with Ta results in a low ohmic contact resistance R_C. Johnson D W et al. [69] used a Cl_2 plasma etch to recess the whole AlGaN barrier layer in the source/drain region and deposit a Ta/Al/Ta layer with optimized thickness. The achieved R_C can be as low as 0.2 Ω·mm with the annealing temperature lower than 600 °C. Maximum drain current is found to be 236 mA/mm at V_{DS} = 10 V. Maximum transconductance and R_{ON} were 55 mS/mm and 41.3 Ω·mm, respectively. Considering the recess ohmic contact technique increased the fabrication complexity, in 2021, Fan et al. [79] developed a non-recessed Ti-Al alloys Au-free scheme instead of Ti/Al-based multilayers. By optimizing the ratio of Ti in Ti_xAl_y and annealing temperature, the low R_C of 0.063 Ω·mm can be obtained with Ti_5Al_1 under the optimal annealing condition. The EDS and TEM results as shown in Ref. [79] show that the Al outdiffusion can be suppressed and Ti can deeply diffuse into the AlGaN barrier layer by using Ti_xAl_y alloy. The extracted N atoms increase with an increasing Ti ratio of Ti_xAl_y. Therefore, a higher Ti ratio Ti_xAl_y is beneficial in forming a highly doped n-type AlGaN region, resulting in a much lower ohmic contact R_C. Furthermore, their group proposed an Au-contained $Ta_{0.83}Al_{0.17}$/Au metal scheme to realize a low ohmic contact of 0.14 Ω·mm at 900 °C [83]. Unlike the spike mechanism in the Ti/Al/Ni/Au method, the Ta-Al-Au mixed more uniformly with each other so that Au-dominated alloy fully penetrated the AlGaN barrier layer for electron path formation. Thus, the R_C could be decreased drastically.

Au-free ohmic contact formation can also be realized by intentional ion implantation in the AlGaN/GaN within the source/drain region [84]. In addition, selective regrowth of the source/drain region after the partial AlGaN etching can also realize Au-free ohmic contact formation [85,86]. However, these methods require strict process conditions, increase the fabrication cost, and make the fabrication process more complex. Therefore, only few universities and institutions investigate these techniques.

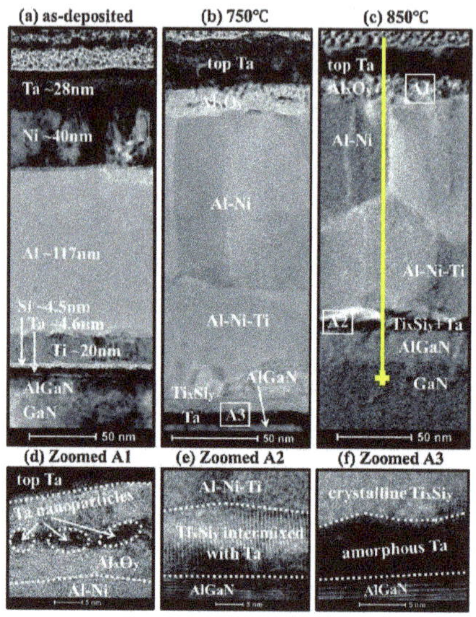

Figure 9. Cross sectional high-resolution TEM images of (**a**) as-deposited, (**b**) 750 °C-annealed, (**c**) 850 °C-annealed non-gold ohmic contacts. (**d**,**e**): zoomed-in TEM images of Ta/Ni interface (A1) and metal/AlGaN interface (A2) of 850 °C-annealed non-gold contact in (**c**), respectively, and (**f**): zoomed-in TEM image of metal/AlGaN interface (A3) of 750 °C -annealed non-gold (Ta/Si/Ti/Al/Ni/Ta) ohmic contact in (**b**). Reprinted with permission from Ref. [82]. Copyright 2013, IOP science.

The AlGaN/GaN HEMTs devices' electrical characteristics, with the existing advanced Au-free ohmic contact technique, can be summarized in Table 2. The results show that AlGaN/GaN HEMTs with Au-free ohmic contacts exhibit similar electrical characteristics comparable to Au-contained AlGaN/GaN HEMTs. The surface morphology of Au-free ohmic metals is more flat than the Au-contained ones due to the smaller root mean square (RMS) surface roughness.

Table 2. Survey of literature data on GaN HEMTs using different Au-free ohmic contact schemes.

Au-Free Schemes	$\mathbf{R_C}$ **(Ω·mm)**	$\mathbf{I_{dmax}}$ **(mA/mm)**	$\mathbf{R_{ON}}$ **(Ω·mm)**	**Specific On-State Resistivity (mΩ/cm²)**	$\mathbf{G_m}$ **(mS/mm)**	**Root-Mean-Square (RMS) Surface Roughness (nm)**	**Ref.**
Ti/Al/W	0.358	422	9.7	/	73.2	/	[33]
Ta/Al/Ta	0.2	263	41.3	/	55	/	[69]
Ti/Al/Ti/TiN	0.21	/	/	/	/	/	[71]
Ta/Si/Ti/Al/Ni/Ta	0.24	830	/	/	250	/	[72]
Ti/Al/Ni/Pt	0.6	700	/	0.77	140	4.6	[73]
Ti/Al/TiN	0.62	/	/	/	/	/	[74]
Ti/Al/NiV	0.13	210	/	2.3	60	/	[76]
Ti/Al/Ti/W	0.41	350	/	/	/	7	[78]
Ti_xAl_y	0.063	/	/	/	/	/	[79]
Ti/Al/Ni/Ti	/	400	15.6	6.39	/	7.9	[80]

3.2. *Low-Temperature Au-Free Ohmic Contact Technology*

The benchmark of R_C values against annealing temperatures of state-of-the-art Au-free ohmic contacts for the AlGaN/GaN HEMT can be seen in Figure 10, including recess and non-recess schemes [33,69–82]. It can be seen that low R_C values (<0.3 Ω·mm) always occur with high annealing temperature (>600 °C) conditions. Only a few methods meet the low R_C and low annealing temperature requirements. Thermal budget is another issue that should be considered in AlGaN/GaN HEMT source/drain ohmic contact formation.

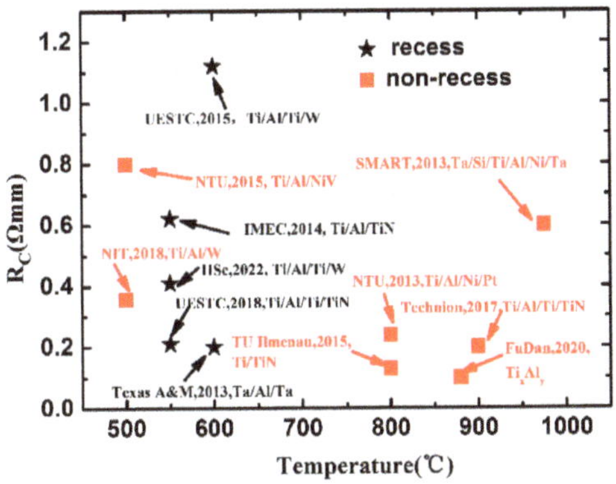

Figure 10. Comparison of R_C as a function of annealing temperature of different Au-free ohmic contact to AlGaN/GaN HEMT from various publications.

A high-temperature atmosphere will lead to the oxidation of AlGaN surfaces and create N defects [34], degrading the dynamic performance of AlGaN/GaN-based HEMTs. Also, TiN protrusions of random sizes take shape during the high-temperature annealing process [35]. Rough surface morphology and unideal ohmic metal edge sharpness can be observed, easily forming short contact between the source/drain metal stack and gate metal electrode [36]. This phenomenon will be more severe for the Q band and Ka band small-sized AlGaN/GaN HEMT. In addition, the high-temperature process is incompatible with the gate-first fabrication technique [37]. Thus, the Au-free low-temperature ohmic contact technique should be investigated and proposed for AlGaN/GaN HEMT. Au-free ohmic contact with AlGaN/GaN HEMT always needs a high temperature for contact formation using traditional RTA. With the existing optimized Au-free metal schemes, the annealing method might be another element which affects the ohmic contact quality. Therefore, the novel annealing technique should be introduced for low-temperature Au-free ohmic contact to AlGaN/GaN HEMT. A novel annealing technique may be employed and investigated for AlGaN/GaN HEMT ohmic contact formation: microwave annealing (MWA) and laser annealing (LA), etc.

MWA: as a novel annealing technique, MWA has special characteristics (including selective heating of specific materials, rapid heating rates and temperature gradients, the elimination of wall effects, and the superheating of solvents) and non-thermal effects [87]. Ohmic metals and 2DEG have strong ability to absorb microwaves. Thus, MWA can be an alternative to AlGaN/GaN HEMT source/drain ohmic contact formation. Using an Au-contained Ti/Al/Ni/Au metal stack, Zhang et al. [88] demonstrated that a low temperature of approximately 600 °C can be used to achieve good quality of ohmic contact (R_C = 0.6 Ω·mm) with an MWA power of 4200 W and heating time of 20min, as shown in Figure 11 [89]. High-resolution atomic force microscope (AFM) results show that the MWA-HEMT source/drain ohmic metal surface, after annealing process, is relatively flat compared to the RTA one. R_C values and root mean square (RMS) surface roughness characteristics for various MWA conditions are compared with the RTA method and shown in Table 3 [90].

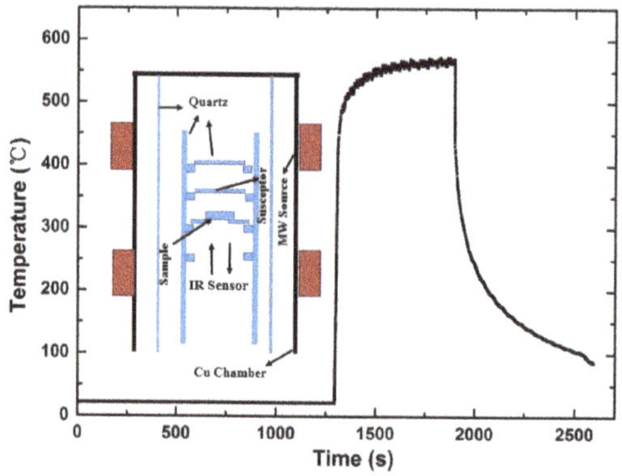

Figure 11. The schematic diagrams of MWA equipment. Reprinted with permission from Ref. [89]. Copyright 2017, IEEE.

Table 3. Comparisons of peak temperature, R_C, and RMS characteristics for various annealing conditions. Reprinted with permission from Ref. [90]. Copyright 2016, IEEE.

Methods	Peak Temperature (°C)	R_C (Ω·mm)	RMS (nm)
MWA, 3500 W, 20 min	579.6	1.16	37.7
MWA, 4200 W, 5 min	579.6	4.57	37.5
MWA, 4200 W,10 min	594	4.02	35.8
MWA, 4200 W, 15 min	597.6	1.44	49.6
MWA, 4200 W, 20 min	597.6	0.6	63.5
RTA, 840 °C, 35 s	840	0.4	101

TEM and EDS analysis are employed to learn the ohmic contact formation mechanism and reasons for surface roughness [88]. From the TEM and EDS results, as shown in Figure 12a,b, a direct current path is also formed in MWA-HEMT, which facilitates the flow of electrons in both directions, indicating that the spike mechanism is still the major mechanism of forming carrier conduction paths. From the EDS results as shown in Table 4, we deduce that less Ni-Au blocks were formed under a low-temperature atmosphere, which accounts for the superior surface morphology formation.

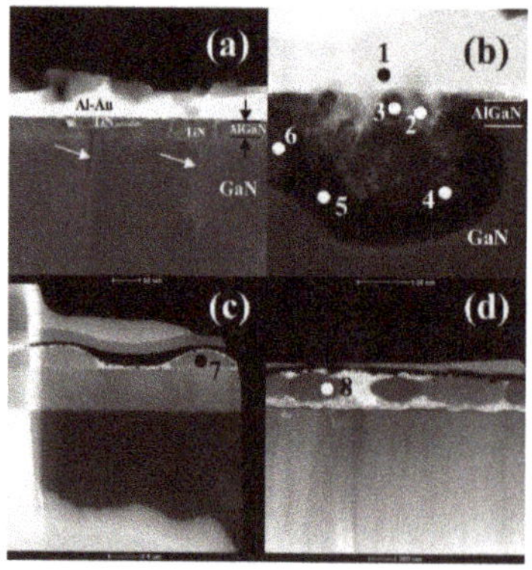

Figure 12. (**a**) TEM image of the post-annealing structure by MWA where the arrows markings indicate dislocations. (**b**) Magnified image of the interface. (**c**) TEM image of bulge area of the alloyed ohmic contact treated by RTA. (**d**) TEM image of bulge area of the alloyed ohmic contact treated by MWA. Reprinted with permission from Ref. [88]. Copyright 2015, IEEE.

Table 4. EDS Composition (at. %) from the points labeled in Figure 12. Reprinted with permission from Ref. [88]. Copyright 2015, IEEE.

	N	Al	Ti	Ni	Ga	Au
1		26.37		2.76		70.87
2		9.95	57.37	7.10	3.42	22.16
3			89.30	2.47		8.23
4		8.96	83.24	1.63		6.16
5		97.81	1.59			0.60
6	19.53	20.27	23.75		35.20	1.26
7		48.77		45.49	5.75	
8		51.35		48.65		

The MWA-HEMT also exhibits good gate leakage and output current characteristics as compared to RTA-HEMT. The maximum drain current value reaches 850 mA/mm. The MWA method was also employed to form ohmic contact to AlN/GaN HEMT [89] and InAlN/GaN HEMT [91]. For InAlN/GaN HEMT, Chou et al. [91] fabricated an MWA-HEMT, which exhibits better R_{ON}, subthreshold swing (SS), RF performance, and current gate leakage characteristics than the RTA one. Furthermore, MWA-HEMT ohmic contact has a much smoother surface morphology, which results in high reliability. For AlN/HEMT [89], MWA also delivers better surface morphology and a lower gate leakage current. The ohmic contact formation mechanism was also studied. The experimental results fit well with the field emission (FE) model, as shown in Figure 13.

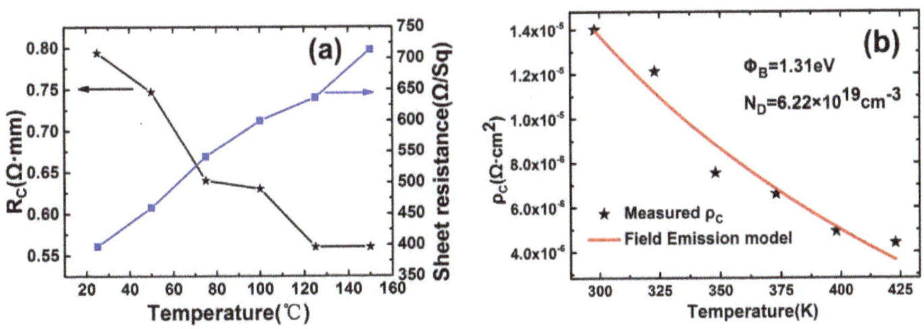

Figure 13. (**a**) Temperature−dependent characteristics of the R_{SH} and R_C of MWA−HEMT. (**b**) Measured (symbols) and modeled (lines) ρ_C of MWA−HEMT as a function of temperature. Reprinted with permission from Ref. [89]. Copyright 2017, IEEE.

The MWA GaN-based HEMTs devices' electrical characteristics compared to the RTA ones can be summarized in Table 5 as follows: the results show that MWA-HEMT exhibits comparable R_C, output current, G_M characteristics, and better R_{ON} and surface morphology. Thus, considering the excellent advantages of MWA, combined with Ti/Al-based, Ta-based, and Ti_xAl_y metal stack schemes, the MWA method is promising to realize low-temperature Au-free ohmic contact formation to AlGaN/GaN HEMT instead of pursuing an optimized record-low temperature for the RTA method.

LA: Tzou A J et al. [92] demonstrated the incorporation of a Si implantation in the GaN HEMTs and non-alloyed ohmic contact process by using the LA method. The surface morphology, contact resistance, on-resistance, and surface trap density can be improved by LA instead of the RTA technique. Ferreyra R A et al. [93] achieved a non-alloy ohmic contact using a Hf/Al/Ti metal stack with the picosecond LA method, featuring the R_C of 0.17 Ω·mm. Without source/drain n^+ GaN regrowth, Sheng's group proposed a laser annealing (LA) for AlGaN/GaN HEMT ohmic contact formation with a Ti/Al/Ni/Au metal stack [94]. The fabricated LA-HEMT exhibits smooth morphology, low contact resistance, high breakdown voltage and low dynamic on-resistance. The wafer is not damaged during the LA period due to the larger energy bandgap of GaN than LA photon energy. By applying a wavelength of 532 nm, 20 ns of a pulse duration, 100 kHz of pulse frequency, and energy density up to 2.99 J/cm^2, the achieved R_C is 2.18 Ω·mm [95]. The ohmic contact formation mechanism by LA has been studied and proposed. The experiment's data fit the thermionic field emission (TFE) model closely. From their results they deduce that the current transport mechanism of the LA sample is dominated by thermionic field emission (TFE). The energy band diagram for the contacts annealed by LA can be seen in Figure 14.

Table 5. GaN-based HEMTs devices' electrical characteristics: MWA vs. RTA [88–91].

Material	Methods	Temperature (°C)	R_C (Ω·mm)	ρ_c ($\times10^{-5}$Ω/cm^2)	RMS (nm)	I_{dmax} (mA/mm)	R_{ON} (Ω·mm)	G_m (mS/mm)
AlGaN/GaN	MWA	579.6	0.6	/	63.5	850	/	/
AlGaN/GaN	RTA	840	0.4	/	101	800	/	/
AlN/GaN	MWA	572.4	0.65	/	29.3	1400	/	270
AlN/GaN	RTA	750	0.4	/	56.5	1320	/	267
InAlN/GaN	MWA	/	/	4.29	6.79	/	4.9	/
InAlN/GaN	RTA	875	/	4.02	115	/	6.1	/

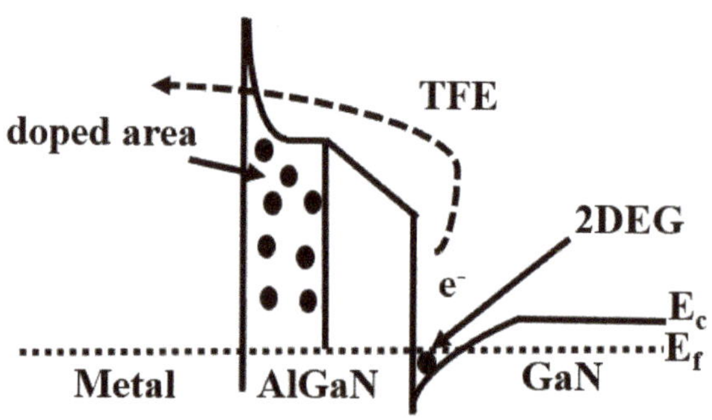

Figure 14. The schematic energy band diagram for the contacts annealed by LA.

The LA method results in a much smoother surface morphology due to less metal diffusion and alloy aggregation for the ultra-short LA time. They also studied the surface damage condition by LA and compared it with the RTA method [96]. The X-ray photoelectron spectroscopy (XPS) results, as shown in Figure 15, show that the oxidation reaction with a high-temperature atmosphere can be eliminated by the LA method. Also, Au-free ohmic contact using an Au-free Ti/Al/Ni/W metal stack combined with the LA annealing technique was proposed and studied. The LA-HEMT exhibits better isolation leakage current, low-density surface states, and suppressed R_{ON} [97].

Instead of a whole device region under a high-power laser pulse, Liu et al. [98] proposed a micro-scale LA approach to form ohmic contact to AlGaN/GaN HEMT. The schematic illustration of the micron-scale laser annealing is shown in Figure 16. Due to the selective source/drain region heating by LA, the gate electrode, which is temperature sensitive, is not thermally affected. Therefore, gate-first devices is realized using this micro-scale LA technique. The extracted R_C is as low as 0.3 Ω·mm when the laser power setting as 0.9 W. The high heating rate tends to form a thicker TiN layer, leaving a high density of N vacancy in AlGaN which makes tunneling much easier. In addition, the LA-HEMT shows better output current and smaller gate leakage.

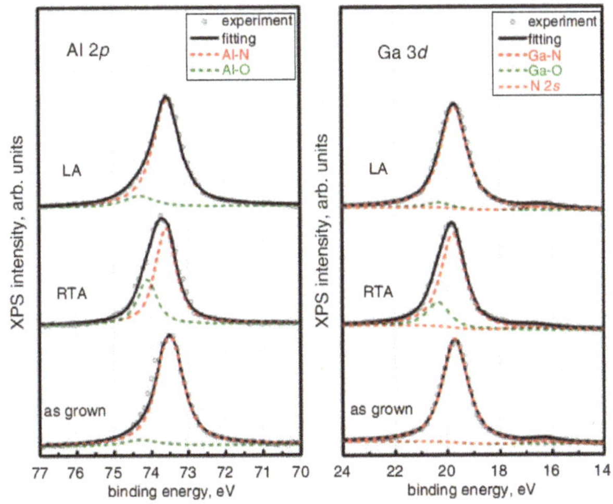

Figure 15. XPS core-level spectra of AlGaN surface as grown, after RTA and after the LA method. Reprinted with permission from Ref. [96]. Copyright 2019, Wiley.

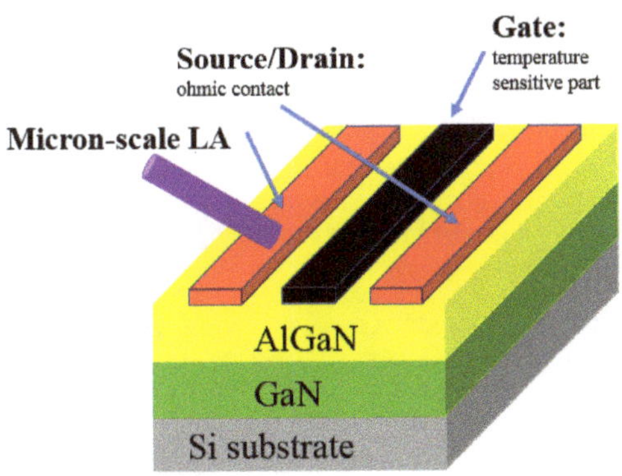

Figure 16. The schematic diagrams of micron-scale LA.

The LA GaN-based HEMTs devices' electrical characteristics compared to the RTA ones can be summarized in Table 6 as follows: from the testing results [94], the laser annealing reveals smooth morphology, low contact resistance, high breakdown voltage, and low dynamic on-resistance for the fabricated devices combined with the existing Au-free metal stack technique; this method is beneficial for gate-first technology and Si-CMOS compatible GaN HEMT technique.

Table 6. GaN-based HEMTs devices' electrical characteristics: LA vs. RTA.

Methods	R_C (Ω·mm)	RMS (nm)	I_{dmax} (mA/mm)	R_{ON} (Ω·mm)	G_m (mS/mm)	Ref.
LA	2.18	16.5	329	14.4	/	[94]
RTA	0.85	/	330	14.3	/	[94]

4. Conclusions

In this work, the conventional AlGaN/GaN HEMT Ti/Al/Ni/Au Au-contained ohmic contact technique was discussed comprehensively, including the selected metal stack, the metal thickness, ohmic recessing the AlGaN layer, surface pre-treatment prior to ohmic metallization, annealing time and temperature, n-type doped in the semiconductor, etc. After that, the existing Au-free ohmic contact was briefly introduced. In addition, to lower the thermal budget for the annealing process, novel MWA and LA annealing techniques were proposed for low-temperature Au-free ohmic contact to AlGaN/GaN HEMT, which is beneficial for gate-first technology and Si-CMOS compatible GaN HEMT technique realization.

Author Contributions: L.-Q.Z.; writing—original draft, investigation, data curation, conceptualization. X.-L.W., W.-Q.M., Z.-Y.W.; writing—original draft, investigation, data curation, conceptualization. Q.X., P.-F.W.; writing—review & editing, visualization, validation. All authors have read and agreed to the published version of the manuscript.

Funding: This work was supported in part by the Doctoral Scientific Research Start-Up Foundation of Henan Normal University under Grant 5101239170008, Henan Province Natural Science Foundation of Youth Fund No. 212300410185, Key scientific research projects of higher education institutions in Henan Province No. 21A510005, 22B510009 and in part by Key Laboratory of Optoelectronic Sensing Integrated Application of Hennan Province.

Institutional Review Board Statement: Not applicable.

Informed Consent Statement: Not applicable.

Data Availability Statement: Not applicable.

Conflicts of Interest: The authors declare no conflict of interest.

References

1. Chen, K.J.; Häberlen, O.; Lidow, A.; lin Tsai, C.; Ueda, T.; Uemoto, Y.; Wu, Y. GaN-on-Si power technology: Devices and applications. *IEEE Trans. Electron Devices* **2017**, *64*, 779–795. [CrossRef]
2. He, J.; Cheng, W.C.; Wang, Q.; Cheng, K.; Yu, H.; Chai, Y. Recent Advances in GaN-Based Power HEMT Devices. *Adv. Electron. Mater.* **2021**, *7*, 2001045. [CrossRef]
3. Mishra, U.K.; Shen, L.; Kazior, T.E.; Wu, Y.F. GaN-based RF power devices and amplifiers. *Proc. IEEE* **2008**, *96*, 287–305. [CrossRef]
4. Palacios, T.; Chakraborty, A.; Rajan, S.; Poblenz, C.; Keller, S.; DenBaars, S.P.; Speck, J.S.; Mishra, U.K. High-power AlGaN/GaN HEMTs for ka-band applications. *IEEE Electron Device Lett.* **2005**, *26*, 781–783. [CrossRef]
5. Pengelly, R.S.; Wood, S.M.; Milligan, J.W.; Sheppard, S.T.; Pribble, W.L. A review of GaN on SiC high electron-mobility power transistors and MMICs. *IEEE Trans. Microw. Theory Tech.* **2012**, *60*, 1764–1783. [CrossRef]
6. Pérez-Tomás, A.; Fontserè, A.; Llobet, J.; Placidi, M.; Rennesson, S.; Baron, N.; Chenot, S.; Moreno, J.C.; Cordier, Y. Analysis of the AlGaN/GaN vertical bulk current on Si, sapphire, and free-standing GaN substrates. *J. Appl. Phys.* **2013**, *113*, 174501. [CrossRef]
7. Liang, Z.; Du, H.; Yuan, Y.; Wang, Q.; Kang, J.; Zhou, H.; Zhang, J.; Hao, Y.; Wang, X.; Zhang, G. Ultra-thin AlGaN/GaN HFET with a high breakdown voltage on sapphire substrates. *Appl. Phys. Lett.* **2021**, *119*, 252101. [CrossRef]
8. Ishida, M.; Ueda, T.; Tanaka, T.; Ueda, D. GaN on Si technologies for power switching devices. *IEEE Trans. Electron Devices* **2013**, *60*, 3053–3059. [CrossRef]
9. Wośko, M.; Szymański, T.; Paszkiewicz, B.; Pokryszka, P.; Paszkiewicz, R. MOVPE growth conditions optimization for AlGaN/GaN/Si heterostructures with SiN and LT-AlN interlayers designed for HEMT applications. *J. Mater. Sci. Mater. Electron.* **2019**, *30*, 4111–4116. [CrossRef]
10. Baliga, B.J. Power semiconductor device figure of merit for high-frequency applications. *IEEE Electron Device Lett.* **1989**, *10*, 455–457. [CrossRef]
11. Tang, Y.; Shinohara, K.; Regan, D.; Corrion, A.; Brown, D.; Wong, J.; Schmitz, A.; Fung, H.; Kim, S.; Micovic, M. Ultrahigh-speed GaN high-electron-mobility transistors with f_T/f_{max} of 454/444 GHz. *IEEE Electron Device Lett.* **2015**, *36*, 549–551. [CrossRef]
12. Su, M.; Chen, C.; Rajan, S. Prospects for the application of GaN power devices in hybrid electric vehicle drive systems. *Semicond. Sci. Technol.* **2013**, *28*, 074012. [CrossRef]
13. Freedsman, J.J.; Egawa, T.; Yamaoka, Y.; Yano, Y.; Ubukata, A.; Tabuchi, T.; Matsumoto, K. Normally-off Al_2O_3/AlGaN/GaN MOS-HEMT on 8 in. Si with low leakage current and high breakdown voltage (825 V). *Appl. Phys. Express* **2014**, *7*, 041003. [CrossRef]
14. Egawa, T. Heteroepitaxial growth and power electronics using AlGaN/GaN HEMT on Si. In Proceedings of the 2012 International Electron Devices Meeting, San Francisco, CA, USA, 10–13 December 2012.
15. Then, H.W.; Radosavljevic, M.; Agababov, P.; Ban, I.; Bristol, R.; Chandhok, M.; Chouksey, S.; Holybee, B.; Huang, C.Y.; Krist, B.; et al. GaN and Si Transistors on 300mm Si (111) Enabled by 3D Monolithic Heterogeneous Integration. In Proceedings of the 2020 IEEE Symposium on VLSI Technology, Honolulu, HI, USA, 16–19 June 2020; pp. 1–2.
16. Asbeck, P.M.; Yu, E.T.; Lau, S.S.; Sullivan, G.J.; Van Hove, J.; Redwing, J. Piezoelectric charge densities in AlGaN/GaN HFETs. *Electron. Lett.* **1997**, *33*, 1230–1231. [CrossRef]

17. Katz, O.; Horn, A.; Bahir, G.; Salzman, J. Electron mobility in an AlGaN/GaN two-dimensional electron gas. I. Carrier concentration dependent mobility. *IEEE Trans. Electron Devices* **2003**, *50*, 2002–2008. [CrossRef]
18. Jessen, G.H.; Fitch, R.C.; Gillespie, J.K.; Via, G.; Crespo, A.; Langley, D.; Denninghoff, D.J.; Trejo, M.; Heller, E.R. Short-channel effect limitations on high-frequency operation of AlGaN/GaN HEMTs for T-gate devices. *IEEE Trans. Electron Devices* **2007**, *54*, 2589–2597. [CrossRef]
19. Ranjan, K.; Arulkumaran, S.; Ng, G.I.; Vicknesh, S. High Johnson's figure of merit (8.32 THz· V) in 0.15-μm conventional T-gate AlGaN/GaN HEMTs on silicon. *Appl. Phys. Express* **2014**, *7*, 044102. [CrossRef]
20. Androse, D.R.; Deb, S.; Radhakrishnan, S.K.; Sekar, E. T-gate AlGaN/GaN HEMT with effective recess engineering for enhancement mode operation. *Mater. Today Proc.* **2021**, *45*, 3556–3559. [CrossRef]
21. Tasker, P.J.; Hughes, B. Importance of source and drain resistance to the maximum f_T of millimeter-wave MODFETs. *IEEE Electron Device Lett.* **1989**, *10*, 291–293. [CrossRef]
22. Li, L.; Nomoto, K.; Pan, M.; Li, W.S.; Hickman, A.; Miller, J.; Lee, K.; Hu, Z.Y.; Bader, S.J.; Lee, S.M.; et al. GaN HEMTs on Si with regrown contacts and cutoff/maximum oscillation frequencies of 250/204 GHz. *IEEE Electron Device Lett.* **2020**, *41*, 689–692. [CrossRef]
23. Shinohara, K.; Corrion, A.; Regan, D.; Milosavljevic, I.; Brown, D.; Burnham, S.; Willadsen, P.J.; Butler, C.; Schmitz, A.; Wheeler, D.; et al. 220GHz f_T and 400GHz f_{max} in 40-nm GaN DH-HEMTs with re-grown ohmic. In Proceedings of the 2010 International Electron Devices Meeting, San Francisco, CA, USA, 6–8 December 2010.
24. Brown, D.F.; Shinohara, K.; Corrion, A.L.; Chu, R.; Williams, A.; Wong, J.C.; Rodriguez, I.A.; Grabar, R.; Johnson, M.; Butler, C.M.; et al. High-Speed, Enhancement-Mode GaN Power Switch with Regrown n+ GaN Ohmic Contacts and Staircase Field Plates. *IEEE Electron Device Lett.* **2013**, *34*, 1118–1120. [CrossRef]
25. Zhang, L.Q.; Huang, H.F.; Liu, X.Y.; Shi, J.S.; Liu, Z.; Zhao, S.X.; Wang, P.F. Two-dimensional device simulation for radio frequency performance of AlGaN/GaN HEMT, Semiconductor Technology International Conference. In Proceedings of the 2015 China Semiconductor Technology International Conference, Shanghai, China, 15–16 March 2015; pp. 1–3.
26. Fontserè, A.; Pérez-Tomás, A.; Placidi, M.; Llobet, J.; Baron, N.; Chenot, S.; Cordier, Y.; Moreno, J.C.; Gammon, P.M.; Jennings, M.R.; et al. Micro and nano analysis of 0.2 Ω mm Ti/Al/Ni/Au ohmic contact to AlGaN/GaN. *Appl. Phys. Lett.* **2011**, *99*, 213504. [CrossRef]
27. Wang, C.; Kim, N.Y. Electrical characterization and nanoscale surface morphology of optimized Ti/Al/Ta/Au ohmic contact for AlGaN/GaN HEMT. *Nanoscale Res. Lett.* **2012**, *7*, 1–8. [CrossRef] [PubMed]
28. Wang, L.; Mohammed, F.M.; Adesida, I. Dislocation-induced nonuniform interfacial reactions of Ti/Al/Mo/Au ohmic contacts on AlGaN/GaN heterostructure. *Appl. Phys. Lett.* **2005**, *87*, 141915. [CrossRef]
29. Shen, B.; Zhou, H.M.; Liu, J.; Zhou, Y.G.; Zhang, R.; Shi, Y.; Zheng, Y.D.; Someya, T.; Arakawa, Y. Ohmic contact and interfacial reaction of Ti/Al/Pt/Au metallic multi-layers on n-$Al_xGa_{1-x}N$/GaN heterostructures. *Opt. Mater.* **2003**, *23*, 197–201. [CrossRef]
30. Mohammed, F.M.; Wang, L.; Adesida, I.; Piner, E. The role of barrier layer on Ohmic performance of Ti/Al-based contact metallizations on AlGaN/GaN heterostructures. *J. Appl. Phys.* **2006**, *100*, 023708. [CrossRef]
31. Piazza, M.; Dua, C.; Oualli, M.; Morvan, E.; Carisetti, D.; Wyczisk, F. Degradation of TiAlNiAu as ohmic contact metal for GaN HEMTs. *Microelectron. Reliab.* **2009**, *49*, 1222–1225. [CrossRef]
32. Roccaforte, F.; Iucolano, F.; Giannazzo, F.; Alberti, A.; Raineri, V. Nanoscale carrier transport in Ti/Al/Ni/Au Ohmic contacts on AlGaN epilayers grown on Si (111). *Appl. Phys. Lett.* **2006**, *89*, 022103. [CrossRef]
33. Yoshida, T.; Egawa, T. Improvement of Au-Free, Ti/Al/W Ohmic Contact on AlGaN/GaN Heterostructure Featuring a Thin-Ti Layer and Low Temperature Annealing. *Phys. Status Solidi A* **2018**, *215*, 1700825. [CrossRef]
34. Ganchenkova, M.G.; Nieminen, R.M. Nitrogen vacancies as major point defects in gallium nitride. *Phys. Rev. Lett.* **2006**, *96*, 196402. [CrossRef]
35. Islam, Z.; Haque, A.; Glavin, N. Real-time visualization of GaN/AlGaN high electron mobility transistor failure at off-state. *Appl. Phys. Lett.* **2018**, *113*, 183102. [CrossRef]
36. Gong, R.; Wang, J.; Liu, S.; Dong, Z.; Yu, M.; Wen, C.; Cai, Y.; Zhang, B. Analysis of surface roughness in Ti/Al/Ni/Au Ohmic contact to AlGaN/GaN high electron mobility transistors. *Appl. Phys. Lett.* **2010**, *97*, 062115. [CrossRef]
37. Saadat, O.I.; Chung, J.W.; Piner, E.L.; Palacios, T. Gate-first AlGaN/GaN HEMT technology for high-frequency applications. *IEEE Electron Device Lett.* **2009**, *30*, 1254–1256. [CrossRef]
38. Rickert, K.A.; Ellis, A.B.; Kim, J.K.; Lee, J.L.; Himpsel, F.J.; Dwikusuma, F.; Kuech, T.F. X-ray photoemission determination of the Schottky barrier height of metal contacts to n-GaN and p-GaN. *J. Appl. Phys.* **2002**, *92*, 6671–6678. [CrossRef]
39. Hirsch, M.T.; Duxstad, K.J.; Haller, E.E. Effects of annealing on Ti Schottky barriers on n-type GaN. *Electron. Lett.* **1997**, *33*, 95–96. [CrossRef]
40. Binari, S.C.; Dietrich, H.B.; Kelner, G.; Rowland, L.B.; Doverspike, K.; Gaskill, D.K. Electrical characterisation of Ti Schottky barriers on n-type GaN. *Electron. Lett.* **1994**, *30*, 909–911. [CrossRef]
41. Lee, M.L.; Sheu, J.K.; Lin, S.W. Schottky barrier heights of metal contacts to n-type gallium nitride with low-temperature-grown cap layer. *Appl. Phys. Lett.* **2006**, *88*, 032103. [CrossRef]
42. Reddy, V.R.; Ramesh, C.K.; Choi, C.J. Structural and electrical properties of Mo/n-GaN Schottky diodes. *Phys. Status Solidi A* **2006**, *203*, 622–627. [CrossRef]

43. Wu, Y.F.; Jiang, W.N.; Keller, B.P.; Keller, S.; Kapolnek, D.; Denbaars, S.P.; Mishra, U.K.; Wilson, B. Low resistance ohmic contact to n-GaN with a separate layer method. *Solid-State Electron.* **1997**, *41*, 165–168. [CrossRef]
44. Schmitz, A.C.; Ping, A.T.; Khan, M.A.; Chen, Q.; Yang, J.W.; Adesida, I. Metal contacts to n-type GaN. *J. Electron. Mater.* **1998**, *27*, 255–260. [CrossRef]
45. Luther, B.P.; Mohney, S.E.; Jackson, T.N.; Asif Khan, M.; Chen, Q.; Yang, J.W. Investigation of the mechanism for Ohmic contact formation in Al and Ti/Al contacts to n-type GaN. *Appl. Phys. Lett.* **1997**, *7*, 57–59. [CrossRef]
46. Ping, A.T.; Khan, M.A.; Adesida, I. Ohmic contacts to n-type GaN using Pd/Al metallization. *J. Electron. Mater.* **1996**, *25*, 819–824. [CrossRef]
47. Fan, Z.; Mohammad, S.N.; Kim, W.; Aktas, Ö.; Botchkarev, A.E.; Morkoç, H. Very low resistance multilayer Ohmic contact to n-GaN. *Appl. Phys. Lett.* **1996**, *68*, 1672–1674. [CrossRef]
48. Wang, L.; Mohammed, F.M.; Adesida, I. Differences in the reaction kinetics and contact formation mechanisms of annealed Ti/Al/Mo/Au Ohmic contacts on n-Ga N and AlGaN/GaN epilayers. *J. Appl. Phys.* **2007**, *101*, 013702. [CrossRef]
49. Luther, B.P.; Mohney, S.E.; Jackson, T.N. Titanium and titanium nitride contacts to n-type gallium nitride. *Semicond. Sci. Technol.* **1998**, *13*, 1322. [CrossRef]
50. Kim, J.K.; Jang, H.W.; Lee, J.L. Mechanism for Ohmic contact formation of Ti on n-type GaN investigated using synchrotron radiation photoemission spectroscopy. *J. Appl. Phys.* **2002**, *91*, 9214–9217. [CrossRef]
51. Van Daele, B.; Van Tendeloo, G.; Ruythooren, W.; Derluyn, J.; Leys, M.R.; Germain, M. The role of Al on Ohmic contact formation on n-type GaN and AlGaN/GaN. *Appl. Phys. Lett.* **2005**, *87*, 061905. [CrossRef]
52. Mohammed, F.M.; Wang, L.; Selvanathan, D.; Hu, H.; Adesida, I. Ohmic contact formation mechanism of Ta/Al/Mo/Au and Ti/Al/Mo/Au metallizations on AlGaN/GaN HEMTs. *J. Vac. Sci. Technol. B Microelectron. Nanometer Struct. Processing Meas. Phenom.* **2005**, *23*, 2330–2335. [CrossRef]
53. Qin, Z.X.; Chen, Z.Z.; Tong, Y.Z.; Ding, X.M.; Hu, X.D.; Yu, T.J.; Zhang, G.Y. Study of Ti/Au, Ti/Al/Au, and Ti/Al/Ni/Au ohmic contacts to n-GaN. *Appl. Phys. A Mater. Sci. Processing* **2004**, *78*, 729–731. [CrossRef]
54. Jacobs, B.; Kramer, M.; Geluk, E.J.; Karouta, F. Optimization of the Ti/Al/Ni/Au ohmic contact on AlGaN/GaN FET structures. *J. Cryst. Growth* **2002**, *241*, 15–18. [CrossRef]
55. Kong, X.; Wei, K.; Liu, G.; Liu, X. Role of Ti/Al relative thickness in the formation mechanism of Ti/Al/Ni/Au Ohmic contacts to AlGaN/GaN heterostructures. *J. Phys. D Appl. Phys.* **2012**, *45*, 265101. [CrossRef]
56. Feng, Q.; Li, L.M.; Hao, Y.; Ni, J.Y.; Zhang, J.C. The improvement of ohmic contact of Ti/Al/Ni/Au to AlGaN/GaN HEMT by multi-step annealing method. *Solid-State Electron.* **2009**, *53*, 955–958. [CrossRef]
57. Yu, H.; McCarthy, L.; Rajan, S.; Keller, S.; Denbaars, S.; Speck, J.; Mishra, U. Ion implanted AlGaN-GaN HEMTs with nonalloyed ohmic contacts. *IEEE Electron Device Lett.* **2005**, *26*, 283–285.
58. Brown, D.F.; Williams, A.; Shinohara, K.; Kurdoghlian, A.; Milosavljevic, I.; Hashimoto, P.; Grabar, R.; Burnham, S.; Bulter, C.; Willadsen, P. W-band power performance of AlGaN/GaN DHFETs with regrown n^+ GaN ohmic contacts by MBE. In Proceedings of the International Electron Devices Meeting, Washington, DC, USA, 5–7 December 2011.
59. Yan, W.; Zhang, R.; Du, Y.; Han, W.; Yang, F. Analysis of the ohmic contacts of Ti/Al/Ni/Au to AlGaN/GaN HEMTs by the multi-step annealing process. *J. Semicond.* **2012**, *33*, 064005. [CrossRef]
60. Buttari, D.; Chini, A.; Meneghesso, G.; Zanoni, E.; Moran, B.; Heikman, S.; Zhang, N.Q.; Shen, L.; Coffie, R.; DenBaars, S.P.; et al. Systematic characterization of Cl_2 reactive ion etching for improved ohmics in AlGaN/GaN HEMTs. *IEEE Electron Device Lett.* **2002**, *23*, 76–78. [CrossRef]
61. Recht, F.; McCarthy, L.; Rajan, S.; Chakraborty, A.; Poblenz, C.; Corrion, A.; Speck, J.S.; Mishra, U.K. Nonalloyed ohmic contacts in AlGaN/GaN HEMTs by ion implantation with reduced activation annealing temperature. *IEEE Electron Device Lett.* **2006**, *27*, 205–207. [CrossRef]
62. Wang, C.; He, Y.; Zheng, X.; Zhao, M.; Mi, M.; Li, X.; Mao, W.; Ma, X.; Hao, Y. Low ohmic-contact resistance in AlGaN/GaN high electron mobility transistors with holes etching in ohmic region. *Electron. Lett.* **2015**, *51*, 2145–2147. [CrossRef]
63. Fujishima, T.; Joglekar, S.; Piedra, D.; Lee, H.S.; Zhang, Y.; Uedono, A.; Palacios, T. Formation of low resistance ohmic contacts in GaN-based high electron mobility transistors with BCl_3 surface plasma treatment. *Appl. Phys. Lett.* **2013**, *103*, 083508. [CrossRef]
64. Bright, A.N.; Thomas, P.J.; Weyland, M.; Tricker, D.M.; Humphreys, C.J.; Davies, R. Correlation of contact resistance with microstructure for Au/Ni/Al/Ti/AlGaN/GaN ohmic contacts using transmission electron microscopy. *J. Appl. Phys.* **2001**, *89*, 3143–3150. [CrossRef]
65. Boudart, B.; Trassaert, S.; Wallart, X.; Pesant, J.C.; Yaradou, O.; Théron, D.; Crosnier, Y.; Lahreche, H.; Omnes, F. Comparison between TiAl and TiAlNiAu ohmic contacts to n-type GaN. *J. Electron. Mater.* **2000**, *29*, 603–606. [CrossRef]
66. Mohammad, S.N. Contact mechanisms and design principles for nonalloyed ohmic contacts to n-GaN. *J. Appl. Phys.* **2004**, *95*, 4856–4865. [CrossRef]
67. Wang, L.; Mohammed, F.M.; Adesida, I. Formation mechanism of Ohmic contacts on AlGaN/GaN heterostructure: Electrical and microstructural characterizations. *J. Appl. Phys.* **2008**, *103*, 093516. [CrossRef]
68. Vertiatchikh, A.; Kaminsky, E.; Teetsov, J.; Robinson, K. Structural properties of alloyed Ti/Al/Ti/Au and Ti/Al/Mo/Au ohmic contacts to AlGaN/GaN. *Solid-State Electron.* **2006**, *50*, 1425–1429. [CrossRef]

69. Johnson, D.W.; Lee, R.T.; Hill, R.J.; Wong, M.H.; Bersuker, G.; Piner, E.L.; Kirsch, P.D.; Harris, H.R. Threshold voltage shift due to charge trapping in dielectric-gated AlGaN/GaN high electron mobility transistors examined in Au-free technology. *IEEE Trans. Electron Devices* **2013**, *60*, 3197–3203. [CrossRef]
70. Shriki, A.; Winter, R.; Calahorra, Y.; Kauffmann, Y.; Ankonina, G.; Eizenberg, M.; Ritter, D. Formation mechanism of gold-based and gold-free ohmic contacts to AlGaN/GaN heterostructure field effect transistors. *J. Appl. Phys.* **2017**, *121*, 065301. [CrossRef]
71. Zhang, J.; Kang, X.; Wang, X.; Huang, S.; Chen, C.; Wei, K.; Zheng, Y.; Zhou, Q.; Chen, W.; Zhang, B.; et al. Ultralow-Contact-Resistance Au-Free Ohmic Contacts with Low Annealing Temperature on AlGaN/GaN Heterostructures. *IEEE Electron Device Lett.* **2018**, *39*, 847–850. [CrossRef]
72. Arulkumaran, S.; Ng, G.I.; Vicknesh, S.; Wang, H.; Ang, K.S.; Kumar, C.M.; Teo, K.L.; Ranjan, K. Demonstration of submicron-gate AlGaN/GaN high-electron-mobility transistors on silicon with complementary meta-oxide-semiconductor-compatible non-gold metal stack. *Appl. Phys. Express* **2012**, *6*, 016501. [CrossRef]
73. Liu, Z.; Sun, M.; Lee, H.S.; Heuken, M.; Palacios, T. AlGaN/AlN/GaN high-electron-mobility transistors fabricated with Au-free technology. *Appl. Phys. Express* **2013**, *6*, 096502. [CrossRef]
74. Firrincieli, A.; De Jaeger, B.; You, S.; Wellekens, D.; Van Hove, M.; Decoutere, S. Au-free low temperature ohmic contacts for AlGaN/GaN power devices on 200 mm Si substrates. *Jpn. J. Appl. Phys.* **2014**, *53*, 04EF01. [CrossRef]
75. Jatal, W.; Baumann, U.; Tonisch, K.; Schwierz, F.; Pezoldt, J. High-frequency performance of GaN high-electron mobility transistors on 3C-SiC/Si substrates with Au-free ohmic contacts. *IEEE Electron. Device Lett.* **2015**, *36*, 123–125. [CrossRef]
76. Tham, W.H.; Bera, L.K.; Ang, D.S.; Dolmanan, S.B.; Bhat, T.N.; Tripathy, S. $Al_xGa_{1-x}N$/GaN MISHEMTs with a Common Gold-Free Metal-Stack for Source/Drain/Gate. *IEEE Electron Device Lett.* **2015**, *36*, 1291–1294. [CrossRef]
77. Zhang, J.; Huang, S.; Bao, Q.; Wang, X.; Wei, K.; Zheng, Y.; Li, Y.; Zhao, C.; Liu, X.; Zhou, Q.; et al. Mechanism of Ti/Al/Ti/W Au-free ohmic contacts to AlGaN/GaN heterostructures via pre-ohmic recess etching and low temperature annealing. *Appl. Phys. Lett.* **2015**, *107*, 262109. [CrossRef]
78. Niranjan, S.; Rao, A.; Muralidharan, R.; Sen, P.; Nath, D.N. Performance Comparison of Au-Based and Au-Free AlGaN/GaN HEMT on Silicon. *IEEE Trans. Electron Devices* **2022**, *69*, 1014–1019. [CrossRef]
79. Fan, M.Y.; Yang, G.Y.; Zhou, G.N.; Jiang, Y.; Li, W.M.; Jiang, Y.L.; Yu, H.Y. Ultra-low Contact Resistivity of <0.1 Ω·mm for Au-free TixAly Alloy Contact on Non-recessed i-AlGaN/GaN. *IEEE Electron Device Lett.* **2020**, *41*, 143–146.
80. Gao, S.; Liu, X.; Chen, J.; Xie, Z.; Zhou, Q.; Wang, H. High breakdown-voltage GaN-based HEMTs on silicon with Ti/Al/Ni/Ti ohmic contacts. *IEEE Electron Device Lett.* **2021**, *42*, 481–484. [CrossRef]
81. Lee, H.S.; Lee, D.S.; Palacios, T. AlGaN/GaN high-electron-mobility transistors fabricated through a Au-free technology. *IEEE Electron Device Lett.* **2011**, *32*, 623–625.
82. Li, Y.; Ng, G.I.; Arulkumaran, S.; Kumar, C.M.M.; Ang, K.S.; Anand, M.J.; Wang, H.; Hofstetter, R.; Ye, G. Low-contact-resistance non-gold Ta/Si/Ti/Al/Ni/Ta Ohmic contacts on undoped AlGaN/GaN high-electron-mobility transistors grown on silicon. *Appl. Phys. Express* **2013**, *6*, 116501. [CrossRef]
83. Fan, M.Y.; Jiang, Y.; Yang, G.Y.; Jiang, Y.L.; Yu, H.Y. Very-Low Resistance Contact to 2D Electron Gas by Annealing Induced Penetration Without Spikes Using TaAl/Au on Non-Recessed i-AlGaN/GaN. *IEEE Electron Device Lett.* **2020**, *41*, 1484–1487. [CrossRef]
84. Nomoto, K.; Tajima, T.; Mishima, T.; Satoh, M.; Nakamura, T. Remarkable reduction of on-resistance by ion implantation in GaN/AlGaN/GaN HEMTs with low gate leakage current. *IEEE Electron Device Lett.* **2007**, *28*, 939–941. [CrossRef]
85. Guo, J.; Li, G.; Faria, F.; Cao, Y.; Wang, R.; Verma, J.; Gao, X.; Guo, S.; Beam, E.; Ketterson, A.; et al. MBE-Regrown Ohmics in InAlN HEMTs with a Regrowth Interface Resistance of 0.05 Ωmm. *IEEE Electron Device Lett.* **2012**, *33*, 525–527. [CrossRef]
86. Seo, H.C.; Chapman, P.; Cho, H.I.; Lee, J.H.; Kim, K. Ti-based nonalloyed Ohmic contacts for $Al_{0.15}Ga_{0.85}N$/GaN high electron mobility transistors using regrown n^+-GaN by plasma assisted molecular beam epitaxy. *Appl. Phys. Lett.* **2008**, *93*, 102102. [CrossRef]
87. Lee, Y.J.; Cho, T.C.; Chuang, S.S.; Hsueh, F.K.; Lu, Y.L.; Sung, P.J.; Chen, H.C.; Current, M.I.; Tseng, T.Y.; Chao, T.S.; et al. Low-temperature microwave annealing processes for future IC fabrication—A review. *IEEE Trans. Electron Devices* **2014**, *61*, 651–665. [CrossRef]
88. Zhang, L.Q.; Shi, J.S.; Huang, H.F.; Liu, X.Y.; Zhao, S.X.; Wang, P.F.; Zhang, D.W. Low-temperature ohmic contact formation in GaN high electron mobility transistors using microwave annealing. *IEEE Electron Device Lett.* **2015**, *36*, 896–898. [CrossRef]
89. Zhang, L.Q.; Liu, Z.; Zhao, S.X.; Lin, M.Z.; Wang, P.F. Low-Temperature Ohmic Contact Formation in AlN/GaN HEMT Using Microwave Annealing. *IEEE Trans. Electron Devices* **2017**, *64*, 1385–1389. [CrossRef]
90. Liu, Z.; Zhang, L.Q.; Shi, J.S.; Liu, X.Y.; Zhao, S.X.; Huang, H.F.; Wang, P.F. Mechanism of ohmic contact formation in AlGaN/GaN high electron mobility transistors using microwave annealing. In Proceedings of the 2016 13th IEEE International Conference on Solid-State and Integrated Circuit Technology (ICSICT), Hangzhou, China, 25–28 October 2016; pp. 1050–1052.
91. Chou, L.I.; Peng, L.Y.; Wang, H.C.; Chiu, H.C.; Wang, H.T.; Chiang, D.L.; Chyi, J.I. InAlN/GaN HEMT using microwave annealing for low temperature ohmic contact formation. In Proceedings of the 2017 IEEE Compound Semiconductor Integrated Circuit Symposium (CSICS), Miami, FL, USA, 22–25 October 2017; pp. 1–3.
92. Tzou, A.J.; Hsieh, D.H.; Chen, S.H.; Li, Z.Y.; Chang, C.Y.; Kuo, H.C. Non-thermal alloyed ohmic contact process of GaN-based HEMTs by pulsed laser annealing. *Semicond. Sci. Technol.* **2016**, *31*, 055003. [CrossRef]

93. Ferreyra, R.A.; Suzuki, A.; Kazumoto, T.; Ueda, D. n^{++} GaN regrowth technique using pico-second laser ablation to form non-alloy ohmic contacts. *IEEE Electron Device Lett.* **2017**, *38*, 1079–1081. [CrossRef]
94. Hou, M.; Xie, G.; Sheng, K. Improved device performance in AlGaN/GaN HEMT by forming ohmic contact with laser annealing. *IEEE Electron Device Lett.* **2018**, *39*, 1137–1140. [CrossRef]
95. Hou, M.; Xie, G.; Sheng, K. Mechanism of Ti/Al/Ni/Au ohmic contacts to AlGaN/GaN heterostructures via laser annealing. *Chin. Phys. B* **2019**, *28*, 037302. [CrossRef]
96. Hou, M.; Xie, G.; Sheng, K. Low surface damage during ohmic contact formation in AlGaN/GaN HEMT by selective laser annealing. *Electron. Lett.* **2019**, *55*, 658–660. [CrossRef]
97. Hou, M.; Xie, G.; Guo, Q.; Sheng, K. Protection of isolated and active regions in AlGaN/GaN HEMTs using selective laser annealing. *Chin. Phys. B* **2021**, *30*, 097302. [CrossRef]
98. Liu, Z.; Chen, D.; Wan, L.; Li, G. Micron-scale annealing for ohmic contact formation applied in GaN HEMT gate-first technology. *IEEE Electron Device Lett.* **2018**, *39*, 1896–1899. [CrossRef]

MDPI
St. Alban-Anlage 66
4052 Basel
Switzerland
www.mdpi.com

Crystals Editorial Office
E-mail: crystals@mdpi.com
www.mdpi.com/journal/crystals

www.ingramcontent.com/pod-product-compliance
Lightning Source LLC
Chambersburg PA
CBHW040837250926
51850CB00007BA/1343
9783036586243